Ergebnisse der Plasmaphysik und der Gaselektronik

Band 6

Transport Properties of Dense Plasmas

Ergebnisse der Plasmaphysik und der Gaselektronik

Band 6

Herausgegeben von
Robert Rompe und Max Steenbeck†

Transport Properties of Dense Plasmas

by

W. Ebeling, V. E. Fortov, Yu. L. Klimontovich, N. P. Kovalenko, W. D. Kraeft, Yu. P. Krasny, D. Kremp, P. P. Kulik, V. A. Riaby, G. Röpke, E. K. Rozanov, M. Schlanges

With 60 Pictures and 2 Tables

Akademie-Verlag · Berlin
1983

Authors:

W. Ebeling, Department of Physics of the Humboldt-University, Berlin

V. E. Fortov, Institute of Chemical Physics, Academy of Sciences of the USSR, Chernogolovka, Moscow Region

Yu. L. Klimontovich, Faculty of Physics, Moscow State University, Moscow

N. P. Kovalenko, Department of Theoretical Physics, Odessa State University, Odessa

W. D. Kraeft, Department of Physics/Electronics of the Ernst-Moritz-Arndt-University, Greifswald

Yu. P. Krasny, Department of Theoretical Physics, Odessa State University, Odessa

D. Kremp, Department of Physics of the Wilhelm-Pieck-University, Rostock

P. P. Kulik, Moscow Institute of Electronic Technique, Moscow

V. A. Riaby, Moscow Institute of Electronic Technique, Moscow

G. Röpke, Department of Physics of the Wilhelm-Pieck-University, Rostock

E. K. Rozanov, Moscow Institute of Electronic Technique, Moscow

M. Schlanges, Department of Physics of the Wilhelm-Pieck-University, Rostock

Erschienen im Akademie-Verlag, DDR-1086 Berlin, Leipziger Str. 3–4

Lizenznummer: 202 · 100/430/83
Printed in the German Democratic Republic
Gesamtherstellung: VEB Druckerei „Thomas Müntzer", 5820 Bad Langensalza
LSV 1195
Bestellnummer: 763 172 3 (5570/6)
04400

Preface

The idea to write this volume arose in May, 1979 on the "Annual Seminar on Statistical Physics" in Lähnwitz near Güstrow, a nice resort place of the Pädagogische Hochschule Liselotte Herrmann Güstrow, where most of the authors were present. The participants of this seminar agreed in the general opinion that the transport properties of charged particle systems starting from ionized gases and ending with liquid metals should be interpreted in a unique way. The basic method of such a unified theory which is still in "status nascendi" nowadays, should be the statistical physics of systems with Coulombic interactions.

The first two chapters of this volume are devoted to mainly experimental aspects in investigating high density plasmas. After transport theory is presented in chapter 3., two special problems are considered being of interest in the theory of high density plasmas: the problem of bound states in kinetic theory (chapter 4.) and the theory of liquid metals (chapter 5.), which may be considered as weakly coupled nonideal plasmas. Besides transport properties also thermodynamic properties of dense Coulomb systems have been discussed in chapters 1. and 5. The theory of thermodynamic properties of plasmas has been given in a previous volume (Nr. 5) of this series.

Three of us (W.E., W.D.K., and G.R.) have taken the difficult task to organize this volume and to fit the chapters one to another, without changing the ideas and the style of the authors of the individual chapters. So we have restricted ourselves to unify some notations and to add some remarks (1.6. and 5.7.) on the cross connections between the chapters, and some references. Formulae are numerated within each chapter separately and citations refer to the chapter on consideration. We want to devote this volume to the beautiful place in Lähnwitz expressing the thanks of the participants to the organizers of the regular Lähnwitz-Seminars and especially to Prof. K. Henneberger and his coworkers.

By the name of the authors

Werner Ebeling, Berlin,

Wolf-Dietrich Kraeft, Greifswald,

Gerd Röpke, Rostock

Contents

1. Experiment and Physical Models of a Strongly Nonideal Plasma

V. E. Fortov

1.1. Introduction

Under the conditions of strong interparticle interaction, the physical properties of plasma, the most abundant state of matter in nature, have always attracted the attention of physicists of various specialities. A nonideal plasma is realized and determines the physical properties of a substance over a wide range of the phase diagram, spreading from conducting solids and liquids up to weakly ionized gases, including the phase boundaries of boiling and melting, as well as the region of metal-dielectric transition in disordered systems [1—3]. In recent years, the high-pressure plasma has aroused also practical interest due to the construction of some modern energetic arrangements and projects, the operation of which is based on pulsed high local concentration of energy in dense media, viz. powerful MHD-generators, power plants [4], propelling devices [5] with a gas-phase nuclear reactor, in arrangements of plasma-chemical industrial technology [6]. A strongly nonideal plasma appears on detonation of chemical [6] and nuclear [7] explosives, on explosive evaporation of metallic pinch liners and magnetocumulative generators [8], under the action of powerful shock waves, concentrated laser radiation, electron and ion beams on a condensed substance [9]. At hypersonic motion of spacecrafts in the dense atmosphere of giant planets [10] and in many other cases. A sharp need in the physical characteristics of coupled plasmas arises when realizing the idea of pulse thermonuclear fusion by laser, electron, or ion compression by explosion of spherical targets [11]. To solve these important problems, comprehensive information is necessary on thermodynamical, electrophysical and optical characteristics of a plasma over a wide region of the phase diagram from the strongly compressed condensed state up to ideal degenerate and Boltzmann gases, involving phase boundaries and the neighbourhood of the critical point.

The physical properties of plasmas are drastically simplified at extremely high pressures and temperatures, when the kinetic energy considerably exceeds that of interparticle interaction, and the models of homogeneous degenerate and Boltzmann ideal plasma can be applied with assurance [2, 3]. A weak interparticle interaction can then be taken into account by the methods of the modern perturbation theory within the framework of quasiclassical or classical approaches to the self-consistent field methods. Here important results were obtained by the scientists from GDR — see the monograph [1] and the chapters 3. and 4. of this book. In a nonideal plasma, the energy of interaction is comparable or exceeds the kinetic energy of particle motion, which hinders the application of asymptotic methods [1, 2] to such systems and presents considerable difficulties in the choice of pseudopotentials of the electron-ion interaction when using parameterless Monte-Carlo methods [12]. Besides, for a strongly interacting plasma, no adequate kinetic equation has been so far formulated for the calculation of the electrophysical properties [13], while the standard classification of elementary processes of light absorption and emission is also invalid here.

A considerable progress achieved nowadays in understanding of physical processes and mathematical description of equilibrium and kinetic characteristics of strongly coupled plasma, became possible due to both the development of theoretical methods and a rapid progress of experimental methods for the production and diagnostics of a high-pressure plasma, which permits to establish the actual boundaries for the applicability of the asymptotic description, and which also are the base for creating the physical models of dense plasma. The main difficulties in the production of a nonideal plasma are due to a necessity of considerable local energy concentrations, which leads to high pressures and temperatures considerably above the thermostrength limits of structural materials of the devices. This makes it necessary to carry out experiments in a forced pulsing regime at a high power level. In the experiments with a nonideal plasma, two ways of its generation have been most widely used nowadays, viz. the electric methods based on Joule dissipation of energy, and the dynamic methods which employ the effects of viscous dissipation of kinetic energy at the front of intense shock waves. The results obtained by the electric method which became classical in the plasma physics are generalized in [14—17] and chapter 2. of this book. In the present review we will discuss the basic physical results obtained during the last ten years mainly by the dynamical methods, which made it possible to produce plasma of extremely high parameters in a wide region of the phase diagram, and we will also consider the theoretical models proposed for its description.

1.2. Dynamical Methods in Dense Plasma Physics

Dynamical methods [18, 19, 20], which play a dominant role in experimental physics of high pressures, are based on the cumulation of energy in a substance under study either as a result of viscous dissipation in the shock front which when propagating over the substance bring about its compression, acceleration and irreversible heating up, or as a result of the adiabatic change in pressure of the medium. These methods are not coupled with the flow of powerful electric currents; they are characterized by high purity and homogeneity of the volume under investigation, by absence of electric and magnetic fields which hinder the diagnostics and lead to the development of various instabilities in plasma and by high reproducibility of the results [6]. Due to these factors together with the possibility of obtaining extremely high parameters and of applying specific methods of the diagnostics, dynamical methods become a convenient means for the investigation of physical properties of plasmas under extreme conditions.

In figure 1.1 the possibilities of different versions of dynamic experiments are presented when investigating plasma. Use of various ernergy sources, viz. compressed [21, 22] and electrically heated [23] gas, chemical [18—20, 24—32] and nuclear explosives [7], powerful laser [33] and neutron [7] beams to excite shock waves has

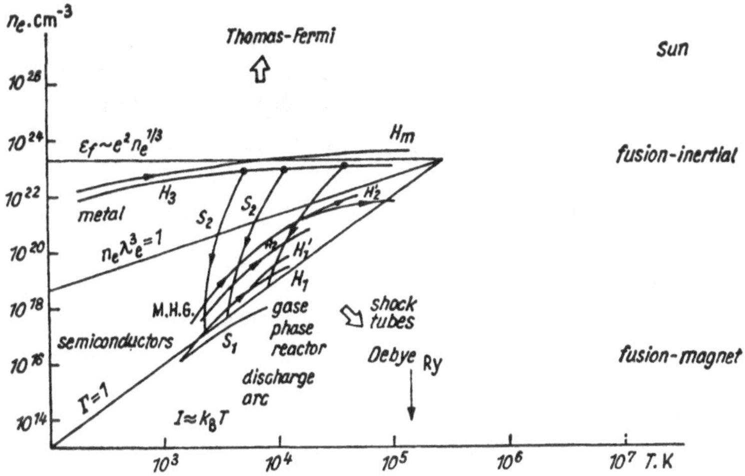

Fig. 1.1. Regions of nonideal plasma existence. Wavy curve $I \approx k_B T$ shows the beginning of ionization. Pointers show the directions of the interaction reduction and simplification of the plasma physical properties description. Experiment: H_1, H_1' — cesium vapour compression by the incident and reflected shock waves, S_1 — isentropic compression of cesium, H_2, H_2' — noble gases compression by the incident and reflected shock waves; shock compression of solid H_3 and porous H_m metals, S_2, S_3 — isentropic expansion of shock compressed metals

permitted to produce a nonideal plasma of different elements over a wide and continuously changing range of densities from solid to gas, at maximum pressures of ten millions of atmosphere and maximum temperatures of tens of electron volts.

The method of shock compression is the most efficient for investigation of a nondegenerate gas-like plasma resulting from the dynamic compression of high-pressure gases the initial states of which are in the neighbourhood of the saturation curve (cesium [21, 22], noble gases [24—29, 31, 33]), or even under supercritical conditions (xenon [27, 32]). While registering the states of single H and double H' compression, one manages to obtain a highly heated plasma of cesium (curves H_1, H_1') and noble gases (H_2, H_2') of supercritical parameters over a wide range of pressures (up to 11 GPa) and temperatures (T up to 10^5 K) and to gain access from the side of the "gas" phase to the region of a condensed state — the maximum density of xenon plasma [32] comes to 4.5 g/cm^3, which is 1.5 times larger than the density of solid xenon and higher than the crystallographic density of aluminium. Adiabatic compression [34, 35] of saturated cesium vapours leads to lower heating-up of plasmas where interaction of a charge with neutrals prevails. It is essential that the dynamic methods permit not only to realize the state of a nondegenerate plasma with extremely high energy concentrations over a wide range of the phase diagram, but under these conditions to fulfil detailed thermodynamic, electrophysical and optical measurements used for verifying the extrapolation properties of the asymptotic theories, as well as for creation of physical models of a nondegenerate "gas-like" plasma.

Compression of metals by intense shock waves generated by detonation of condensed explosives [18, 19, 30] transfers a substance into a state with pressures of 0.5 up to

1 TPa and temperatures of several thousands of degrees when the metal is melted, and a disordered electron-ionic plasma is realized with a degenerate or partly degenerate electron component (the lower part of the curves H_3 and H_m at fig. 1.1). These data are the base for semi-empirical models of the equation of state [3, 18].

The use of powerful explosions [36—41], pulsed neutron [7] and laser [33] radiation increases the plasma pressures up to tens of TPa and is the base for extrapolation tests of the results of quasi-classical theories [1, 42] (the upper parts of the curves H_m, H_3).

The adiabatic expansion method [30] based on plasma generation by adiabatic expansion (curves S_2) of the condensed substance, pre-pressed and irreversibly heated at the front of intensive shock waves (H_3), permits to study the intermediate states between solid and gas. Thus, it has been possible to investigate the properties of metals over a wide range of the phase diagram from the strongly compressed condensed state up to the ideal gas, including the region of degenerate and Boltzmann plasmas, near-critical and two-phase states, as well as the region of the "metal-dielectric" transition. These results form the base of wide-range equations of state [3, 30] which perform a semi-empirical description of all the four states of matter.

It is important that under the conditions of dynamic experiments, it is possible not only to obtain extremely high plasma parameters, but also to realize a sufficiently full specific diagnostics of the arising states, which is based on the use of the laws of mass, momentum and energy conservation at the front of the shock wave [43]

$$\frac{V_H}{V_0} = \frac{D - u}{D}; \qquad P_H = P_0 + \frac{Du}{V_0};$$
$$E_H - E_0 = \frac{1}{2}(P_H - P_0)(V_0 - V_H) \tag{1.1}$$

or at isentropic flow in the Riemann wave:

$$V_S = V_H + \int_P^{P_H} \left(\frac{\mathrm{d}u}{\mathrm{d}P}\right)^2 \mathrm{d}P; \qquad E_S = E_H - \int_P^{P_H} P \left(\frac{\mathrm{d}u}{\mathrm{d}P}\right)^2 \mathrm{d}P. \tag{1.2}$$

Use of these relationships reduces the problem of determining the calorimetric equation of state $E = E(P, V)$ to measuring of kinematic parameters of the shock wave movement (phase D and group u velocities) and contact surfaces, i.e. to recording of distances and times, which can be performed with high accuracy. The temperature of plasma can be either directly measured in the experiment together with the kinematic parameters, or calculated from the measured dependence $E = E(P, V)$ by numerical integration of the linear inhomogeneous differential equation of the second principle of thermodynamics [43, 44]

$$(P + E'_P)\frac{\partial T}{\partial P} - E'_P\frac{\partial T}{\partial V} = T. \tag{1.2a}$$

By the method of characteristics follows:

$$\frac{\mathrm{d}P}{\mathrm{d}V} = -\frac{P + \left(\frac{\partial E}{\partial V}\right)_P}{\left(\frac{\partial E}{\partial P}\right)_V},$$

$$\frac{\mathrm{d}T}{\mathrm{d}V} = -\frac{T}{\left(\frac{\partial E}{\partial P}\right)_V}.$$

The accuracy of the solution obtained, depending on the experimental errors and those of the initial data, was estimated using the Monte Carlo method [44] by modelling of a probable structure of the measuring process on the computer.

1.3. Thermodynamics of a Nonideal Plasma

Plasmas near equilibrium may be described by three characteristic lengths, the Landau radius

$$l = e^2/k_B T,$$

the Debye radius

$$r_D = \varkappa^{-1} = (k_B T/8\pi n_e e^2)^{1/2},$$

and the thermal de Broglie wave lengths of electrons and ions

$$\Lambda_e = h(2\pi m_e k_B T)^{-1/2}; \qquad \Lambda_i = h(2\pi m_i k_B T)^{-1/2}.$$

In the following we shall consider the parameter

$$l/r_D = e^3(8\pi n_e)^{1/2}(k_B T)^{-3/2}$$

as a measure of non-ideality and the parameter

$$n_e \Lambda_e^3$$

as a measure of degeneracy.

The first experiments investigating the effect of nonideality on the thermodynamical properties of plasma were performed with cesium [21, 22] in the pneumatical heated shock tube over the range of pressures $P \approx (0.15-20)$ MPa, temperatures $T \approx (2.6-20) \cdot 10^3$ K, electron concentrations $n_e \approx (5 \cdot 10^{15} - 5 \cdot 10^{19})$ cm^{-3}, where Coulomb interaction is strong $l/r_D \approx 0.2$ up to 2.2 and determines the physical properties of plasma with developed ionization.

In the experiments on shock compression of heavy noble gases [25, 28, 29] higher parameters of plasma are obtained, P up to 3.4 GPa, T up to $6 \cdot 10^4$ K, and densities $\varrho \approx 0.4$ g/cm^3, $n_e \approx 3 \cdot 10^{21}$ cm^{-3} that approach the density of condensed xenon, while in the experiments [32] $\varrho \approx 4.5$ g/cm^3 exceeded this density. Under these conditions the typical intermolecular distances in plasmas [29] equaled 60–70 nm and were comparable to the dimensions of ions and atoms in the ground state; the largest parameters of non-ideality $l/r_D \approx 5$ are close to the maximum possible values for a non-degenerate plasma.

The experimental data for noble gases [25, 28, 29] relate to single-acting ($\alpha_{Ar} \leqq 0.7$) and double-acting ($\alpha_{Xe} \leqq 1.8$) ionization and permit to advance into the region of a plasma of solid-state densities from the side of a "gas-like" plasma.

For the interpretation of the thermodynamical properties of the plasma we shall use here the "chemical" model of the plasma [1, 3]. In the framework of the chemical model the free energy F is represented as a function of the numbers of free electrons N_e and of free ions and neutrals belonging to the k-th species N_k. In a certain approximation F may be represented as a sum of the ideal, the Coulombic and the hard sphere contribution

$$F(V, T, N_e, N_k) = k_B T \left\{ N_e \ln\left(\frac{n_e \Lambda_e^3}{2e}\right) + \sum_k N_k \left(\frac{n_k \Lambda_k^3}{e\, Q_k}\right)\right\} + \Delta F_C + \Delta F_{HS},$$
$$e = 2.718282 \ldots \tag{1.3}$$

Here $n_e = N_e/V$ and $n_k = N_k/V$ denote the densities of the corresponding species and further Q_k the internal partition functions of the species k. In order to avoid divergencies we shall use the Brillouin-Planck-Larkin partition function [1]

$$Q_k = \sum_n g_n(e^{-\beta E_n} - 1 + \beta E_n) ; \qquad \beta = 1/k_B T ,$$

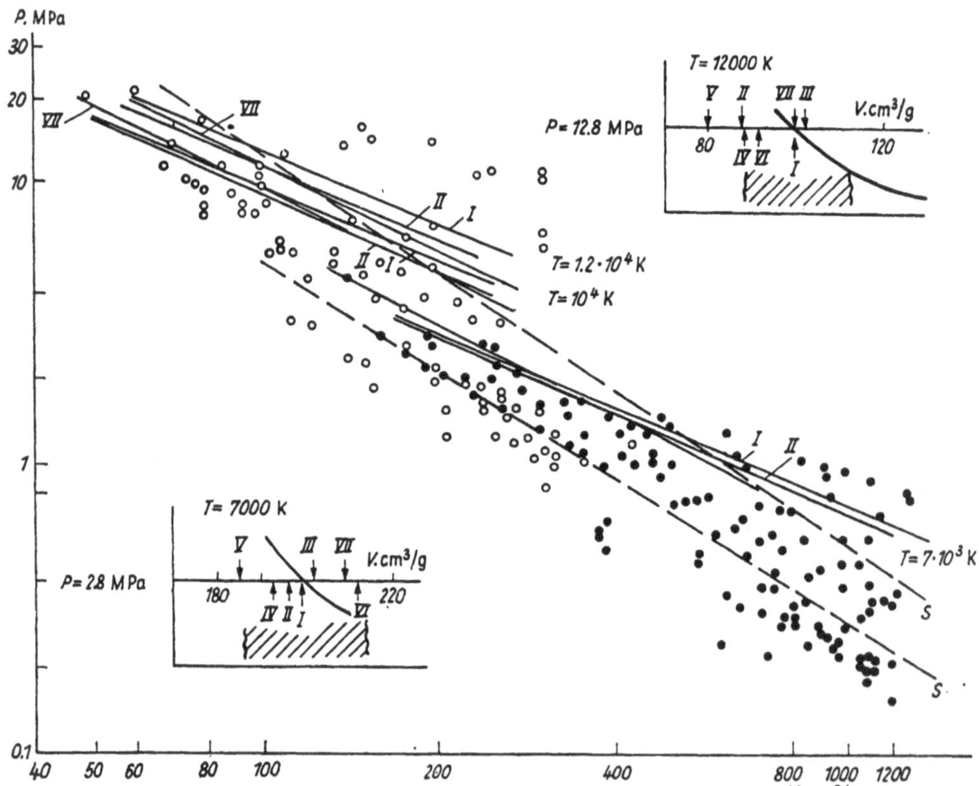

Fig. 1.2. Results of shock compression of cesium plasma [21, 22], T — isotherms, S — isentropes, ○ — incident, ● — reflected shocks

where E_n represents the energy levels and g_n its weight factor. The Coulombic part of the free energy is in the lowest approximation given by the Debye expression for the canonical ensemble which is given by

$$\Delta F_{\mathrm{C}} = -k_{\mathrm{B}} T V \varkappa^3 / 12\pi \tag{1.4}$$

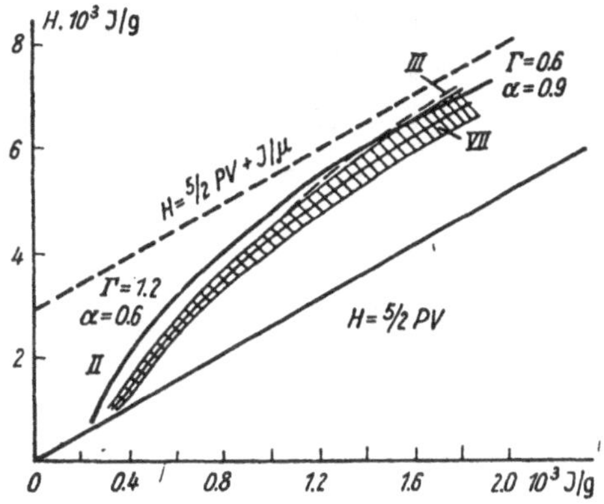

Fig. 1.3. Isochore of cesium plasma $V = 200$ cm³/g. The error corridor is shaded

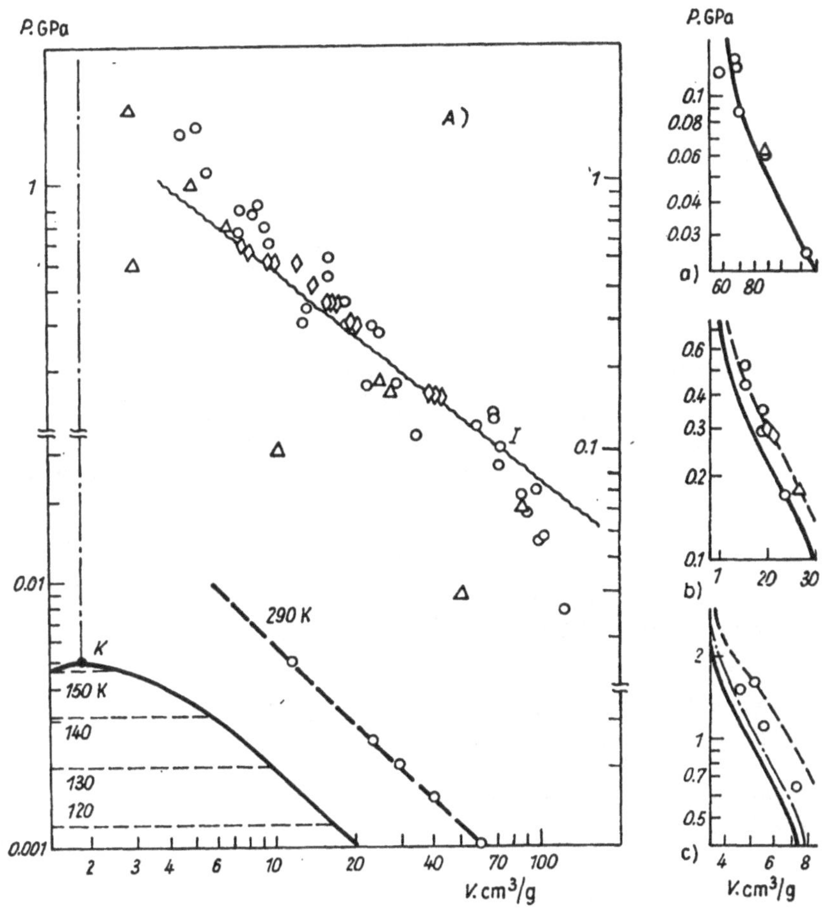

Fig. 1.4.A)

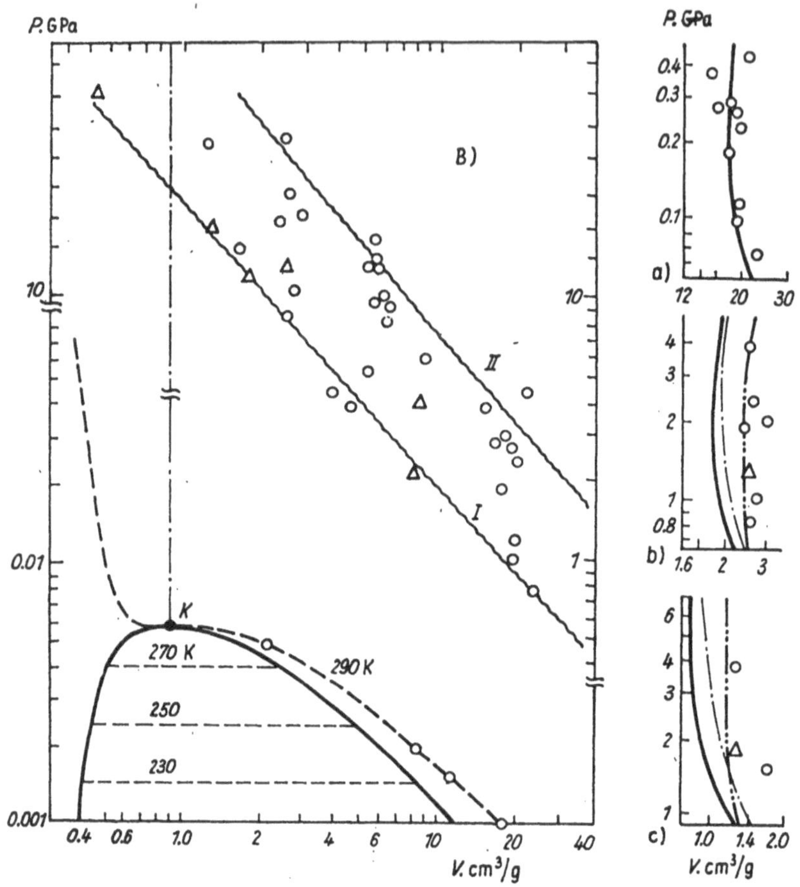

Fig. 1.4.B)

Fig. 1.4. Phase diagrams of argon (A)) and xenon (B)). The boundaries of the two-phase region are marked, the critical point is K. Dashed-isotherms of the initial states, dash-dot — the isochore $V = V_{cr}$, lines-boundary of single I ($\chi_{Ar^+} = \chi_{Ar}$, $\chi_{Ar} = n_{Ar}/n_{tot}$) and double II ($\chi_{Xe^{++}} = \chi_{Xe^+}$) ionization. Points — shock wave experiments. On the right is shown a comparision of the experimental shock adiabates with the results of the model calculations. Solid curves — allowance for the interaction of the charged particles in the ring approximation [3, 4, 46], dash-dot — additional allowance for the interaction of the atoms in the approximation of the second virial coefficient [46a, 69], dashed lines — calculation using the "confined" atom model (1.3), (1.6) up to (1.8) double dash-dot — allowance for the interaction of the charged particles in the model (1.10), (1.11) with pseudopotential (1.9). On A): initial pressures: a) $P_0 = 0.1$ MPa, b) $P_0 = 0.5$ MPa, c) $P_0 = 2$ MPa; on B): a) $P_0 = 0.1$ MPa, b) $P_0 = 1$ MPa, c) $P_0 = 2$ MPa

($\varkappa$ — reciprocal Debye radius). There exist now many improvements of this expression, among them also the ring approach in the grand canonical ensemble given below [1, 3].

In all the thermodynamic registrations performed up to now [20, 21, 25, 28, 29] one can notice a tendency toward decreasing of the measured (index I on figs. 1.2—1.4) internal energy as compared to the traditional [3, 4] plasma calculation by the ring Debye approach in the grand canonical ensemble of statistical mechanics (index II on figs. 1.2, 1.3):

$$\Delta F_{\mathrm{C}} = -\frac{\varkappa^3}{24\pi} - \sum_a n_a \ln\left(1 + \frac{Z_a^2 \tilde{\gamma}(\xi)}{2}\right); \; \varkappa = \frac{2\sqrt{\pi}e}{\sqrt{k_{\mathrm{B}}T}} (\sum_a Z_a^2 \xi_a)^{1/2}; \; \tilde{\gamma} = \frac{e^2}{k_{\mathrm{B}}T}\varkappa,$$

$$Z = \sum g_n(\mathrm{e}^{-\beta E_n} - 1 + \beta E_n), \qquad \beta = (k_{\mathrm{B}}T)^{-1} \tag{1.5}$$

(ξ_a fugacity). In figures 1.2, 1.3 the results of measurements at the cesium plasma equation of state are compared with a number of other plasma approximations, viz. the model of an ideal plasma $\Delta F = 0$, $Z = 2$ — III and Q according to (1.5) — IV, the Debye theory in the canonical ensemble according to (1.4) — V [3] a pseudopotential Monte-Carlo model [12] — VI. A relative contribution of other models to the equation of state is analysed in [45]. The analysis of those results show [4, 21, 22, 29] that though the most of the theoretical models agree within the limits of measuring accuracy with the experimental isotherms of a cesium plasma, those models cannot provide a description both of thermal and caloric data. Due to the fact that larger compressions were obtained for a plasma of noble gases, the given effect manifested itself more distinctively here and the theoretical models in this case contradict not only caloric, but thermal measurements as well.

Interpretation of the set of the experimental data has shown [3, 4] that in a strongly compressed plasma a short-range repulsion is realized which can not be described by standard plasma theories [2, 45]. In this case, though by introducing modifications for the interaction, one can achieve agreement in the thermal data, such a procedure leads to a still greater discrepancy in the caloric information. A common description of thermal and caloric data is possible with a considerable modification of the contribution of bound states — the effect which is ignored by most plasma thermodynamic theories [1, 3, 4, 45].

Independently of this conclusion, the experimental data definitely indicate that the theories, which give correction for nonideality not exceeding the corrections of the ring Debye approximation in the grand canonical ensemble, have good extrapolation properties into the strongly nonideal region [4, 45a].

At typical plasma temperatures $T \approx (10 \text{ up to } 60) \cdot 10^3$ K, the main type of short-range interaction is repulsion, for which an estimation was given within the framework of the second virial coefficient [46] using the Lennard-Jones potential with parameters chosen on the base of experiments at low temperatures when an overwhelming number of molecules is in the energy ground state. Such a procedure, making the agreement with experiment much better (fig. 1.4) at low pressures, does not eliminate a systematic difference at high compressions, pointing to the presence of an additional repulsion mechanism in the system. The fact is that at high temperatures and pressures typical of the plasma experiments discussed here, a larger part of atoms and ions is in excited states, the parameters of short-range repulsion of which exceed the corresponding characteristics in the ground state.

Within the plasma high density region considered, the average interparticle distances are comparable with typical sizes of atoms and ions, which together with strong Coulomb interaction of free charges may result in considerable distortion of discrete energy levels. To describe this effect, a model not traditional for plasma physics has been developed [29, 47] where the influence of the medium on the bounded electrons was described by the effective potential

$$U(r) = \begin{cases} -Ze^2/r & r < r_c \\ \infty & r > r_c \end{cases} \tag{1.6}$$

In searching for eigenvalues of energy and wave functions, a version of the Hartree-Fock method was used [47], where the system of integrodifferential equations for the radial part of the wave function $f_{nl}(r)$ was solved numerically (use is made of atomic units $\hbar = 1$, $m_e = 1$, $e = 1$):

$$\left[\frac{d^2}{dr^2} + 2V_{nl}(r) - \frac{l(l+1)}{r^2} - \varepsilon_{nl}\right] f_{nl}(r) = \int_0^{r_c} G_{nl}(r, r') f_{nl}(r') \, dr + \sum_{n' \neq n} \varepsilon_{nl,n'l} f_{n'l}(r) \tag{1.7}$$

where $V_{nl}(r)$ is a self-consistent potential involving interaction of electrons with the nucleus and with each other. The integral in the right-hand part of (1.7) is a non-local part of the potential or an exchange term in the calculation of which (approximation of the "complete exchange") no additional simplifications are made. $\varepsilon_{nl,n'l}$ and ε_{nl} are non-diagonal factors and eigenvalues determined from the conditions

$$f_{nl}(0) = 0; \qquad f_{nl}(r_c) = 0;$$
$$\int f_{nl}(r) \cdot f_{n'l}(r) \, dr = \delta_{nn'}.$$

The system (1.7) was solved on the electronic computer at various values of r_c, which defines a discrete spectrum of the coupled plasma (fig. 1.5). The equilibrium value of the parameter r_c was obtained from the condition of the free energy minimum (1.3) $\frac{\partial F}{\partial r_c} = 0$ depending on r_c through atomic and ionic partition functions as well as the correction to the free energy which takes into account the interaction of hard spheres

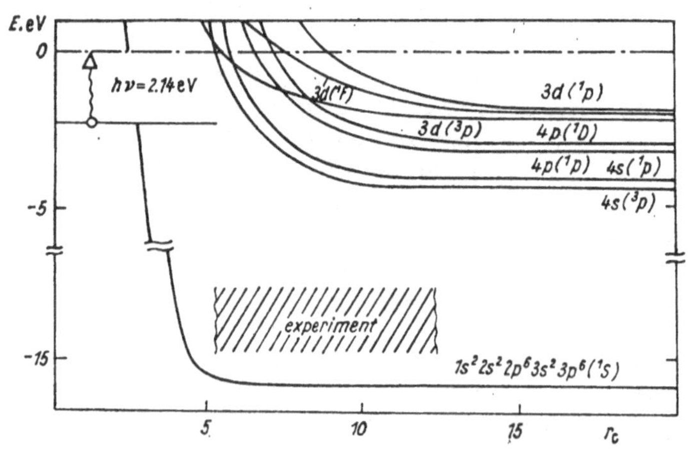

Fig. 1.5. Quantum-mechanical calculation of the energy spectrum of compressed argon in the framework of the "confined" atom model (1.3), (1.6) up to (1.8). Experimental region [25, 28, 29, 31] is shaded, $h\nu = 2.14$ eV — the energy of the light quants, registered in experiment [31]; r_c in atomic units ($\hbar = 1$, $m_e = 1$, $e = 1$)

in the form [48]

$$\Delta F_{HS} = Nk_BT\frac{4-3y}{(y-1)^2}y\,; \qquad y = \frac{4\pi}{3}r_c^3N \tag{1.8}$$

approximating the results of numerical calculations by the molecular dynamics method.

This thermodynamically enclosed model is in fact the synthesis of purely plasma ideas on ionization equilibrium and a model of hard spheres taken from the theory of simple fluids. In contrast to cell models of solids [3], the above approximation is made within the boundaries of a quasichemical way of description with an explicit account of translational degrees of freedom of separate particles and distinctive division of electrons into free and bound. From the correlations of this model (index VII) with experiments (figs. 1.2–1.4) one can see that the model of a "confined" atom correctly reflects the experimentally revealed tendency towards the decrease of plasma compressibility, overestimating, in some way, the repulsion effect at near-critical densities, where, as in the theory of fluids [3], the model of "soft" spheres is more adequate.

A characteristic property of the quasichemical method of description for a strongly non-ideal plasma is the condition of dividing charges into free and bound. Therefore different ways of separating the effects, treated as distortion of the excited states contribution, may be considered as an exhibition of a quantum nature of the electron-ion interaction at short distances. An approximate account of this effect in the chemical plasma model is realized by introduction of an effective pair electron-ion pseudopotential [1, 12], and appropriate corrections for the interaction under the conditions of strong compression are no longer a function of only Coulomb nonideal parameters, but are defined by a specific electronic structure of a particular chemical element.

Under the conditions of strong nonideality, an exact form of pseudopotential is undetermined and can be constructed in a semiempirical manner, as in the case of the one-component plasma [50]

$$\Phi_{ei}^* = -\frac{Z_ie^2}{r}(1-\mathrm{e}^{-r/\varkappa'})\,; \qquad \Phi_{ee}^* = \frac{e^2}{r}\,; \qquad \Phi_{ij}^* = \frac{Z_iZ_je^2}{r}\,. \tag{1.9}$$

$\varkappa'$ is a free parameter. Following [50], a binary correlation function can be presented as

$$F_{\alpha\beta}(r) = 1 \pm A\,\frac{\mathrm{e}^{-qr}-\mathrm{e}^{pr}}{r} \equiv 1 \pm \Psi_0\,\mathrm{e}^{-\nu r}\frac{\operatorname{sh}\omega r}{\omega r}\,, \tag{1.10}$$

where in order to find free parameters A, p, q (or Ψ_0, ν, ω), the conditions of screening [50, 51] and an approximate relation between the amplitude of a screening cloud Ψ_0 and the depth of the pseudopotential are used in this model:

$$n\int[F_+(r)-F_-(r)]\,\mathrm{d}V = 1\,,$$

$$n\int[F_+(r)-F_-(r)]\frac{r^2}{r_\mathrm{D}^2}\,\mathrm{d}V = 3\,,$$

$$-\Psi_0 \cong -\ln F_+ \cong \beta\,[\Phi_{ei}^*(0)-\Delta\mu_e-\Delta\mu_i]\,. \tag{1.11}$$

To choose the main parameter of the model, the depth of the pseudopotential, use is made of the experimental data on shock-wave compression of cesium plasmas [21, 22], which show that the best description of this experiment can be obtained on putting $\Phi_{ei}^*(0)$ equal to the value of energy which divides particles into free and bound $-k_B T$. Comparison of the model (1.9) up to (1.11) with the results of shock compression of noble gases points to a reasonable description of this experimental information.

Thus, two independent groups of experiments on plasmas of cesium and noble gases show that within the boundaries of the model under consideration, the efficient depth of the pseudopotential must be significantly less than the value recommended in [12], calculated with the use of wave functions of an isolated hydrogen atom.

An exceedingly interesting region for applying plasma representations is the hypermegabar pressure range realized by compression of solids by shock waves of extreme intensities. According to the ideas of [3, 42], plasma properties are drastically simplified at extremely high pressures and densities, when internal electron shells of atoms and ions turn out to be "diluted" and a quasiuniform distribution of the electron density is realized inside Wigner-Seitz elementary cells. In this case a quasiclassical approach to the self-consistent field method is valid [42], when the description is carried out in terms of the average density of the electron plasma component, rather than by a quantummechanical language of wave functions and eigenvalues. These concepts form the base of the Thomas-Fermi model [42] the region of applicability of which is defined by the smallness of exchange, correlation and shell effects, which leads to a general estimation of the lower boundary for the applicability of this model $P \gg P_b \approx \frac{e^2}{a_0^4}$ ($a_0 = \frac{\hbar^2}{me^2}$ is the Bohr electron radius), corresponding to extremely high pressures $P \gg 30$ GPa. The first extrapolation extimates [37] of limits of the applicability of the quasiclassical description, based on the results of fixation of relative metal compressibility, have given the values of boundary pressures of 15 up to 20 TPa. Retreatment of these data performed thereupon in [40] by the use of a more specific version (where quantum and exchange corrections are taken into account) of a shock adiabate of lead has permitted the authors [40] to make a conclusion on the applicability of quantum statistical models for a cold ($T = 0$ K) substance starting from pressures of ≈ 30 TPa, and for $T \approx 10$ eV — from ≈ 5 TPa.

These conclusions, however, are inconsistent with later experiments [41, 52] where strong deviations from the Thomas-Fermi model were recorded over the pressure range of 1 up to 16 TPa. These differences are due to so-called oscillation effects [42, 54] which are not taken into consideration by the simplest [53] models, they depend on the discrete energy spectrum and lead to sharp inmonotonies of thermodynamic functions (and even to phase transitions [42]) at pressures $P > 30$ TPa. Oscillations in the region of superhigh pressures, which are due to a shell structure also result from more accurate quantum mechanical calculations by the method of augmented plane waves [55] and different versions of the Hartree-Fock-Slater method [56, 57]. Therefore the problem of the development of methods for the theoretical description of a superdense plasma, and particularly, the problem of the actual limit of the applicability of quasiclassical methods remains unsolved to a great extent and attracts now an increased attention.

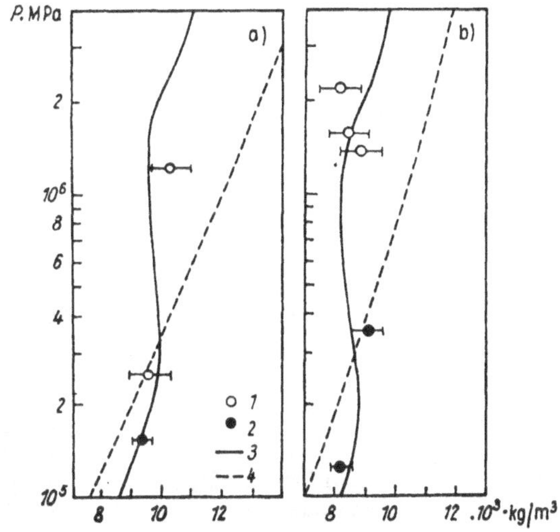

Fig. 1.6. Compression of porous copper by intense shock waves [39, 60]. Solid curves — the Hugeniots, calculated by the plasma chemical model, dashed — Thomas-Fermi theory with the quantum and exchange corrections [53]

Figure 1.6 shows the results of experiments on compression of porous copper by intense shock waves [39] which make it possible to obtain extremely high concentrations of thermal energy ($\varepsilon \approx 0.75$ MJ/cm³) of a superdense ($n_e \approx 2 \cdot 10^{23}$ cm^{-3}) plasma of high (P up to 2.09 TPa) pressure. At maximum temperatures attained, $T \approx 2.5 \cdot 10^5$ K, degeneration of electrons is removed, $n_e\lambda_e^3 \approx 0.7$, the ionization rate factor of such a plasma reaches five, while short-range and Coulomb interaction ($\Gamma \approx 2$) is strong. To describe these data, a "chemical" plasma model (1.3) was used in [58] with account for effects of degeneration and Coulomb nonideality following (1.4) and description of the contribution of short-range action by the model of hard spheres (1.8). To estimate the packing of these spheres y use was made in (1.8) of the results of quantum mechanical calculations of compressed ions by (1.7) with a potential (1.6), fig. 1.7. It can be seen that a "chemical" plasma model reasonably describes the plasma properties at high temperatures over the TPa pressure range, where detectable discrepancies are observed between the experiment and quasiclassical theories.

The experiments on adiabatic expansions of metals precompressed by intense shock waves [30, 71—74] give a unique possibility of studying the behaviour of a superdense plasma, starting from a strongly compressed condensed state with a typical pressure of several megabars, where ions are greatly disordered and electrons are degenerated up to a quasinonideal Boltzmann plasma and expanded metallic vapour, including curves of high-temperature melting, boiling and neighbourhood of a critical point. Such experiments permit to cover an exceedingly wide range of parameters (four orders in pressure and two in density, fig. 1.1) where as the plasma expands, various slightly investigated physical processes occur in it, viz. degeneration of electrons is removed and their energetic spectrum is radically reconstructed, recombination of the plasma and disordering of the ion component take place, transition "metal-dielectric" is realized in the disordered system, a plasma with strong interparticle interaction, complicated as to the structure, appears. Presence of strong interpartical interactions

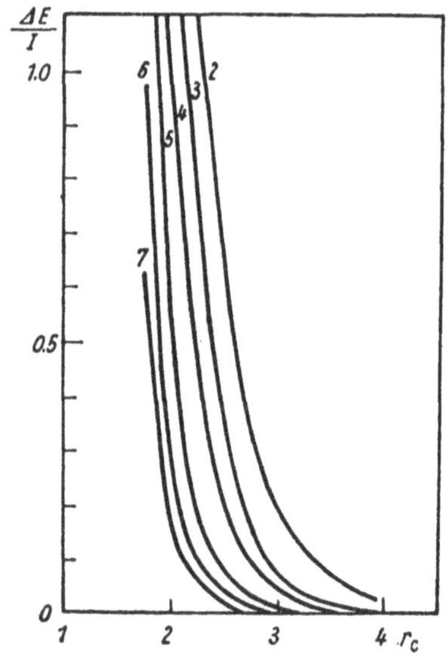

Fig. 1.7. Relative variation of the ground state energy of copper ions (ionization degrees are shown near the curves) by the model (1.3), (1.6) up to (1.8). I — the ionization potentials; r_c in atomic units

hinders a successive theoretical description of plasma properties over such a wide range of parameters, putting in the prime front the heuristic models [2, 3, 67, 75, 76] which describe separate effects in the isolated parts of the phase diagram, as well as semi-empirical models, which generalize in a unified way the experimental and theoretical information over the wide range of parameters [3].

The main qualitative result of most heuristic models is an indication of a possible loss of thermodynamic stability and appearence of new exotic phases in a nonideal plasma, which should considerably distort a usual form of the phase diagram of a matter. The data of statical measurements now available [14—17], as well as the results on shock-wave and isentropic compression [18—41] and expansion of metals [30, 61—64] show the absence of detectable thermodynamic and gasodynamic anomalies which might be interpreted as plasma phase transformations. We shall emphasize in this connection that just in the parameter range uder investigation (fig. 1.1) the phase anomalies discussed in [2, 3, 65—67] are most likely, since increase in temperature and decrease in density of a Boltzmann plasma, as well as increase in pressure of a degenerated plasma, lead to a relative decrease in the nonideality effects.

The experimental data [18, 19, 30, 60—64] are the base for constructing of semi-empirical equations of state [30, 68] which describe in the unified analytical form the whole set of statical and dynamical data available now in solid, liquid and plasma phases, which reproduce the effects of melting, evaporation, thermal ionization, and pressure ionization and take into account the removal of degeneration of the electron gas and at superhigh pressures and temperatures have a correct asymptotics to classical (the theory by Debye-Hückel) and quasiclassical (the theory by Thomas-Fermi) theoretical models (figs. 1.8, 1.9). In the semiempirical equations of state [30] the representation of the free energy is used

$$F(V, T) = E_x(V) + F_e(V, T) + F_r(V, T)\,, \tag{1.12}$$

where $E_x(V)$ determines the part of interaction at $T = 0$ K, and F_e and F_r are the thermal contributions of electrons and ions.

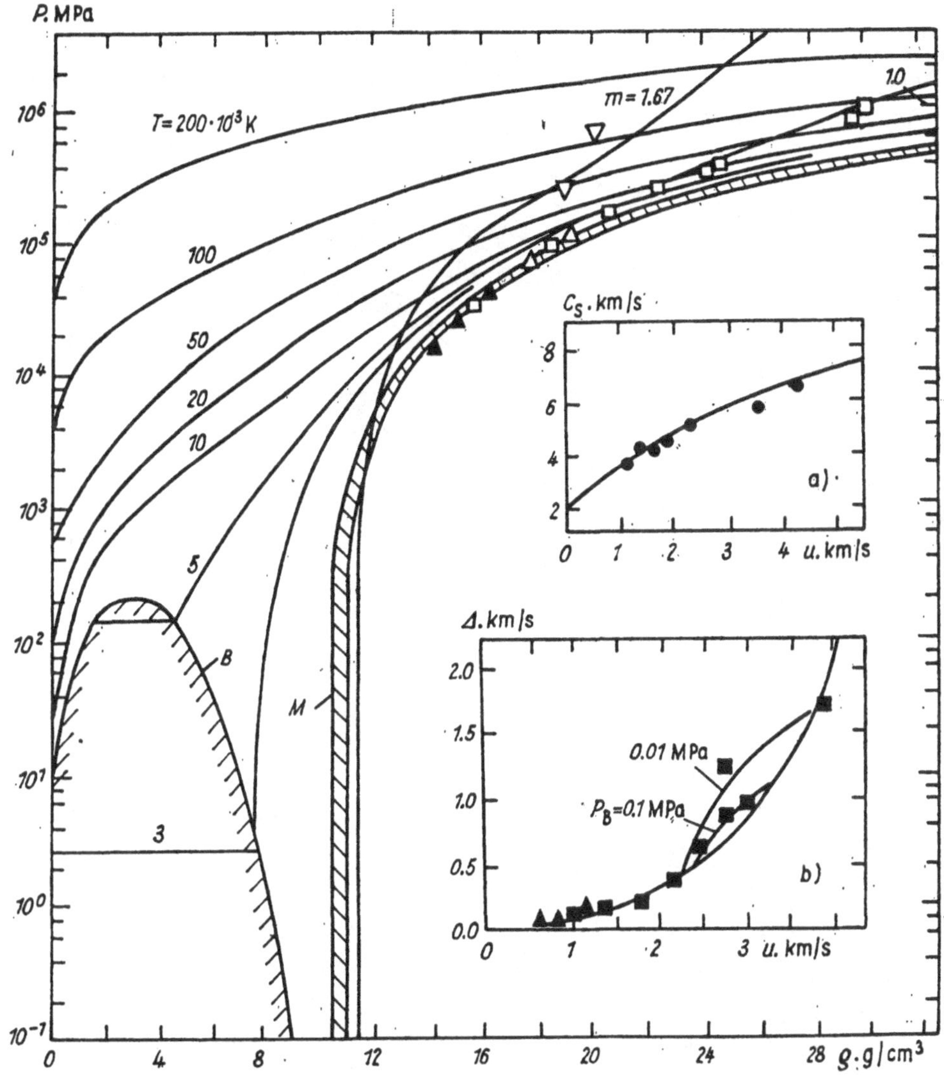

Fig. 1.8. Phase diagram of lead. A semiempirical equation of state was used (1.13) up to (1.15) [30]. H_m — Hugeniots by the different porosity $m = \varrho_0/\varrho_{00}$, T — isotherms, M, B — phase boundaries of melting and boiling, two-phase regions are shaded. a) — sound velocity (C_s) on the Hugeniot. b) — deviation from the "doubling velocity" rule

"Cold" pressure $P_x(V)$ was chosen as a power series. To determine its coefficients, use was made of experimentally measured metal properties in normal conditions, at $P \approx 0.1$ TPa and the data of a quasiclassic theory at high compressions $\sigma = \varrho/\varrho_0 \approx 2$ up to 200. The contribution of thermally excited electrons was described by the term

$$F_e(V, T) = -C_e(\sigma, T)\, T \ln\left[1 + \frac{B_e(T)\, T}{2C_{ei}}\, e^{-\gamma_e(\sigma, T)}\right], \tag{1.13}$$

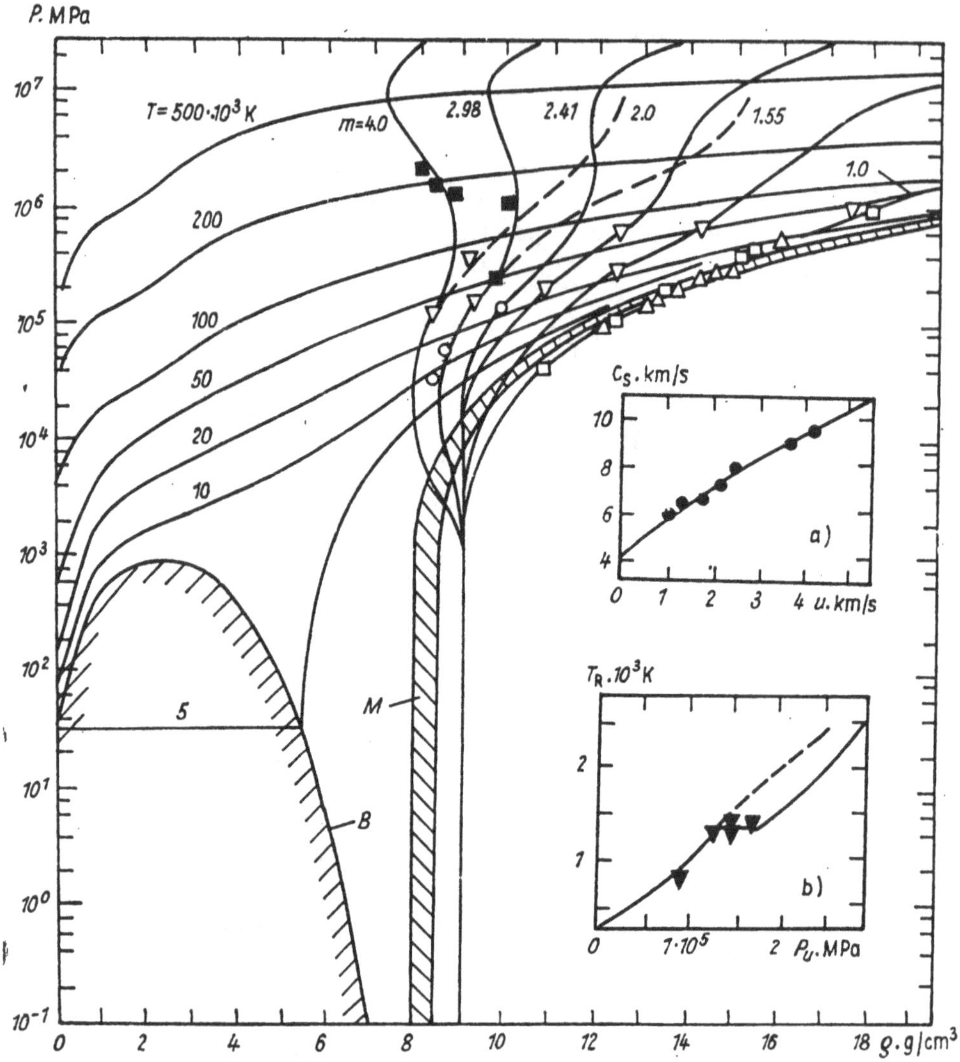

Fig. 1.9. Phase diagram of copper. The notations see fig. 1.8. Dashed curves — the Hugeniots from [60]; T_R — temperature in the isentropic wave

where the selection of generalized analogues of the electron thermal capacity coefficient $B_e(T)$, the Grüneisen electron coefficient γ_e and thermal capacity of the electron gas C_e provide fulfilment of the asymptotics of the degenerate electron gas in the solid region

$$F_e(V, T) = -\tfrac{1}{2}\beta_0 T^2 \sigma^{-\gamma_e}$$

and of asymptotics of Boltzmann plasma, at high temperatures

$$F_e(V, T) = C_{ei} T \ln (\sigma^{2/3} T) .$$

In selecting a functional form of B_e, γ_e, C_e, account was taken of the results of quantum-mechanical calculations of the energy spectrum of a solid by the cell model [69], the ionization effects on heating and compression of a substance, as well as corresponding experimental data.

Contribution of ions to thermal dynamics of a solid phase was chosen in the form

$$F_r^{(s)}(V, T) = 3RT \ln\left(\frac{\theta\sigma}{T}\right), \tag{1.14}$$

where the functional dependence of the typical oscillation frequency provided a quantitative description of the experiment and a correct asymptotics for the Grüneisen coefficient $\gamma = \dfrac{\mathrm{d}\ln\theta}{\mathrm{d}\ln\sigma} \to 2/3$ at $\sigma \to 0$ and $\sigma \to \infty$. For a disordered ion system, it was assumed that

$$F_r^{(l)} = \frac{3RT}{2}\left[1 + \frac{\sigma T_T}{(\sigma + \sigma_T)(T + T_T)}\right] \ln \frac{\theta_a(\sigma, T)}{T} + F_m(V, T)\,, \tag{1.15}$$

where $F_m(V, T)$ also describes the details of the melting process while the constants σ_T, T_T and the functions $\theta_a(\sigma, T)$ are introduced to describe the experimental data and provide the correct gas asymptotics on heating and expansion of plasmas. A set of constants and functions in the equation of state (1.12) up to (1.15) was chosen from the condition for the best description of experimental and theoretical informations available.

Experimental and theoretical information available is given in [30] and permits to carry out the calculations of physical processes [9] and gas-dynamic [70] phenomena in plasma over the whole region of the phase diagram (figs. 1.8, 1.9).

1.4. Electrical Conductivity of a High Pressure Plasma

Electric conductivity is one of the most indicative and easily observed characteristics of the plasma that determines its dissipative heating and the interaction with the electromagnetic field. The major part of the experimental data obtained now are presented in figure 1.10, where the results of the measurements are treated by a simple thermodynamic model, in which e.g. a ring Debye approximation in a grand canonical ensemble of statistical mechanics is used [3, 4]. The non-ideality parameters achieved in the first experiments [67] did not exceed $l/r_D \approx 0.2$ up to 0.5 and the measured values of conductivity are in agreement [4, 79] with the calculations based on the kinetic equations or on the methods of time-dependent correlation functions [4, 13]. With increasing non-ideality there arise considerable complexities both with the substantiation of the initial kinetic equations and with mathematical methods of their solutions. In particular, in view of the strong Coulomb interaction in a dense plasma it is impossible to separate unambiguously the characteristic times of the elementary processes in plasma, and in general, the time evolution of the system under the influence of an external field is no longer described by a Markov process [13]. A special problem is the correct calculation of the bound states in a partly ionized plasma [80] (see also section 3.3.7. and chapter 4.) because of the difficulties in formulating the adequate kinetic equations and of the lack of the transport cross sections for the semi-empirical methods to be used. The calculation of the kinetic properties therefore, are of an extrapolative character for the nonideal plasma and are needed to be checked up experimentally.

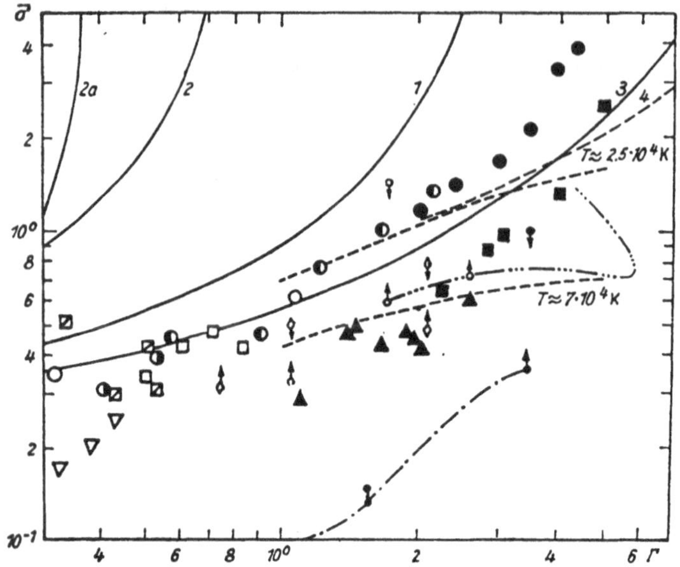

Fig. 1.10. Dependence of the Coulomb component of the electric conductivity $\bar{\sigma} = (1 \pm z)^{1/2} \cdot \gamma^*(z) \dfrac{\sigma}{\omega_P} \Gamma$ $\left(\omega_P = \sqrt{\dfrac{4\pi n e^2}{m_e}}\right.$; $\gamma^*(z)$ — electron-electron correction term$\Big)$ on the non-ideality parameter. Theories: 1 — [72], 2 — [81], 2a — [73], 3 — modified Ziman approach (1.15) up to (1.20), dashed-pseudopotential (1.21) calculations for different temperatures, 4 — [74, 75]. Experiment: ◐, ●, ■, ▲ — argon, xenon [26, 27, 32] and ◇ cesium [76] shock compression, other points were obtained from arcs and discharges [4, 27], —·—·—·— "slow" electrical explosion of cesium [77], —···—···—···— — isobaric heating of cesium [14, 78]

Figure 1.10 shows a number of such extrapolations. Curve 1 is obtained by a numerical integration of the Fokker-Planck kinetic equation (the Spitzer theory [72]). By applying the diagram technique (taking into account ring and ladder fragments) to the equations of motion for the time-dependent Green's functions in [81] a kinetic equation was obtained in which the effective two particle interaction in the first order of l/r_D has been considered (curve 2). The collective effects in the Coulomb interaction were taken into account in [73] on the basis of the Fokker-Planck equation for the one-particle distribution where the collision integral contains all moments of the distribution function (curve 2a).

The methods of shock compression enables to carry out the measurements [23, 26, 27, 32, 76] of the static conductivity ($\omega \ll \omega_P = [4\pi n_e e^2/m_e]^{1/2}$) of the plasma in a wide range of non-ideality parameters from $l/r_D \approx 0.3$, where differences between the asymptotic theories are not large and there is a considerable amount of experimental data [67], up to the region of extremely high parameters of the Coulomb interaction $l/r_D \approx$ ≈ 5 up to 10, where most of the theoretical approximations are in disagreement and where the results of the experiments give a base for building the physical models of electron transport in a dense disordered substance [82—84]. It is essential that this range of parameters is studied for plasmas of different elements (cesium [76, 35], air [23], neon, argon and xenon [26, 27, 32]) for which together with the electrophysical properties some thermodynamic and gasdynamic measurements were made (chap. 1.3.), permitting to explore more definitely the physical conditions in which this plasma exists.

The maximal parameters of the plasma (P up to ≈ 11 GPa, $n_e \approx 10^{22}$ cm^{-3}, $T \approx 10^4$ up to 10^5 K), were obtained in experiments with xenon (fig. 1.11) where one

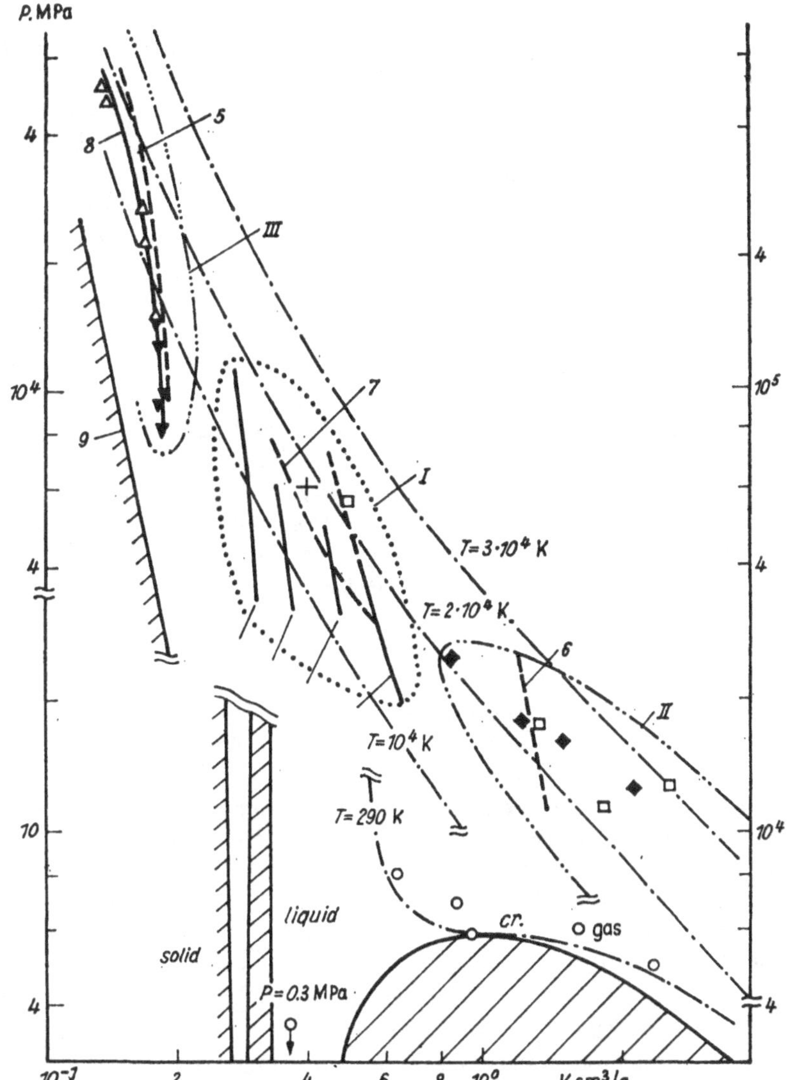

Fig. 1.11. Phase diagram of xenon. Calculated shock adiabats: 1 — $P_0 = 5$ MPa, 2 — $P_0 = 6$ MPa, 3 — $P_0 = 7$ MPa, 4 — $P_0 = 8$ MPa, 5 — $P_0 = 0.3$ MPa, $T_0 = 165$ K, 6 — $P_0 = 2$ MPa, 7 — secondary compression shock adiabat — [32], 8 — bound model calculations [85], 9 — isotherm $T = 0$ K. Thermodynamic measurement data: 10 (+) — "reflection" method [32], 11 (□) — [28], 12 (△) — [9, 85], measurement of electrical conductivity: 13 (◆) — [26], 14 (▼) — [9, 85]

succeeded in covering supercritical densities by the experiments and intracing in a continuous manner the behaviour of the electric conductivity from the state of low density II being described by the dilute plasma models [79—84] up to the region of solid state densities produced by a shock compression of liquid xenon where thermodynamics is described by the band theories of solids [9, 85] and the electric conductivity is in agreement with the semiconductor models [74, 75, 85].

The experiments performed now (fig. 1.10) point to the lowering of the measured values of the electric conductivity in comparison with the Spitzer theory [72] while some theories [73, 81] predicted conductivity rising with increasing the Coulomb non-ideality. The quantitative divergence between different groups of the experimental points (fig. 1.10) are caused both by the peculiarities of the behaviour of the low temperature plasma as well as by the actual disagreements of the initial data and the difficulties of extracting the Coulomb component of the electrical conductivity in a weakly ionized plasma. The latter circumstance is mostly characteristic of the alkaline metal plasmas [14, 76, 78] where the contribution of neutrals is especially high due to large cross sections of the electrons' scattering on atoms in the ground and excited states [86] and due to the insufficient degree of ionization of the plasma studied. This facts lead to the uncertainty of extracting the Coulomb component from the data of the low temperature measurements [14, 76, 78] reaching hundreds of percents — see arrows in figure 1.10.

The use of the technique of high explosive shock waves made it possible [26, 27, 32] to obtain a plasma with a high ionization degree for which the problem of extracting the Coulomb components is absent, simplifying essentially the interpretation of the electrophysical results, reducing the uncertainty in the thermodynamic composition and cross sections of the non-Coulomb elementary processes. The data obtained by the dynamic methods can be conventionally separated into "low temperature" with $T \leqq 2 \cdot 10^4$ K [26, 32] and "high-temperature" ones with $T \geqq 2 \cdot 10^4$ K [27]. The low temperature points correspond here to extremely high densities $\varrho \approx 4$ g/cm^3 which are close to the verge of degeneration of the electronic component of the plasma. Here a strong Coulomb interaction is realized and it arises a nonideal plasma with a considerable single and double ionization in the high temperature region. It can be seen (fig. 1.10) that the results obtained with different gases (Ar, Xe, Ne, air) are in agreement and permit us to trace the influence of the Coulomb interaction on the electric conductivity in a wide and continuously varying range of parameters of the non-ideality $l/r_D \approx 0.3$ up to 10.

The experimentally recorded reduction of the nonideal plasma conductivity in comparison with the data of the asymptotic theories can be due to [87] the increase of the cross sections of the Coulomb interaction as compared to those of theoretical estimates which predict at $l/r_D \geqq 1$ anomalously small ($\leqq 10^{-8}$ cm) screening radii. There seems [26, 87] to occur in the nonideal plasma a renormalization of interaction and the radius of the charge correlation decreases in compressing the plasma to values less than the average interparticle distance which is similar to the case in metals and doped semi-conductors [74, 75]. The corresponding model [74, 75] used to describe the electron scattering at the impurity centres of the semiconductors are denoted by the index 4 in figure 1.10.

To describe the electric conductivity of the nonideal plasma on the basis of the experimental data available a theoretical model has been worked out [84] in which the ion correlations were described in the Ziman approximation [88] used in the theory of liquid metals and semiconductors and the charge scattering was calculated in the Born approximation with the screened Coulomb potential. In the approximation of the linear response for the electric conductivity it holds true ([88] and chaps. 3. and 5.)

$$\sigma = \frac{4e^2}{3\sqrt{\pi}\, m_e} \xi_e \int_0^\infty \frac{x^{3/2} e^{-x} dx}{\nu_{ea}(x) + \sum \tilde{\gamma}(\xi_i)\, \nu_{ei}(x)}, \tag{1.15}$$

where ξ_e — activity connected in the framework of the ring Debye approximation with the electron concentration through the correlation [4]:

$$\xi_e = n_e \left(1 + \frac{\tilde{\gamma}(\xi_e)}{2}\right)^{-1},$$

where γ is the root of the equation:

$$\tilde{\gamma}^2 = \left(\frac{e^2}{k_B T}\right)^3 4\pi \left\{ \sum_k \left[n_k Z_k^2 \Big/ \left(1 + Z_k^2 \frac{\tilde{\gamma}}{2}\right)\right]\right\}. \tag{1.16}$$

The frequency of the electron collision with ions is calculated in the Ziman approximation [88] (see also chaps. 3. and 5.):

$$\nu_{ei}(\varepsilon) = \frac{\sqrt{2}}{16\pi n_e m_e^{1/2} \varepsilon^{3/2}} \int_0^{\frac{4\varepsilon}{Z_j e^2}} |u_j(q)|^2 a_i(q)\, q^3\, dq. \tag{1.17}$$

In the Debye approximation for the binary electron-ion correlation function:

$$g_{ij}(r) = 1 - \frac{Z_i Z_j e^2}{k_B T r} e^{-\varkappa_0 r},$$

$$\varkappa_0^2 = 4\pi e^2 (k_B T)^{-1} \left[\sum Z_i^2 n_i + n_e\right]; \tag{1.18}$$

for the structure factor $a_i(q)$ and formfactor of a pseudopotential $u_j(q)$ we have

$$a_i(\boldsymbol{q}) = 1 + \sum_{n_j} n_j \int e^{i\boldsymbol{q}\boldsymbol{r}} \left[g_{ij}(\boldsymbol{r}) - 1\right] d^3\boldsymbol{r} = 1 - \frac{4\pi e^2 z_i \sum n_k Z_k}{k_B T (q^2 + \varkappa_0^2)}, \tag{1.19}$$

$$u_j(\boldsymbol{q}) = n_j \int e^{i\boldsymbol{q}\boldsymbol{r}} u_j(\boldsymbol{r})\, d^3\boldsymbol{r} = \frac{4\pi Z_j n_j e^2}{q^2 + \varkappa_0^2}. \tag{1.20}$$

The relations obtained have a correct Spitzer asymptotics in the limit $l/r_D \to 0$, possessing acceptable extrapolation properties in the highly nonideal region of the parameters which do not have non-physical discrepanices and depict satisfactorily "low temperature" (with $T < 2 \cdot 10^4$ K) experiments up to the experimentally high values of l/r_D (fig. 1.10).

A characteristic particularity of the experiments with nonideal plasmas is the relatively low temperature $T \leqq 4 \cdot 10^4$ K as these experiments were planned to obtain a considerable Coulomb non-ideality which drops as the temperature increases. Use of cumulative explosive shock tubes and the effects of shock wave reflection from the condensed targets enabled us [27, 32] to advance in the field of extremely high plasma heating $T \approx 10^5$ K, where xenon plasma is highly ionized ($Z_i \approx 3$), to be in conditions of strong Coulomb interaction $l/r_D = 1$ up to 5. Conforming to the earlier results obtained in the "low temperature" region with the rise of shock wave velocity (or heating of the plasma) the electric conductivity of the plasma increases slowlier than it can be expected on the base of usual [4,72] plasma models of the electron transport (fig. 1.12).

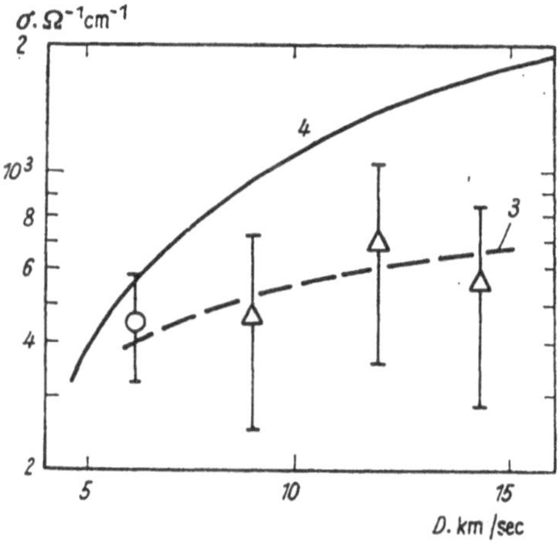

Fig. 1.12. Dependence of electric conductivity of xenon plasma on the shock front velocity D at $P_0 = 0.1$ MPa. Experimental data: 1 (△) — [27], 2 (○) — [26], calculations: 3 — (1.21) up to (1.23), 4 — [72]

This curious behaviour pof the plasma displays most typically in Figure 1.10, witnessing the differences of the Coulomb component of the electric conductivity of the plasma — the dimensionless electric conductivity of the high temperature plasma turns out to be less than that of the low temperature one at the same values of the Coulomb non-ideality. This result is in a qualitative disagreement with the models [89, 90] extremely simplifying the description of close interactions in the plasma by means of the forced restriction of the minimum electron scattering distance by the electron de Broglie wave length and these models predict an opposite dependence of the electric conductivity on the temperature.

The analysis of the electron scattering processes under experimental conditions [27] shows that short range interactions play a noticeable role as the Coulomb scattering amplitude $f_C \approx (e^2/k_B T)$ proves to be comparable with the typical sizes of the xenon ions of about 0,4 nm and the electron during its scattering with the ion can penetrate into the environment of the nucleus occupied by the bound electrons. The potential of a electron-ion interaction at short distances is stronger than the external Coulomb ion potential, resulting in an increase of the electron-ion cross section in comparison with the pure Coulomb one and, hence, this fact gives the relative decreasing of the electric conductivity of the plasma. The evaluation of this effect in the dilute plasma [91] points out that a noticeable contribution arises only at extremely high temperatures while the calculations for the dense single-ionized plasma [92] lower noticeably the values of temperature.

To describe this effect a model [27] was proposed which takes into account the influence of the non-coulombic character of electron scattering at small distances on their mobility in a dense plasma. The conductivity was calculated in the Frost approximation [4] and the transport cross sections used in the calculations were numerically found by the method of "potential" waves. The interaction of the electron with the ion was described by the effective pair potential [91] taking into account the presence of the ion structure:

$$\varphi_{ej} = -\left[\frac{Z_j}{r} + \frac{(Z - Z_j)}{r} e^{-\tilde{\beta} r}\right] e^{-r/r_D}; \quad \tilde{\beta} = \frac{1.8 Z^{4/3}}{(Z - Z_j)} \tag{1.21}$$

where Z, Z_j the charges of the nucleus and of the ion respectively, radius r and the Debye radius r_D are measured here in atomic units (a.u., $\hbar = 1$, $m_e = 1$, $e = 1$). At small radii the potential (1.21) coincides with the Thomas-Fermi potential but at $r \gg \tilde{\beta}^{-1}$ it turns into the Debye potential. This potential was used in numerical solutions of the Schrödinger equation for the radial part of the wave function

$$\frac{d^2 f_{kl}}{dr^2} - \left[2\varphi_{ej} + \frac{l(l+1)}{r^2} - k^2\right] f_{kl} = 0 \tag{1.22}$$

at different energies $k^2/2$ (k — the value of the electron momentum) and the orbital momenta l. The solution to the equation (1.22) was patterned to radii r, where the scattering potential $\varphi_{ej}(r)$ is much less than the centrifugal one $l(l+1)/r^2$ and where the solution (1.22) has the form [93]:

$$f_{kl} = rA[\cos(\delta_l)\, j_l(kr) - \sin(\delta_l)\, n_l(kr)]$$

with A — the normalization constant, $j_l(kr)$ and $n_l(kr)$ — the spherical Bessel and Neumann functions [93]. The solution is valid up to r_0 where $f_{kl}(r_0) = 0$, in this case the scattering phase is determined by the expression

$$\operatorname{tg} \delta_l = \frac{j_l(kr_0)}{n_l(kr_0)},$$

the spherical function is calculated by the recurrence formula

$$n_{l+1}(kr_0) = \frac{(2l-1)}{kr_0} n_l(kr_0) - n_{l-1}(kr_0).$$

The function was found by the method of changing l at fixed r_0. The transport cross sections of the electron-ion scattering was calculated using the scattering phase shifts (used up to 50 phases) in correspondence with [94]:

$$Q_{ej} = \frac{4\pi}{k^2} \left| \sum_{l=0} \{(2l+1)\sin^2\delta_l - (2l+2)\exp[i(\delta_{l+1} - \delta_l)]\sin\delta_l \sin\delta_{l+1}\} \right|. \tag{1.23}$$

The results of these calculations for the potential (1.21) and for the Debye one ($r_D = 7.75$ a.u., $Z_j = 1$) are given in figure 1.13 where, in addition to them, a number of traditionally used approximations are presented. It is seen that in the region of energies of electrons 0.7 a.u. making a major contribution to the electric conductivity of plasma at $T \approx 7 \cdot 10^4$ K the presence of a stronger than Z_j/r interaction increases the electron cross-section about two times, while the use of the Born approximation with the potential (1.21) gives very over-estimated values. One should notice a satisfactory accordance of the exact calculation of (1.23) with the interpolation formula [95]. The results of the electric conductivity calculation for the model used here at $T \approx 2.5 \cdot 10^4$ K

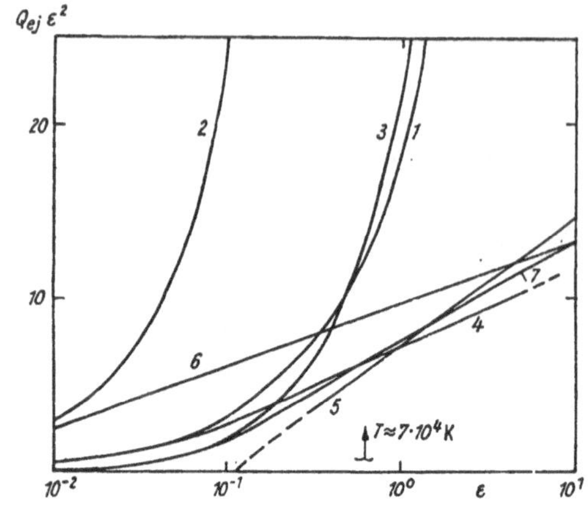

Fig. 1.13. Transport electron-ion cross-section at $r_D \approx 7.75$ atomic units as a function of electron energy. Calculations: for the potential (1.21): 1 — on (1.22)—(1.23), 2 — Born approximation [91], 3 — interpolation [91]; for the Debye potential: 4 — on (1.22)—(1.23), 5 — quasiclassical asymptotics [72], 6 — Born approximation [90], 7 — interpolation [95] for $T \approx 7 \cdot 10^4$ K; atomic units are used

and $T \approx 7 \cdot 10^4$ K show (fig. 1.10) that the model of the "ion core" (1.21) up to (1.23) enables a reasonable description of the effect of "lamination" of the electric conductivity isotherms being displayed in a high temperature plasma.

1.5. Optical Properties of a Highly Compressed Plasma

The solution of many present-day problems of radiation gasdynamics and plasma physica call for detailed informations on the optical properties of a high-density plasma in which the effects of interparticle interaction play an essential role. An investigation of the radiation of such a plasma is also of definite fundamental interest, as it makes it possible to trace the influence of non-ideality on the dynamics and on the energy spectrum of a dense disordered plasma.

The theoretical ideas developed up to now [96] concern dilute plasmas where the elementary processes can be easily separated and the influence of deviations from ideal conditions reduces to a broadening of the spectral lines and to shifts of the photorecombination thresholds. The region of increased densities has become the object of investigations only most recently (see Refs. [23, 97—99]) and the degree to which it has so far been studied is obviously insufficient. The data available here correspond to non-ideality parameters $l/r_D \leqq 0.2$ up to 0.3, $n_e \leqq 10^{18}$ cm^{-3} and are to a great degree of a contradictory character in many respects but indicate nevertheless that the continuum radiation exceeds the calculated value when the plasma density is increased. At the same time the plasma screening [100] and the strongly fluctuating microfields [101] can noticeably distort the near-threshold region of the spectrum, by decreasing the intensity of the recombination radiation of the dense plasma [98, 99, 102]. It should be remarked that a more detailed investigation [116] does not predict a decrease of the intensity in the nearthreshold region.

Of considerable interest is the radiation of the high density air plasma where the effects of interaction between particles play an essential role and new mechanisms of the radiative electron transitions begin to display. Most uncertain here is the problem of the recombinational electron-atom radiation with the formation of negative

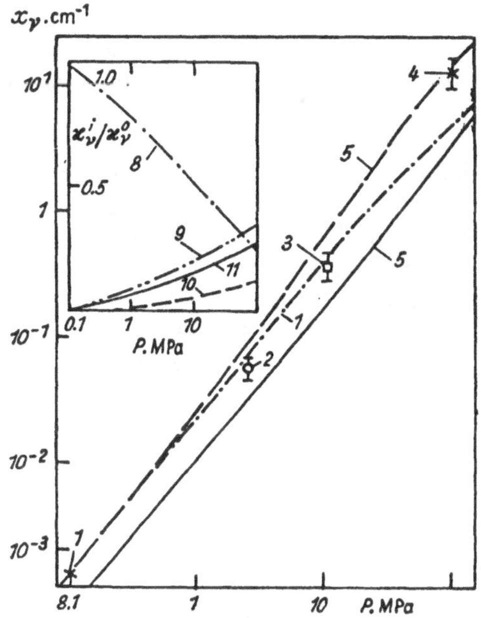

Fig. 1.14. Dependence of air-plasma optical absorption coefficient on the pressure [105]. Experiment: 1 — [107], 2 — [108], 3 — [23], 4 — [105]. Calculations: 5 — [106], 6 — allowance of photodisintegration with cross-section from the experiments at normal pressure, 7 — destruction of the negative ions by microfields is taken into account, 8 — relative component of the electron-ion continuum, 9 — free-free transitions in the fields of neutrals, 10 — molecules, 11 — O^-

nitrogen ions and the account of the influence of the microfields. In the experiments made up till now [23, 103, 104] the maximum pressures of an air plasma did not exceed 150 bars, being not sufficient to treat unambiguously the display of the plasma non-ideality and to single out correctly the cross sections of the photodecay of negative nitrogen ions as well.

The use of explosive dynamic methods [104a, 105] has made it possible to increase the pressure of the plasma by nearly one order of magnitude and to carry out a quantitative analysis of the radiation mechanisms of the many component plasma of air. The results obtained (see fig. 1.14) are compare dwith theoretical calculations [106] taking into account the absorption in the molecular bonds, the photoionization of atoms N and O, free-free electron transitions in the field of ions N^+ and O^+ and in those of neutrals N_2, N and O and the photoionization of O^-. A relative contribution to the absorption coefficient of these mechanisms at $\lambda \approx 490$ nm, $T = 13000$ K is given in dependence on the pressure in figure 1.14a). There can be observed in the experiments a noticeable excess of the estimated absorption coefficients over the calculations, which is beyond the limits of the errors of measurements and uncertainties being brought in by some non-accuracy in calculating the plasma composition and by the information on the cross sections of elementary radiation processes. The largest contribution to the calculated absorption coefficient is made by "free-free" transitions in the fields of neutrals (40—50%), by photoionization of atoms (25%) and photo detachment of O^- (20—30%). The cross sections of the latter two processes are known sufficiently well and were repeatedly checked up experimentally.

The discrepancy observed between the theory and experiments may be recognized by a supposition of the availability in the plasma of unstable negative nitrogen ions $N^-(^3P)$ with the concentration ≈ 2 up to $3 \cdot 10^{17}$ cm^{-3} and with the cross section of photodecay in the visible region of the spectrum $f \approx 1$ up to $3 \cdot 10^{-17}$ cm^2. The appearing decrease of the cross section of photodecay of N^- in comparison with the cross sections (curve 7 in fig. 1.14) found on the basis of measurements at atmosphere pressure is interpreted [105] as a decrease of equilibrium concentration of unstable negative ions due to the influence of the plasma microfields.

The spectrum region near the threshold of ionization has been attracting recently an increased attention as there is predicted here [98] a nonrealization of the part of highly excited states joining the continuous spectrum of the dense plasma. From these stands there were interpreted, in particular, the results of the registration of the mercury discharge spectrum [99]. In the dynamic experiments the near-threshold region of the spectrum was thoroughly studied in a hydrogen plasma on an electric discharge shock tube [23] where at $n_e \approx 8 \cdot 10^{17}$ cm^{-3}, $T \approx 14{,}500$ K no essential anomalies in radiating plasma were observed (fig. 1.15) and the shift of the threshold

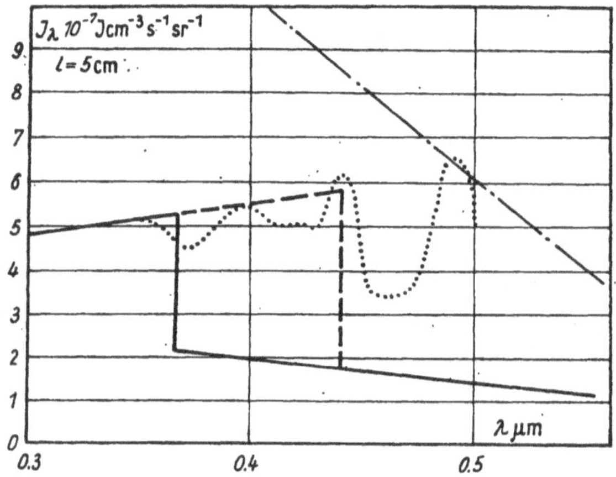

Fig. 1.15. Optical radiation of the hydrogen (30%) and argon (10%) plasma [23]. ——— Calculations where free-free, bound-free transitions in fields of ions and photoattachment of electrons to H^- and the argon recombination continuum are taken into account; — — — ionization edge shift by the Inglis-Teller formula [96]; —·—·— black-body radiation at $14.5 \cdot 10^3$ K; · · · · · · experiments [23]

was described by the Inglis-Teller formula [96], see also [116]. The effects of the plasma non-ideality were recorded more conveniently in experiments with argon [31] where a wide and continuously varying range of parameters from the previously investigated states [97] with $n_e \leqq 10^{18}$ cm^{-3} and $l/r_D \leqq 0.3$ to the extremely high parameters $n_e \approx 2 \cdot 10^{20}$ cm^{-3} and $P \approx 500$ MPa was realized, where strong $l/r_D \approx 1.6$ nonideality gives rise to qualitative change of the radiation characteristics of the plasma (fig. 1.16). Curves 1, 2 in figure 1.16 correspond to the traditional methods of calculation of the optical properties of a low temperature plasma [96]. Absorption in free-free transitions of electrons in the fields of the ions is described by the Kramers formula [96, 112] — curve 1. The calculation of the plasma screening results in a decrease of the absorption coefficient (curve 3). To obtain the total absorption coefficient (curve 2) it is necessary, in addition to the free-free transitions, to take account of the photoionization of the bound states. For the chosen frequency $h\nu = 2.14$ eV, the main contribution to the absorption is made by photoionization from the states $3p^5 3d$, for which the correspond-

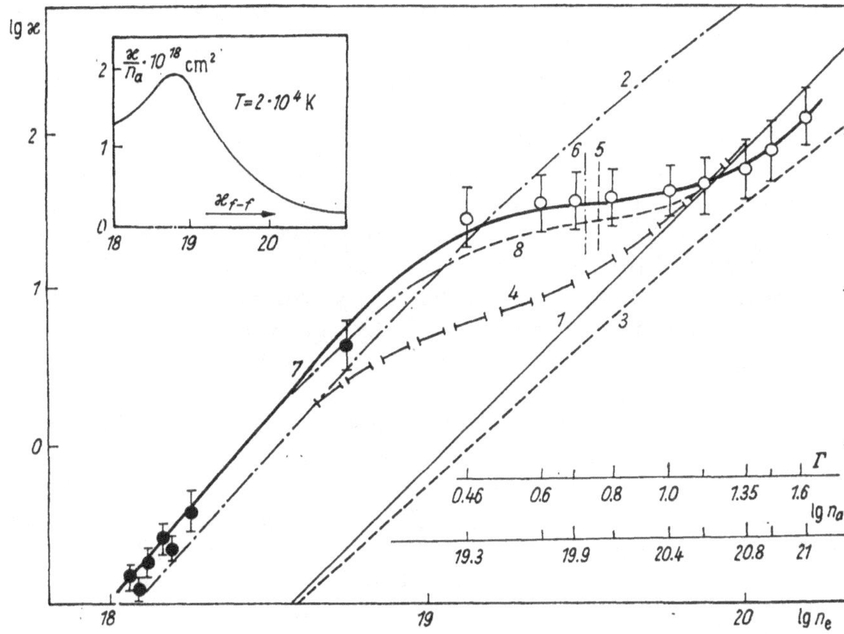

Fig. 1.16. Optical absorption coefficient of shock compressed argon plasma at $T = 20\,000$ K. Open points — measurements from [31], solid points — other measurements (see [31]), 5—6 — estimation of $\varkappa$ on the basis of [110, 111]. Calculations: 1 — free-free transitions, 2 — calculations with photoionizations, 3 — free-free transitions with screening taken into account, 4 — simplified version of the "confined" atom model (1.3), (1.6) up to (1.8) with cross-sections from [109], curve 7 — connects together experimental points obtained by the various measurements, refered to a frequency $\nu = 5.17 \cdot 10^{14}\ \mathrm{s}^{-1}$, and to $T = 2 \cdot 10^4$ K, 8 — calculation from the model (1.3), (1.6) up to (1.8) with cross-sections, calculated by (1.25), (1.27) up to (1.31)

ing cross sections were calculated in [109] by the quantum-defect method. The experimental data in the region of low densities are seen to be in agreement with the numerous measurements performed earlier and support the tendency that the radiation being registered exceeds the calculated values. With further increase of plasma compression the specific absorption coefficient per atom (see the inset of fig. 1.16) decreases monotonically and $\varkappa$ turns out to be smaller by one order of magnitude in comparison with the results of the traditional plasma calculation, and close to the values determined from free-free transitions only. The plasma "transparentization" effect observed in the experiments is treated in [31] as a consequence of the deformation of the electron energy spectrum of atoms if they are compressed by the surrounding particles. Regardless of the actual models used for the description (fluctuating microfields, screening by charges [100, 113]), the interparticle interaction makes the intra-atomic potential to be short-ranged, resulting in a finite number of discrete energy levels and in a successive transition of the highly excited states to the continuum with increasing plasma density. This vanishing of the part of the excited energy levels excludes the mechanism of photoionization absorption from these states and consequently leads to the observed decrease of the absorption coefficient [31].

For an interpretation of this effect in [31] we used the "confined atom" model (1.3), (1.6) up to (1.8) which describes thermodynamics of a highly compressed argon plasma (sect. 1.3). The simplest version of this model applicable to the photoionization calculation consists [31] in using unperturbed cross sections of the bound-free transitions from [109] with the thresholds shifted to the region of small frequencies in accordance with the energy levels deformed by the interparticle interaction (fig. 1.5). It is seen (fig. 1.16, index 8) that this model reflects qualitatively the tendency revealed in the experiment [31] to the "transparentization" of the plasma if being compressed.

A more developed model of the optical properties of a dense plasma [114] is based not only on taking into account the change of the energy spectrum of the plasma but also on the calculation of a change of magnitude and frequency dependences of photoionization cross sections with plasma compression.

To estimate the photoionization cross sections of compressed atoms we have used the Hartree-Fock method [47] in which the state of the atom in the LS-coupling approximation is replaced by the configuration with the total orbital and spin moments of the system. In the Hartree-Fock method being a one-electron approximation the given electron is usually considered to move in the central field of the nucleus and the remaining electrons (self-consistent field). For noble gases and alkali metals each configuration is corresponding to one term (one pair L and S) while in the complex atoms containing more than one electron of partially filled shells one configuration produces several terms. In the method [47] the initial state of the atom is characterized by the total momentum L and S, the initial state of the optical electron is characterized by major quantum and orbital numbers n, l, the final state of the optical electron (in a continuous spectrum) is characterized by the angular momentum l', energy ε. The system which consists of the electron of a continuous spectrum and of an atomic core is described by the total momenta L', S', the atomic core is depicted by L_1, S_1. The process of photoionization under the conditions when the atomic core is not changed is symbolically written as a transition of the kind:

$$S_1L_1nlLS \to L_1S_1\varepsilon l'L'S' .$$

In the dipole approximation the cross section of the photoionization has the form [115] (in atomic units):

$$\sigma(\omega) = \frac{4\pi^2}{3}\alpha\omega \sum_{l'=l\pm1} \sum_{l} (2l+1) \begin{Bmatrix} lLL_1 \\ L'l1 \end{Bmatrix} l_{\max} |M_{\mathrm{bf}}|^2 \tag{1.24}$$

where α is the fine structure constant, ω the radiation frequency, $\begin{Bmatrix} lLL_1 \\ L'l1 \end{Bmatrix}$ — $6j$ the Wigner coefficient [115], $l_{\max}$ the maximum of the numbers l, l'. The sum according to L' is the sum according to all total angular momenta of the optical electron + atomic core system which are possible if the atomic core has a total angular momentum L and the optical electron of the continuum l'. The matrix element of the transition in $\boldsymbol{p}$-representation has the form [115]:

$$M_{\mathrm{bf}}^{p} = \frac{1}{\omega}\int f_{\mathrm{b}}(r)\left[\frac{\mathrm{d}f_{\mathrm{f}}(r)}{\mathrm{d}r} \pm l_{\max}\frac{f_{\mathrm{f}}(r)}{r}\right]\mathrm{d}r\,, \tag{1.25}$$

in $\boldsymbol{r}$-representation

$$M^r_{\mathrm{bf}} = \int f_{\mathrm{b}}(r)\, r f_{\mathrm{f}}(r)\, \mathrm{d}r \tag{1.26}$$

where $f(r) = rR(r)$, and $R(r)$ is the radial part of the one-electron wave function in the central field approximation. It should be underlined that both forms of the matrix element have the same value if $f_{\mathrm{b}}(r)$ and $f_{\mathrm{f}}(r)$ are eigenfunctions of one and the same Hamiltonian.

In calculating wave functions of the bound and free electrons we have used the "freezed core" approximation in which was taken into account the influence of inner electrons on the optical one in the discrete and continuous spectra, including the volume interaction and taking no account of the reverse influence. The wave function of the electron of the discrete spectrum $f_{\mathrm{b}}(r)$ was calculated in the confined atom approximation (1.6), (1.7), (1.8). In the continuum the function $f_{\mathrm{f}}(r)$ was determined from solving the equation

$$\frac{\mathrm{d}^2 f_{kl}}{\mathrm{d}r^2} + V_l(r)\, f_{kl} + k^2 f_{kl} + \int f_{kl}(r')\, G_{kl}(r')\, \mathrm{d}r' = 0 \tag{1.27}$$

where $V_l(r)$ is the potential which has the Coulomb asymptotics at infinity. Integration in the exchange integral was limited by the value r_{c} where the wave functions of the inner electrons obey the condition [3]:

$$f_{\mathrm{b}}(r_{\mathrm{c}}) = 0\,, \tag{1.28}$$

k^2 is connected with the radiation frequency by the conservation of energy principle

$$\omega = -\varepsilon_{nl} + \frac{k^2}{2} \tag{1.29}$$

where ε_{nl} is the energy of an optical electron in the bound state. The first boundary condition for the equation (1.27) has the form

$$f_{kl}(0) = 0 \tag{1.30}$$

at $r \to \infty$ the wave function has the asymptotics

$$f_{kl} \approx \sqrt{\frac{2}{\pi}}\, \frac{1}{[k^2 + V_l(r)]^{1/4}} \sin \int\limits_0^r \sqrt{k^2 + V_l(r')}\, \mathrm{d}r'\,. \tag{1.31}$$

In solving numerically the equation (1.27) the yield on the asymptotics was checked up so that the successive three maxima of the function

$$f_{kl}(r)\, [k^2 + V_l(r)]^{1/4}$$

should have the same value at the fixed exactness.

The matrix elements were calculated for the compressed argon atoms in $\boldsymbol{p}$- and $\boldsymbol{r}$-representation. Figure 1.17 shows a cross section of the argon atom photoionization from the state $1s^2 2s^2 2p^6 3s^2 3p^5 4p^1 - {}^1D$ calculated in the isolated atom approach. It is seen that the cross sections in $\boldsymbol{r}$- and $\boldsymbol{p}$-representation are close to one another. The difference of these cross sections increases with compression which is seen from figure 1.18 where is shown the same cross section at the atom radius $r_{\mathrm{c}} = 8$ a.u. It is to be noted that the use of the matrix element in the $\boldsymbol{p}$-representation is more justified for two reasons. On the one hand it is known that by employing the Hartree-Fock

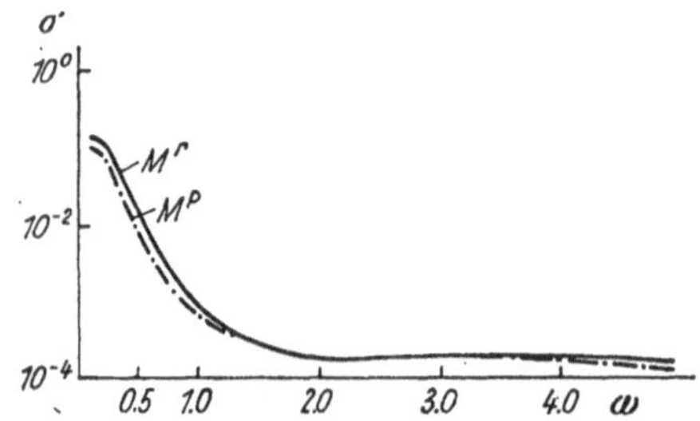

Fig. 1.17

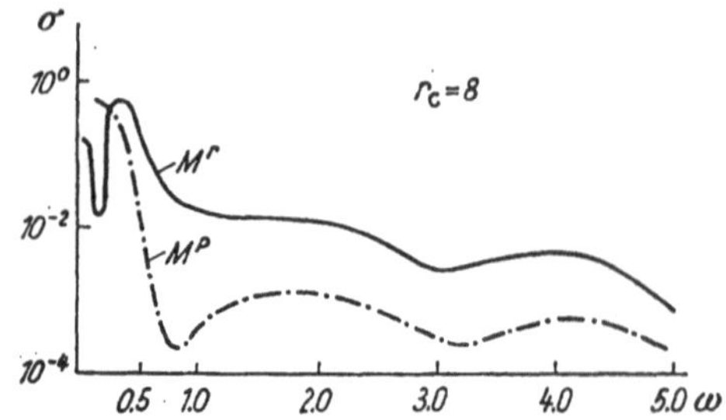

Fig. 1.18

Fig. 1.17. Photoionization cross-section of an "isolated" ($r_c = \infty$) argon atom from $4P^1(^1D)$ state; atomic units are used

Fig. 1.18. Photoionization cross-section of a "confined" ($r_c = 8$) argon atom from $4P^1(^1D)$ state; atomic units are used

method the cross section, calculated in $\boldsymbol{p}$-representation gives a better accordance with the experimental data than the cross section in $\boldsymbol{r}$-representation. On the other hand the approximation (1.6) for the initiation of compressing the atom in the plasma has an influence first of all on the tail of the wave function $f_b(r)$. The major contribution to the electron scattering in $\boldsymbol{p}$-representation is made by the central region of radii and in $\boldsymbol{r}$-representation it comes particularly from by the values near the boundary of the atom, overstating the influence of the compression on the wave functions of the discrete spectrum. The comparison of this model with the experiment (fig. 1.16) shows the shift of the photoionization thresholds in the region of low densities and the increase of the corresponding cross sections on the compression of the plasma to ensure the quantitative description of the regularities observed in the experiment.

1.6. Remarks

A theory of properties of dense Coulomb systems should unify the description of dense ion plasmas, liquid metals (see chap. 5.), heavily doped semiconductors, strongly excited semiconductors, impurities in metals etc. Connections are shown for instance in the theory of electrical conductivity (sect. 1.4), see also chapter 3. and [117]. Especially, the incorporation of structure factor effects and the use of scattering phase shifts well known from liquid metal theory seems to be of great importance, see the discussion in section 5.7. and further references given there.

In the present chapter, the "confined atom" model (eqs. (1.6) up to (1.8)) has been introduced to obtain a good coincidence between theory and experiment, for instance in thermodynamical (sect. 1.3.) and optical properties (sect. 1.5.) of a high dense plasma. In this model, different energy levels merge at different parameter values r_c in the continuum (see fig. 1.5). Another description of this effect has been given recently using consequent many body theory and Green's function technique [118, 121].

As pointed out in section 1.4. a special problem is the correct calculation of the bound states in a partly ionized plasma. For attempts to describe the influence of bound states on kinetic properties in a plasma see section 3.3.7. and chapter 4. and references given there. Further effects on the conductivity are discussed in chapter 3.

Recently the optical properties of a highly compressed plasma were derived from a quantum statistical treatment of the plasma. Such a approach has been formulated in [122—124] where expressions for the dielectric function of a partially ionized plasma have been derived. Within this approach the shift [124] and width of spectral lines in high dense plasmas can be found; furtheron, it has been shown that no decrease of intensity in the near-threshold region occurs [116]. Of course, improvements of recent theory of plasma spectroscopy as refered, for instance, in [125, 126] may be obtained in a systematic manner from quantum statistical many body theory.

2. The Electrical and Thermal Conductivities of Highly Non ideal Alkali Metal Plasmas

P. P. Kulik, V. A. Riaby, E. K. Rozanov

2.1. Introduction

In this chapter two main problems of nonideal plasma physics are considered: the problem of the Coulomb logarithm and the transition from thermal to pressure ionization. Clues to treat these problems have been obtained from experiments with impulse capillary isobaric discharge (ICID) in alkali metal (Li, Na, K, Cs) vapours yielding information about electrical (σ) and radiative thermal (λ) conductivities in the range of non-ideality parameter $0.1 \leqq \gamma \leqq 2$ ($\gamma = e^2 n_e{}^{1/3}/k_B T$). The Coulomb electrical conductivity σ_c is shown to be submitted to a double quantum effect, leading generally to an nonmonotonous dependence of the nondimensional electrical conductivity $\sigma_c^* = \sigma_c e^2 m_e^{1/2}(k_B T)^{-3/2}$ on temperature. Low temperature data for $\sigma(T, P)$ and $\lambda(T, P)$ reveal the character of the metal-dielectric transition in a dense plasma.

The investigation of nonideal plasma transport properties deals mainly with two problems: the so-called problem of the Coulomb logarithm and the transition from thermal ionization to pressure (density) ionization (metal-dielectric transition in dense plasmas).

2.1.1. The Problem of the "Coulomb Logarithm"

This problem arises whenever of any transport property of partially or fully ionized plasmas is calculated in the range of Coulomb non-ideality $\gamma = e^2 n_e^{1/3}/k_B T \gtrless 10^{-1}$. The electrical conductivity, for example, of a coulombic system, obeying the Boltzmann statistics, can be expressed in the form

$$\sigma_c = \frac{(k_B T)^{3/2}}{e^2 m_e^{1/2}} \sigma_c^*(\Lambda) \tag{2.1}$$

where the non-dimensional function $\sigma_c^*(\Lambda)$ takes account of the collective character of the coulombic interactions. Its argument Λ is usually expressed as the ratio of two linear dimensions, determining the area of coulombic interactions: b_{max} the electrostatic shielding radius (for ideal plasma $b_{max} = r_D = \sqrt{k_B T/8\pi n_e e^2}$); b_{min} the distance of closest approach (for classical plasma $b_{min} \approx e^2/k_B T$). According to strict ideal plasma theories (in particular Spitzer-Härm theory [1]), for $\gamma \ll 1$,

$$\sigma_c^* \approx \ln^{-1} \Lambda \tag{2.2}$$

where $\ln \Lambda$ is the so-called "Coulomb logarithm".

Attempts to calculate σ_c using formally equations (2.1)—(2.2) for nonideal plasma ($\gamma > 10^{-1}$) lead to the well-known coulombic divergence. Indeed, with increase of γ, $\Lambda \sim \gamma^{-3/2}$ decreases so that $\ln \Lambda \to 0$ and $\sigma_0 \to \infty$, which is a nonsense. This fact shows that the Debye theory of electrostatic shielding cannot be applied to nonideal plasma.

It should be emphasized that for a classical plasma σ_c^* depends only on γ. It means that isobars $\sigma_c^*(T, \gamma)_P$ and isotherms $\sigma_c^*(P, \gamma)_T$ should form an universal dependence $\sigma_c^*(\gamma)$. This situation is the same for any Coulomb plasma property [2].

Various theoretical approaches to this problem have been undertaken. They are based on model Coulomb potential shielded either at the mean interparticle distance $n_e^{-1/3}$ [2, 3] or at a distance proportional to r_D [4, 5], on dynamic shielding theory [6], on the model of localized electronic states [7, 8], on computer game calculations [9, 10]. All these approaches show the similarity of the classical plasma Coulomb properties not only for ideal, but also for nonideal plasmas. This conclusion is of particular interest and makes the results of measurements of nonideal plasma electrical conductivity especially important, since they allow to deduce values of any other Coulomb plasma property provided the level of plasma non-ideality is known.

The problem grows more sophisticated when quantum effects begin to influence Coulomb interactions. It happens, for instance, at high temperatures when

$$e^2/k_B T \leqq \lambda\!\!\!^{-}_e \tag{2.3}$$

where $\lambda\!\!\!^{-}_e = \hbar/\sqrt{2\pi m_e k_B T}$ is the de Broglie mean electronic wave length. In this case instead of $e^2/k_B T$ for $b_{\min}$ we should take $b_{\min} \approx \lambda\!\!\!^{-}_e$, so that $\Lambda \sim \gamma^{-3/2} T^{-1/2}$. Hence for a so-defined quantum plasma, the isotherms $\sigma_c^*(\gamma)_T$ should be separated for different values of T, the upper isotherms corresponding to higher temperatures [11]. More rigorous theories [12—15] confirm this qualitative conclusion.

When plasma temperature further raises another competing phenomenon appears, due to non-coulombic electron-ion interactions (or, in other terms, to electron scattering on ionic electron shells) [16]. This effect occurs, obviously, when

$$\sup\left(\frac{e^2}{k_B T}, \lambda\!\!\!^{-}_e\right) \leqq r_0 \tag{2.4}$$

where r_0 is the ionic radius. The resulting electron transport cross-section then exceeds the coulombic one. Thus, at high temperatures, when (2.4) holds, for $\gamma = \text{const}$, $\sigma_c^*(T)_\gamma$ should now decrease with temperature, since the electron-ion cross-section is now larger than the Coulomb cross-section.

The resulting dependence $\sigma_c^*(T)_\gamma$ is schematically presented on figure 2.1. At low temperatures $\sigma_c^*(T)_\gamma \approx \text{const}$ according to the automodel character of σ_c^* for classical plasmas. At higher temperatures $\sigma_c^*(T)_\gamma$ grows with T, since $\Lambda \approx T^{-1/2}$. At very high temperatures $\sigma_c^*(T)_\gamma$ decreases. The amplitude and the position of the maximum depend on r_0. It should be emphasized that for large r_0 (i.e. for Xe, Kr, Ar) it may happen, that the maximum does not appear at all and that $\sigma_c^*(T)_\gamma$ is a monotonously decreasing function. For metals both effects should realize and $\sigma_c^*(T)_\gamma$ should probably be represented by the curve of figure 2.1. Obviously, for hydrogen this dependence should be monotonously increasing.

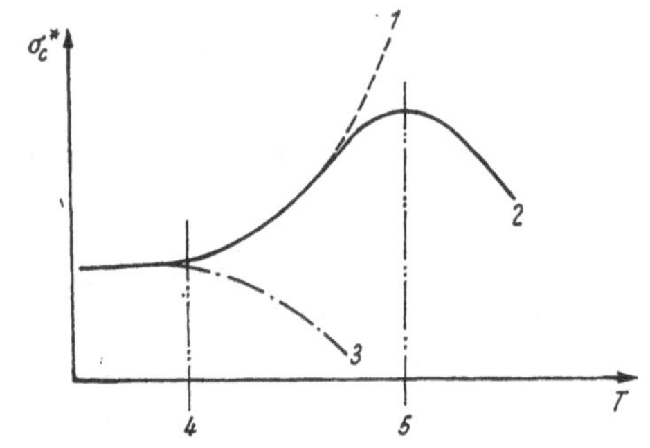

Fig. 2.1. Qualitative behaviour of the non-dimensional coulombic electrical conductivity of highly nonideal plasma subjected to a double quantum effect. 1 — hydrogen, 2 — metal plasma (for moderate, r_0), 3 — inert gas plasma (high, r_0), 4 — $\frac{e^2}{k_B T} \approx \lambda_e$, 5 — $\lambda_e \approx r_0$

2.1.2. The transition from thermal to pressure ionization

At low temperatures, when plasma is weakly ionized the simultaneous increasing of temperature and pressure should be accompanied by two competing ionization mechanisms. When plasma density is low, the common thermal ionization dominates. At large density pressure ionization occurs. By pressure ionization, we mean free electron concentration increasing due to plasma density growth. This ionization mechanism may be caused, for instance, by atomic electron shells overlapping, lowering of the binding energy, electronic levels shift and widening due to interparticle random micro-fields, electron tunnelling, neutral and charged cluster formation.

These density effects influence specifically the behaviour of the electrical and thermal conductivities, which are the subject of the present experimental investigation.

The electrical conductivity. Two isobars $\sigma(T)_P$ are qualitatively presented on figure 2.2, a) for $P_1 > P_c$ and $P_2 < P_c$ (by P_c we eman the critical pressure for the liquid-gas (plasma) transition). The solid lines correspond to the real behaviour of σ, the dashed lines show the hypothetical behaviour assuming that density effects are absent and ideal rarefied plasma equations can formally be applied (for instance the

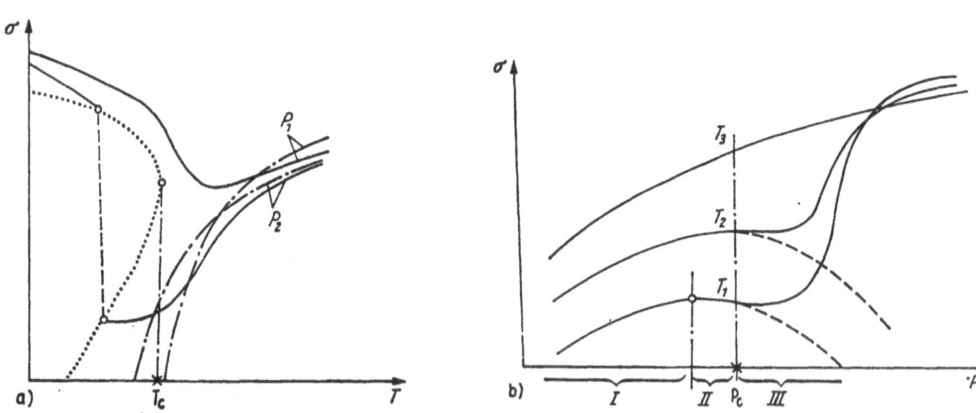

Fig. 2.2. Qualitative behaviour of the nonideal plasma electrical conductivity. a) Isobars $P_1 > P_c > P_2$: dashed lines — ideal plasma; solid lines — nonideal plasma; the dotted line corresponds to phase equilibrium "vapour — liquid". b) Isotherms $T_3 > T_2 > T_1$

Lorentz formula $\sigma_L \sim n_e/n_a Q_{ea} T^{1/2}$). The left part of the diagram corresponds to fully ionized state due to pressure ionization (plasma metallization), the right one corresponds to full thermal ionization.

At full thermal ionization the ideal values of σ are larger than real ones because of Coulomb interactions non-ideality (according to Spitzer formula $\sigma_c \to \infty$ at $\gamma \approx 0.7$). As the temperature decreases, we come to a partially ionized plasma state. Here, the formerly mentioned density effects occur. Most of them provide abrupt exponential increasing of σ when T is lowered. The formation of negative ionic clusters, the presence of excited targets for free electrons, the collective electron interactions, etc. tend to lower σ compared with ideal plasma conductivity. For $P > P_c$ the transition is continuous and the higher the pressure is, the smoother it should be.

Figure 2.2 b) shows the influence of pressure ionization on isotherms $\sigma(P)_T$. At low temperature $T_1 < I/k_B$ (I is the ionization potential) this dependence has three characteristic regions: I — corresponds to complete thermal ionization ($\sigma \approx \ln P$ for ideal plasma); II — plasma is weakly ionized and $\sigma \approx (n_e/n_a) \approx P^{-1/2}$; III — σ abruptly (exponentially) rises with pressure due to pressure ionization. At higher temperatures this transition becomes smoother, and for $T_3 > I/k_B$ the plasma remains completely ionized at all pressures.

The thermal conductivity. In the case of the thermal conductivity a similar transition must take place, but here the physical processes are more complicated. At low temperature, corresponding to liquid metal, the thermal conductivity is due to Fermi electrons and $\lambda_e(T)$ decreases if T grows. At high temperatures

$$\lambda = \lambda_e + \lambda_i + \lambda_a + \lambda_R$$

where λ_R is the radiative thermal conductivity (its high temperature asymptote is the Cramers curve $\lambda_{cr} \approx T^{13/2} n_e^{-2}$) and λ_e, λ_i, λ_a are the electron, ion, and atom components of the molecular thermal conductivity.

When plasma is dense enough, it can happen that $\lambda_R \gg \lambda_e$ ($\lambda_e > \lambda_a \approx \lambda_i$) and that the photon free-path length is markedly lower than the plasma dimension. In this case, the thermal conductivity is determined by the radiative thermal conductivity. Figure 2.3 corresponds to this case, which seems to be the most important and the most frequent in the practice of dense plasma experiments. In the vicinity of the

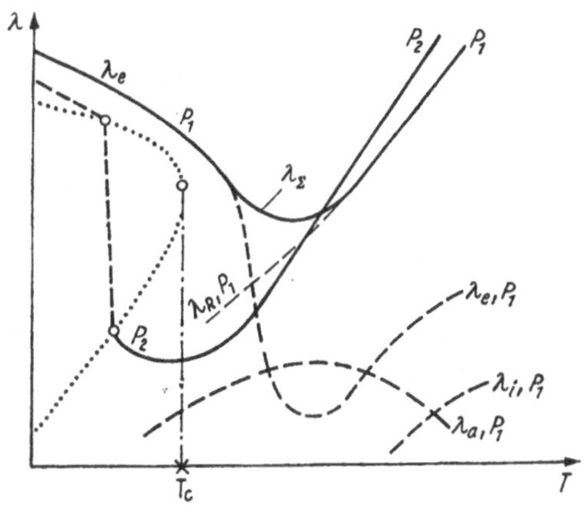

Fig. 2.3. Qualitative behaviour of the plasma thermal conductivity in the form of isobars $P_1 > P_c > P_2$. Solid lines — total conductivity; dashed line — λ components for P_1; dotted line corresponds to phase equilibrium "vapour-liquid"

critical temperature T_c, $\lambda(T)_P$ has a minimum rather similar in character to that of $\sigma(T)_P$. Its depth depends on the behaviour of all components of λ, including λ_R. Therefore, in the range of the transition from thermal to pressure ionization, in addition to density effects determining the electron concentration and mobility, the phenomena of photon emission and absorption have a primary importance for the mechanism of thermal conductivity. These phenomena are related to the effect of enhanced plasma transparency due to the non-existence of upper atomic electron levels in nonideal plasmas [17], the effect of atomic spectrum change when the atom is kept in a finite cell [18], the influence of non-ideality on electronic bremstrahlung [19], etc. The qualitative analysis of these phenomena shows that for higher pressure, especially for supercritical pressures (i.e. $P_1 > P_c$) the isobars $\lambda(T)_P$ should smoothen and their minimum should become shallower (compare the curves for $P_1 > P_c$ and for $P_2 < P_c$ in fig. 2.3).

The available theoretical knowledge on non-ideal plasma physics do not yet allow calculations of σ and λ in the considered range of parameters, even by an order of magnitude.

The present work deals with the experimental determination of $\sigma(T, P)$ and $\lambda(T, P)$ for highly nonideal plasma in a range of parameters, where the discussed phenomena should occur.

As a method of dense plasma generation, we used the impulse capillary isobaric discharge (ICID). Alkali plasmas (Li, Na, K, Cs) are investigated. A method of simultaneous determination of σ and λ is elaborated, allowing to obtain data in the range $P \approx (10-100)$ MPa and $T \approx (4-16) \cdot 10^3$ K.

2.2. The Experimental Impulse Capillary Isobaric Discharge Device (ICID)

The schematic diagram of the device together with the main electrical feeding and measurement circuits are presented on figure 2.4. The impulse discharge is initiated between tantalum electrodes 1 and 2 through the channel of a quartz capillary 3. The cup electrode 2 contains the investigated metal 4, as a rule in solid state. Tungsten probes 5 are hermetically inserted in the capillary. One of them, the nearest to the cathode 2 is thicker and may be used as an auxiliary electrode to trigger the arc. The discharge assembly is settled on the obturator 7 of the high-pressure vessel 8 which can be filled with argon at a pressure up to 100 MPa. The induction pressure-gauge 11 allows to control time-variation of pressure at the capillary exit in the vessel part 12 out of capillary channel.

The electrical feeding of the device is provided by a capacitor bank $C \leqq 7500\ \mu$F connected to electrodes 1 and 2 through an inductor $L = (0.05-0.5)$ H and an inductance-free resistor $R = 0.001\ \Omega$, used for current measurements. The discharge is initiated by a wire (13) electrical explosion or by breakdown between electrodes 2 and 6 using an auxiliary capacitor $C' \ll C$. The initial voltage of both capacitors is up to 1000 V. The main parameters of the device are:

capillary length $l = 15-20$ mm,
capillary channel diameter $d = 1.5-2$ mm,
total discharge duration $\tau \approx 10^{-2}$ s,

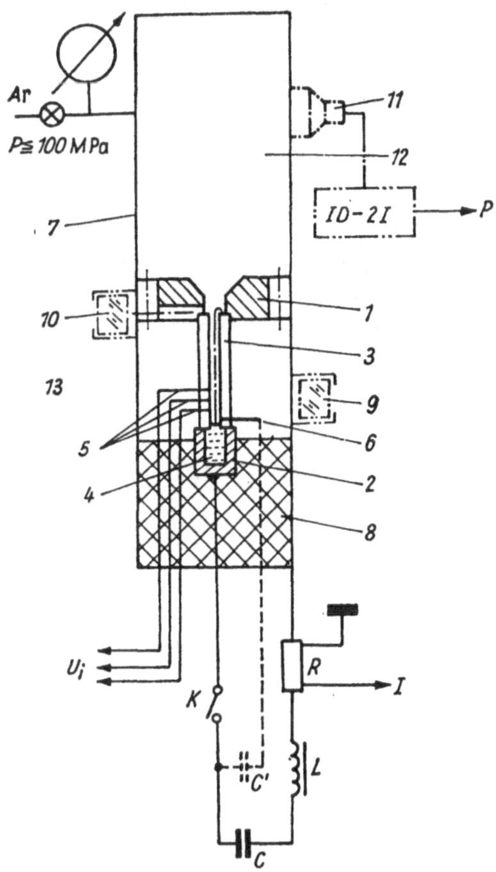

Fig. 2.4. The schematic diagram of the ICID-device and its main electrical circuits

current $I \leqq 1000$ A,
voltage $U \leqq 500$ V,
pressure $P = 10-100$ MPa,
plasma temperature at the surface of the arc column $T_s = 4000-14000$ K,
axis plasma temperature $T_0 \leqq 20000$ K.

2.3. The Model of the Physical Processes in ICID

The present plasma generator can be used for investigations of dense plasma properties if the following requirements to the method of plasma generation and to the design of the device are realized:

(i) The arc discharge is stabilized by the capillary channel wall.
(ii) The discharge is quasi-stationary.
(iii) The arc column is a longitudinally uniform, slowly moving flow in which the static pressure is equal to the outer pressure.
(iv) The plasma flow is a circular cylinder.

(v) The plasma is optically thick so that only surface radiation is possible.

(vi) The medium around the column completely absorbs the plasma radiation, i.e. radiation reflection back to the plasma column is excluded.

(vii) During the discharge, the channel diameter remains constant.

These requirements correspond to the following physical conditions:

(i) The volt-ampere characteristic of the arc should be a raising one:

$$\mathrm{d}E/\mathrm{d}I > 0\,.$$

(ii) The skin-effect should be small, i.e. $\delta/R \gg 1$, where $R = \frac{1}{2}d$ is the radius of the arc column and δ is the well-known depth of the skin-layer. Thereby the electrical field strength E can be considered to be uniform in space, so that the current equation takes the form

$$I = 2\pi E \int_0^R \sigma(T, r)\, r\mathrm{d}r$$

and the electrical conductivity, averaged over the arc cross-section, can be written:

$$\bar{\sigma} = 2R^{-2} \int_0^R \sigma(T, r)\, r\, \mathrm{d}r = \frac{I}{\pi R^2 E}\,. \tag{2.5}$$

The plasma thermal inertia should be small so that the corresponding term in the power balance equation should be negligible compared to the input power:

$$\varrho C_P \frac{\partial T}{\partial \tau} \ll \sigma E^2\,. \tag{2.6}$$

Integrating this equation over the plasma column cross-section it is easy to obtain an approximate condition for the mean temperature variation rate:

$$\frac{\overline{\mathrm{d}T}}{\mathrm{d}\tau} \ll \frac{\bar{\sigma} E^2}{\pi R^2 \overline{\varrho C_P}}\,. \tag{2.7}$$

(iii) The plasma flow velocity should be essentially subsonic ensuring the pressure drop ΔP along the column axis to be small compared with the absolute pressure level P of the gas medium in the vessel:

$$\Delta P \ll P\,.$$

(iv) All the plasma column parameters should depend only on radius.

(v) The whole electrical power input to the plasma column EI should be radially radiated by an optical plasma boundary layer the thickness of which is of the order of magnitude of the photon mean free-path l_λ obeying the condition $l_\lambda/R \ll 1$ in the range of wavelengths corresponding to the main part of radiation heat flux from the column. To a good extend, l_λ can be substituted by the Rosseland mean free-path l_{Ros} so that

$$l_{\mathrm{Ros}} \ll R\,. \tag{2.8}$$

The mentioned optical restrictions allow to determine the boundary temperature of the plasma column using the Stephan-Boltzmann law:

$$T_{\mathrm{S}} = (EI/2\pi R\varkappa)^{1/4} \tag{2.9}$$

where $\varkappa$ is the Stephan-Boltzmann constant.

(vi) The plasma column radiation is partially absorbed by the solid capillary wall together with a possible inside layer of its vapour and also by the gas in the high pressure vessel and by the wall of the vessel. If radiation back to the plasma column is negligible, then the ICID power balance can be expressed using the well-known Ellenbaas-Heller equation

$$\frac{1}{r}\frac{\partial}{\partial r}\lambda\,(T, r)\, r\frac{\partial T}{\partial r} + \sigma(T, r)\, E^2 = 0\,. \tag{2.10}$$

Integrating this equation over r taking account of the boundary conditions

$$T|_{R=r} = T_S = (EI/2\pi R\varkappa)^{1/4}\,,$$
$$\left.\frac{\partial T}{\partial r}\right|_{r=0} = 0\,, \tag{2.11}$$

one obtains the integral equation for the plasma column temperature profile

$$T(r) = T_s + E^2 \int\limits_r^R \frac{\mathrm{d}r}{r\lambda(T, r)} \int\limits_0^r \sigma(T, r)\, r\, \mathrm{d}r \tag{2.12}$$

containing the temperature dependent functions $\lambda(T)$ (the thermal conductivity) and $\sigma(T)$ (the electrical conductivity) to be determined.

(vii) The discharge duration should be short enough for the plasma column radius R to be assumed constant. This condition helps to overcome mathematical difficulties in solving (2.12) to obtain $\lambda(T)$ and $\sigma(T)$ simultaneously from discharge UI-characteristic measurements.

2.4. The Experimental Control of ICID-Model Validity

Earlier methodical researches, mainly [20—22], have shown that the ICID model just described can be considered as a good approximation to the processes in a real experiment on capillary discharges, provided its main parameters are those given at the end of 2.2.

Most important is the question of plasma purity. It was cleared up by direct measurements of the alkali metal mass flow G_M issued from the cup electrode and of the capillary material vapour mass flow G_c. G_M has been found to be by a factor of 40 up to 50 larger than G_c.

G_M has been evaluated both by measurements of the volume of the cathode crater in the case of solid alkali metal filling of the cup electrode and from electrical power drop $I\,\Delta U$ near the cup electrode measured using electrostatic probes and responsible for alkali metal evaporation from the cup electrode. For $d = 2$ mm and $\tau \approx 20$ ms $G_M \approx (30—40)$ mg. This result corresponds to the case of negative polarity of the cup electrode. For the reverse polarity the result is about 25%—30% lower. Therefore, in all final experiments the cup electrode was negative. The probe measurements showed that, in any case, the potential drop ΔU was localized utterly at the cup electrode, ensuring maximum possible mass flow of metal vapours feeding the arc column.

The question to be solved now concerns the G_M distribution in time and space. Time distribution was easily to be controlled by oscillographic registration of $\Delta U(\tau)$. It has turned out to be roughly proportional to EI. Monotonous space distribution can be concluded from the absence of any liquid metal drops. At supercritical pressures this conclusion is obvious. For undercritical pressures large drops (of dimension comparable to the interprobe gap and capillary channel diameter, i.e. $> 0.1d$) should be noticed on oscillographic $\Delta U(\tau)$ recordings. They dont' exist (see figs. 2.5, 2.6). Small drops $< 0.1d$, if they exist really, can easily be shown by simple calculations to evaporate within a space of 1–2 mm ($\approx d$) above the cup electrode (the measured plasma mass flow velocity being ≈ 10 m/s) and so, not to reach the area of probe measurements.

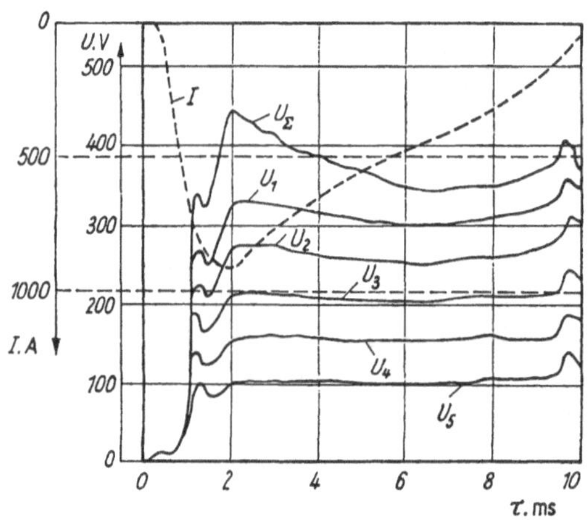

Fig. 2.5. Time sweep of I, U_Σ and electrical potentials of five wall-probes uniformly distributed along the capillary (Cs, $P \doteq 25$ MPa, $d = 1.8$ mm, $l = 20$ mm, interprobe distance $\Delta l = 3$ mm, inverse polarity of the discharge)

The values of G_C have been measured as the quantity of capillary material vaporized during the discharge. Judging by the traces of arc burning on the capillary inner surface, G_C has been found to be uniformly distributed along the capillary inner surface and for discharge energies near 500 J leads to a radius increase $\Delta R \approx 0.005$ mm at $R \approx 1$ mm for the quartz capillary. This material was prefered to glass since, in the last case, the values of ΔR were larger and dark spots appeared on the inner surface of the capillary towards the end of the discharge time. Such spots seem to result from intense chemical reactions between alkali metal and hot (melted) glass and to disturb the arc column symmetry and thus the ICID-model validity. $\Delta R \approx 0.005$ mm corresponds to $G_C \approx 10^{-3}$ g, i.e. to 2%–4% of G_M. So, even if uniformly distributed in the whole plasma column, such a small quantity of admixture should not sensibly influence the investigated plasma properties. In reality quartz vapour enters the plasma flow of $v \approx 10$ m/s forming a thin weakly conducting boundary layer between the metal plasma column and the solid (or melted) capillary wall. Its thickness δ_q is estimated to be small compared to d: $\delta_q/d \lesssim 2\%$.

For a laminar flow $\delta_q \approx L_D \approx \sqrt{\tau D}$, where D is the diffusion coefficient, L_D — diffusion length. For $D \approx 10^{-3}$ cm^2/s and $\tau \approx 10^{-3}$ s (time for a particle to pass through

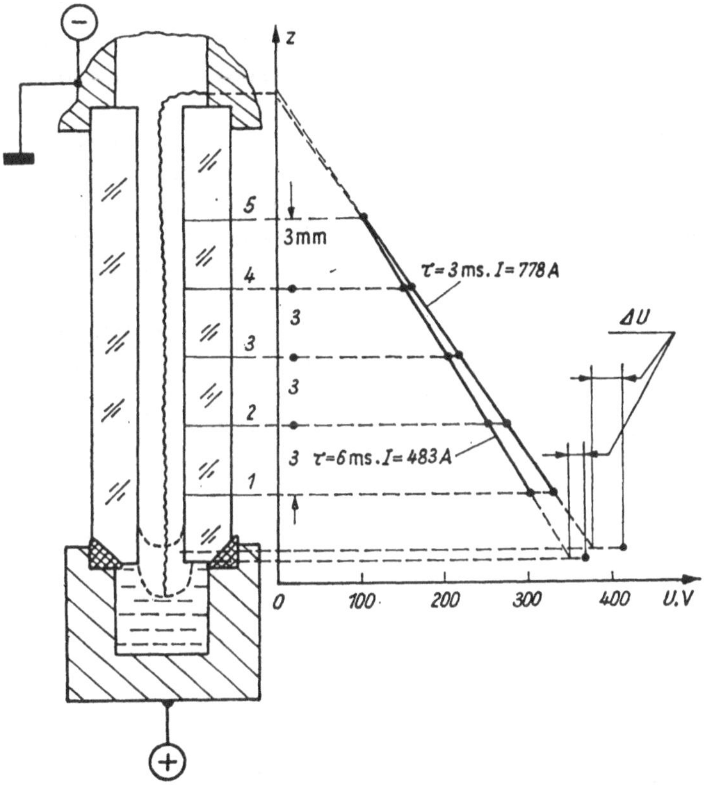

Fig. 2.6. An example of electrical potential distributions for the data of fig. 2.5

a 10 mm long capillary) $\delta_q \approx 10^{-2}$ mm. For a turbulent flow (here the Reynolds number Re $\leqq 10^4$) a pessimistic estimation gives $\delta_q \approx L_T \approx 4 \cdot 10^{-2}$ mm [23], where L_T is the turbulent diffusion length.

In this layer the temperature drops from the plasma surface temperature T_S to the capillary inner wall temperature T_W equal to the saturation (boiling) temperature T_{sat}, which depends on pressure: for quartz $T_{sat} \approx 4000$, 4300 and 4700 K at $P = 10$, 30 and 100 MPa respectively [24]. The quartz vapour layer here is optically thin. Its presence cannot affect the measured values of $\bar{\sigma}$ since $\delta_q \ll R$.

Indirect confirmation of this conclusion follows from several experimental facts. Firstly let us consider the axial electric potential distribution presented as an example on figure 6. This distribution was measured by probes. Within an error of 1 up to 2% it turned out to be linear. It means that $\delta_q \leqq 10^{-2}$ mm at $R \approx 1$ mm. So the presence of this thin layer does not affect the measured values of $\bar{\sigma}$.

A similar conclusion can be deduced from the measured $T_S(z)$-distributions. Measurements were carried out using a one-channel photoelectric micropyrometer for two effective wavelengths (0.65 and 0.63 μm) [20]. No systematic variation of T_S along Z axis was observed within an error of ± 500 K. It proves that the possible narrowing of the plasma column towards the capillary exit due to the growth of the quartz vapour boundary layer is weak so that the temperature T_S is insensitive to it.

Another possibility to check up plasma purity is to analyse the time variation of the UI-characteristics. The heating of a capillary up to the moment when the temperature of the channel inner wall reaches vaporization level takes some 1 up to 2 ms while the arc burning settles during < 0.5 ms [25]. So during 0.5 up to 1.5 ms at the beginning of the process the plasma column is absolutely pure. As a rule this time interval corresponds to the part of the UI-characteristic when the input power EI is increasing. Further, its growth stops and EI begins to decrease. The coincidence of the increasing and decreasing parts of EI-characteristic means that at the first 1 and the last 3 points of this branch of the UI-characteristic the values of the mean electrical conductivity are the same and so do not reveal any plasma impurity. Such a situation is shown on figure 2.7. Here $\tau_1 \approx 1$ ms, $\tau_3 \approx 8$ ms (data registration is finished at $\tau_4 \approx 15$ ms when $T_S \approx 4300$ K). Since, at the point 1, the plasma is surely pure (the vaporization of quartz has not yet begun), and so, at the point 3 where $I_3 \approx I_1$ and $U_{\Sigma 3} \approx U_{\Sigma 1}$, i.e. $T_{S1} \approx T_{S3}$ and $\bar{\sigma}_1 \approx \bar{\sigma}_3$, the plasma should be pure too.

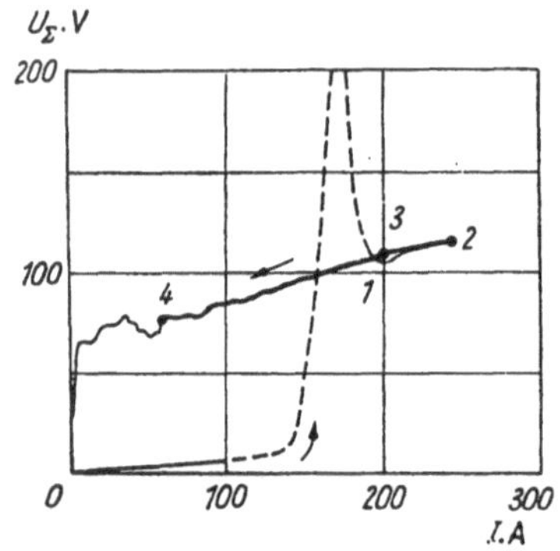

Fig. 2.7. UI-characteristic for Cs at $P = 50$ MPa, quartz capillary: $d = 1.69$ mm, $l = 15.2$ mm, $U_0 = 500$ V

Summarizing, we have come to the conclusion that the ICID plasma column is practically pure and axially uniform. Its dimensions practically coincide with those of the capillary channel.

The next question concerns the pressure. In [20] the pressure was measured at both ends of the capillary and was found to be equal to the pressure in the gas volume N_0 above the capillary. So the pressure drop along the capillary is small compared to the absolute pressure level. For the pressure in the V_0 volume to remain nearly constant during the discharge when the plasma flow enters this volume, the value V_0 must be as large as possible compared to the capillary channel volume. In the present experiments $V_0 \approx 70$ cm^3, so the total relative pressure raise during the discharge does not exceed $\approx 1\%$ for the maximum discharge energy of 1500 J. So the plasma pressure may be assumed to be constant and equal to the static pressure in the vessel.

The optical measurements of T_S in [20] consisted of a plasma radiation flux intensity registration taking account of light absorption in the plasma flow boundary layer and in the quartz. The measured values of T_S for Cs at 50 MPa corresponded to those calculated using (2.9) within an error of ± 500 K for the range $5000 \leqq T_S \leqq 10000$ K. Complementary experiments [21, 22] have shown that the use of formula (2.9) can be extended to the range $4000 \leqq T_S \leqq 12000$ K even for $P = 10$ MPa. In this case measurements were carried out using the hole in the anode 1 and the window 10 shown on figure 2.4. Here the plasma column radiation was free from capillary wall absorption (anode spot radiation was arranged to be covered by plasma column itself). The conclusion about the validity of (2.9) for Cs can be applied to other alkali metals since their optical properties are of the same order of magnitude.

The next problem is to determine the lower temperature limit for which the ICID model is still available and UI-measurements yield correct result for σ and λ. Arguments for such limitations are given by the high speed photography of the plasma column. The frames show that at the end of the discharge, because of thermal inertia of the capillary material, the wall surface temperature T_W becomes larger than T_S. In this case the ICID model is no longer available (see 2.3., condition 6)). Since the saturation temperature of quartz is $T_{sat} = T_W = 4000$ up to 4700 K (depending on pressure) so, these values of the temperature should be assumed to be limiting for T_S. Unfortunately this range of limiting temperatures correspond to $\sigma \approx \sigma_{min}$ (see fig. 2.2a)) so that by means of ICID it seems not to be possible to obtain results in the area of inverse $\sigma(T)$ dependence (i.e. if $\sigma(T)$ decreases with raising temperature at $P = \text{const}$). Another reason for low temperature limitation is that in the case of an inverse $\sigma(T)$ dependence, any arc burning is unstable [26].

Plasma column axial symmetry has been proved directly using probe measurements. For this purpose, the probes were settled along the axis of the capillary at different azimuthal positions. No disturbance was noticed, so the possible probe material vaporization and the formation of a probe vapor track, if it really happened, was insignificant and did not influence the probe measurements within an error of 1%.

The bulk of methodic results allows to conclude that for the present range of parameters and capillary dimensions, the real experimental process corresponds to the ICID model of section 2.3. So the measured UI-(EI-)characteristics contain information about $\sigma(T)$ and $\lambda(T)$ for the investigated metallic plasma.

2.5. The Results of Measurements

The initial capacitor bank voltage was varied in the range of 400 up to 1000 V, the static argon pressure in the working vessel was controlled within an error of ± 1 up to 3%. The quartz capillary diameter was measured individually within an error $\pm 0.5\%$. Most of the discharge EI-characteristics were recorded using a storage electron oscillograph within an error ± 10 up to 40% (up to $\pm 10\%$ in the case of light-beam oscillograph). Synchronized high-speed filming (2000 frames per second) of the probe area through the optical window 9 (fig. 2.4) was used to find the moment τ_1, denoted as point 1 on the UI-characteristic of figure 2.7, after which the interprobe space was uniformly filled with plasma.

For each pressure 8 up to 10 available experiments were provided. The obtained EI-curves were statistically averaged: the mean-square error was about $\pm 8\%$. An example of EI-characteristics for Cs is given on figure 2.8. Similar results have been obtained for Li, Na and K plasmas. Using equations (2.5) and (2.9) these curves were transformed into $\bar{\sigma}_{exp}(T_S)$ dependences (see example for Cs on fig. 2.9). The mean error for $\bar{\sigma}_{exp}$ varies from $\pm 10\%$ at high temperatures to $\pm 15\%$ at low temperatures.

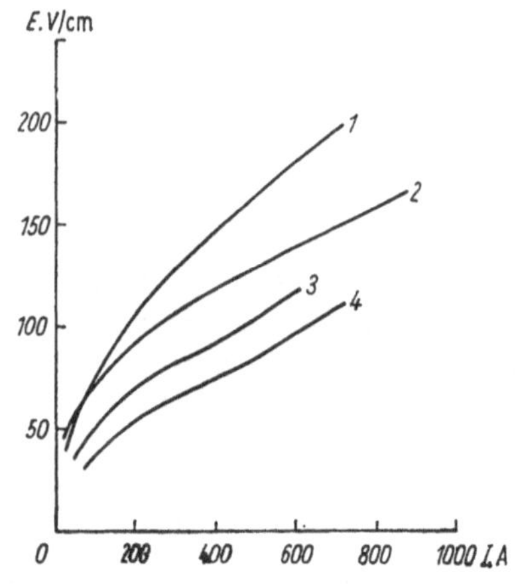

Fig. 2.8. Mean EI-characteristics for Cs at different plasma pressures. 1 — $P = 10$ MPa, $d = 2.08$ mm, 2 — $P = 25$ MPa, $d = 2.12$ mm, 3 — $P = 60$ MPa, $d = 2.09$ mm, 4 — $P = 100$ MPa, $d = 2.14$ mm

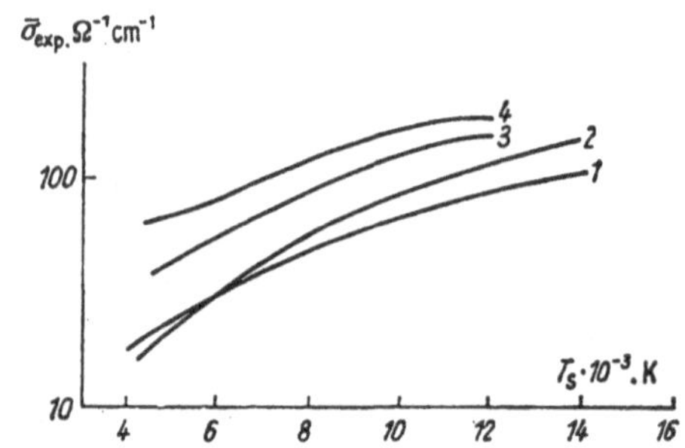

Fig. 2.9. The averaged dependences of the mean electrical conductivity of Cs on the arc column outer surface temperature for different plasma pressures. 1 — $P = 10$ MPa, $d = 2.08$ mm, 2 — $P = 25$ MPa, $d = 2.12$ mm, 3 — $P = 60$ MPa, $d = 2.09$ mm, 4 — $P = 100$ MPa, $d = 2.14$ mm

2.6. Mathematical Processing of the Experimental Data

The purpose of this part of work is to solve the so-called full inverse problem of electrical arc physics, i.e. to extract the functions $\sigma(T)$ and $\lambda(T)$ from the system of integral equations (2.5), (2.11), (2.12).

2.6.1. Existence of the Solution

The existence of a solution for the system of equations (2.5), (2.11), (2.12) is deduced from the optimal equation theory (see for example [27]). Assuming the functions $\sigma(T)$ and $\lambda(T)$ to be driving functions, making linear the functional

$$F_0 = \sum_{i=1}^{k} [\bar{\sigma}_{\exp i}(T_{Si}) - \bar{\sigma}_{\mathrm{th}\, i}(T_{Si})]^2 \tag{2.13}$$

(here $\bar{\sigma}_{\mathrm{th}i}$ is defined by the integral in (2.5)) and using (2.12) to get the equation of state of the system, it is easy to obtain the optimal Hamiltonian which has a maximum for the given driving functions. Such a way is mathematically rigorous but its practical realization is sophisticated [27]. It is simpler to assume the driving functions to be independent arguments and to vary them until F_0 reaches its minimum. This way requires to demonstrate that F_0 has really at least one minimum which is physically reasonable.

So, for this aim, a preliminary model solution was elaborated for the case when the solution $\sigma(T)$ and $\lambda(T)$ is assumed to be known a-priori. For $\sigma(T)$ we used the function obtained in [25] and for $\lambda(T)$, the theoretical curves calculated in [28]. Solving the system of equations (2.5), (2.11), (2.12) we reconstructed the function $\bar{\sigma}(T_S)$. Let us denote the so obtained model function by $\bar{\sigma}_M(T_S)$. Now, using $\bar{\sigma}_M(T_S)$ as an initial function, we solved the system (2.5), (2.11), (2.12) in the inverse direction to obtain $\sigma(T)$ and $\lambda(T)$. Thereby, instead of $\sigma(T)$ we used the power series expansion

$$\sigma_p(T) = \sum_{n=0}^{s} \sigma_n T^n \tag{2.14}$$

the coefficients σ_n of which were to be determined, and instead of $\lambda(T)$, we used the function

$$\lambda_p(T) = \lambda_0 T^m \tag{2.15}$$

where λ_0 and m are two arbitrary parameters, also to be determined. Such a function seems good to approximate the theoretical data of [28] for the radiation thermal conductivity of alkali metal plasmas.

Such a model solution has shown that the so elaborated model function $\bar{\sigma}_M(T_S)$ "remembers" the functions $\sigma(T)$ and especially $\lambda_0(T)$ used for its synthesis: F_0 was calculated for different functions $\lambda'(T)$. It has a minimum when $\lambda'(T) \equiv \lambda_0(T)$ (see fig. 2.10), proving that the elaborated method is in principle available to obtain the physical values of $\lambda(T)$ and $\sigma(T)$ from the experimental data $\bar{\sigma}_{exp}(T_S)$.

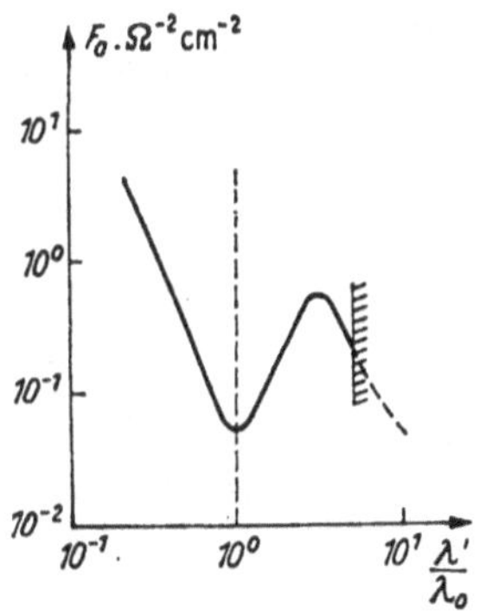

Fig. 2.10. The dependence of the functional F_0 on the ratio λ'/λ_0 obtained by solving the model inverse problem

2.6.2. Solution Singularity

Unfortunately, it is obvious that in addition to the physical solution the system (2.5), (2.11), (2.12) admits at least one trivial mathematical solution $\sigma(T) \equiv \bar{\sigma}_{exp}(T_S)$, $\lambda(T) \equiv \infty$ (in this case $T_i(r) = \text{const} \equiv T_S$). This mathematical solution has no physical meaning: $\lambda = \infty$ should mean that the average photon mean free path is ∞, i.e. much larger than the capillary diameter. This is in contradiction to the conditions of the ICID model. So, the physical solution of the problem must obey the following additional physical conditions (corresponding to the restrictions of the ICID model, given in 2.3.):

— $\sigma > 0$, $\lambda > 0$;
— $\sigma(T)$, $\lambda(T)$ — continuous functions;
— $l_R \ll R$.

The last condition can be defined more precisely according to [29]: $l_R \leqq \frac{1}{3} R$.

2.6.3. Solution Stability, Error Analysis

In our case, the solution stability is defined by the dependence of the divergence of the functions $\sigma(T)$ and $\lambda(T)$ on the scattering of experimental values $\bar{\sigma}_{exp}(T_S)$, i.e. by the relation between the errors of the final results and the initial data. From (5) we see that $\bar{\sigma}$ is proportional to $\sigma(T)$ so $\Delta\sigma(T) = \Delta\bar{\sigma}_{exp}$. To obtain the error in $\lambda_R(T)$ we solved the system (2.5), (2.11), (2.12) for different equidistant curves $\lambda(T)$. We came to the result that for a real error in σ of $\delta_\sigma \approx \pm 10\%$, the corresponding error $\delta_\lambda \approx \approx \pm 50\%$. This conclusion corresponds to a minimal error. The real one, containing additional methodic errors due to the approximate character of the solution of the equations, is somewhat larger.

2.6.4. The Algorithm

The algorithm consists in finding the appropriate probe functions $\sigma_p(T)$ and $\lambda_p(T)$ minimizing the functional F_0. The resulting error depends on the choice of the probe functions.

In preliminary calculations both monotonous and extremal probe-functions $\sigma_p(T)$ were tested (curves 1 and 2 on fig. 2.11) since both kinds of resulting dependences

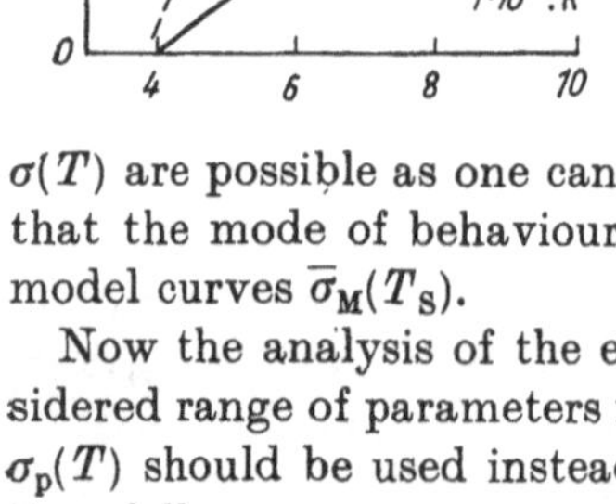

Fig. 2.11. Two kinds of probe-functions $\sigma_p(T)$ and related to them dependences $\bar{\sigma}_M(T)$ obtained using the algorithm of the direct problem. 1, 2 — probe-functions, 1′, 2′ — resulting curves $\bar{\sigma}_M(T)$ respectively

$\sigma(T)$ are possible as one can see from figure 2.2. Curves 1′ and 2′ on figure 2.11 show that the mode of behaviour of the probe functions is concerved in the transformed model curves $\bar{\sigma}_M(T_S)$.

Now the analysis of the experimental dependences $\bar{\sigma}_{exp}(T_S)$ show that in the considered range of parameters no minima are observed. So a monotonous probe function $\sigma_p(T)$ should be used instead of the general power series expansion (14). We chosed it as follows:

$$\sigma_p(T) = \sigma_0 + \sigma_1(T - T')^n . \tag{2.16}$$

Here σ_0, σ_1, T', n are parameters to be determined and $T' < \min(T_S)$. For $\lambda_p(T)$ equation (2.15) was used. So, the solution consists of finding six parameters σ_0, σ_1,

T', n, λ_0 and m. This was achieved by a statistical iteration method described in [30]. The resulting error for $\lambda(T)$ is of a factor of 2, and for $\sigma(T)$ it is about $\pm 30\%$.

2.6.5. "Two-Diameter" Method

Thé above described algorithm allows to determine $\sigma(T)$ and $\lambda(T)$ from experimental isobaric curves $\bar{\sigma}_{exp}(T_S)$ for capillaries of the same diameter. Another alogrithm completing the first one is applicable when isobaric curves $\bar{\sigma}_{exp}(T_S)$ are obtained at different capillary diameters. Such an example is given on figure 2.12 for Li at $P =$

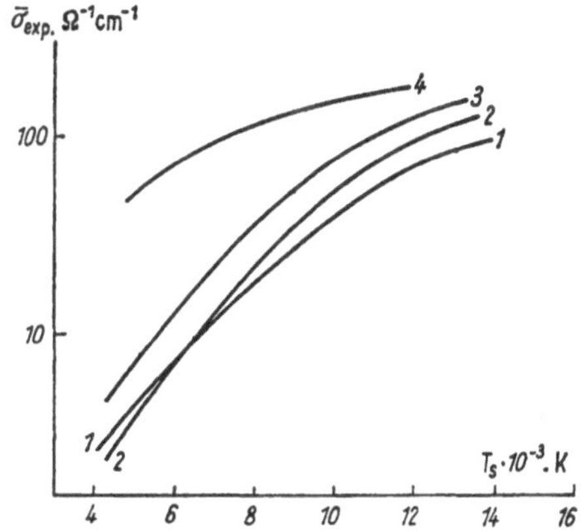

Fig. 2.12. The averaged dependences of the mean electrical conductivity of Li on the arc column outer surface temperature for different plasma pressures. 1 — $P = 10$ MPa, $d = 2$ mm, 2 — $P = 30$ MPa, $d = 1.83$ mm, 3 — $P = 30$ MPa, $d = 2.29$ mm, 4 — $P = 100$ MPa, $d = 2.12$ mm

$= 30$ MPa. It can be seen that the curve $\bar{\sigma}_{exp}(T_S)$ corresponding to $d = 2.29$ mm lays systematically higher than that for $d = 1.83$ mm. This fact, easy to explain from the standpoint of the ICID model, has been used to elaborate a second two-diameter iterating procedure, independent of the first, allowing to obtain $\sigma(T)$ and $\lambda(T)$. Its advantage consists of the absolute singularity of the solution. The resulting error in σ and λ depends on the difference Δd between the two chosen values of the diameters. It is evident that the larger Δd is, the less the error is. Unfortunately, in practice it seems difficult to get correct results at significantly different diameters: at low diameters the radiative thermal conductivity approach is poorer realized (condition 5) of 2.3.), at large ones, the plasma thermal inertia (condition 2) of 2.3.) becomes important. In both cases, the ICID model fails. So the two-diameter method is applicable only for not too different diameters and then the error is significant. Nevertheless in the present work it has been locally applied (in particular for Li at $P = 30$ MPa) and the results obtained by both methods have been found to coincide.

2.7. The Experimental Results and their Discussion

The isobars $\sigma_p(T)$ for Cs, K, Na and Li are presented in figure 2.13. The average error of these data is $\pm 30\%$. The results for the cesium electrical conductivity (fig. 2.13a)) are compared with the experimental data of [31] obtained by radio-frequency measure-

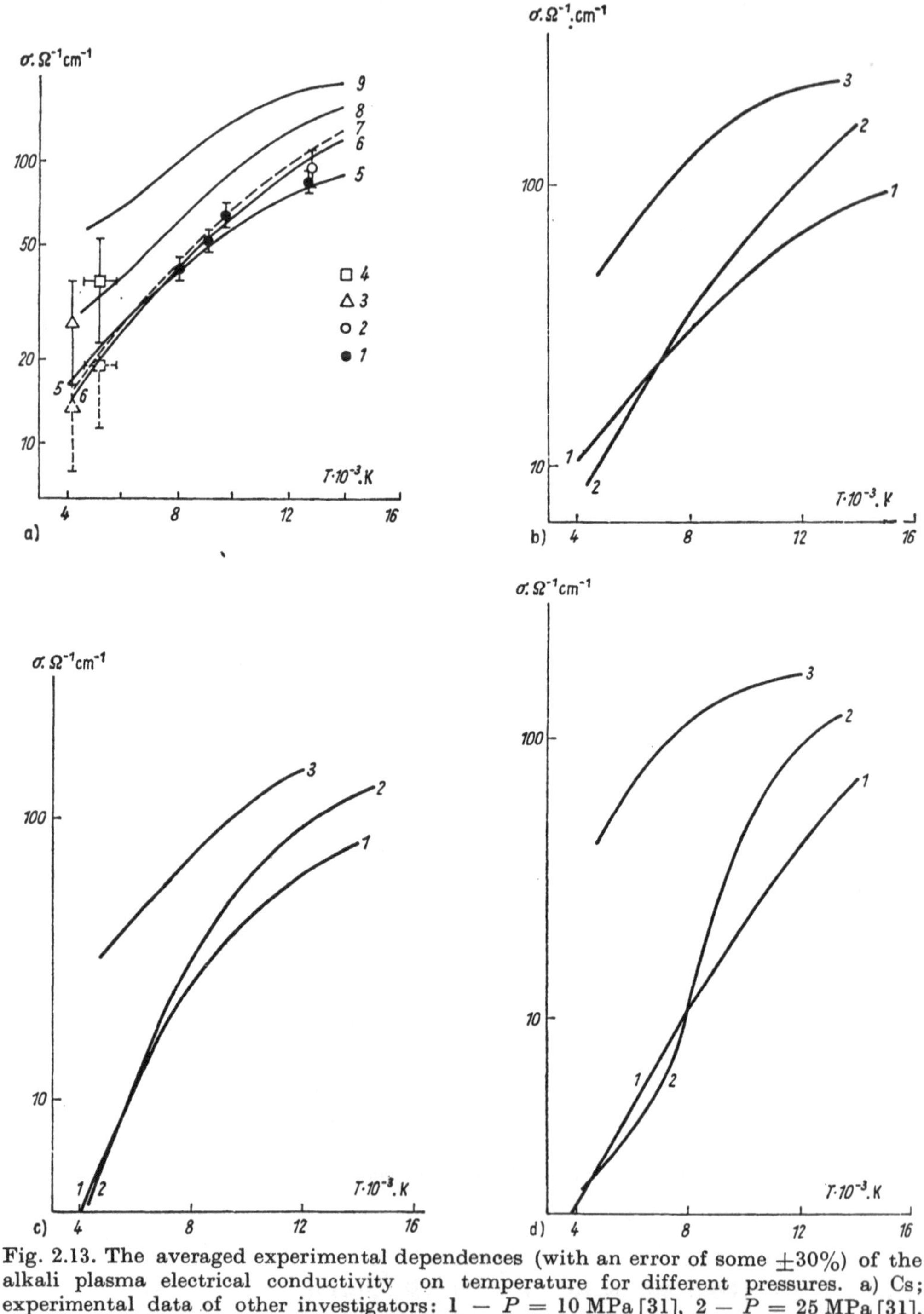

Fig. 2.13. The averaged experimental dependences (with an error of some ±30%) of the alkali plasma electrical conductivity on temperature for different pressures. a) Cs; experimental data of other investigators: 1 — $P = 10$ MPa [31], 2 — $P = 25$ MPa [31], 3 — $P = 10$ MPa [32], 4 — $P = 25$ MPa [32]; the present work: 5 — $P = 10$ MPa, 6 — $P = 25$ MPa, 7 — $P = 30$ MPa (result of interpolation), 8 — $P = 60$ MPa, 9 — $P = 100$ MPa; b) K: 1 — $P = 10$ MPa, 2 — $P = 30$ MPa, 3 — $P = 100$ MPa; c) Na: 1 — $P = 10$ MPa, 2 — $P = 30$ MPa, 3 — $P = 100$ MPa; d) Li: 1 — $P = 10$ MPa, 2 — $P = 30$ MPa, 3 — $P = 100$ MPa

ments in shock-wave experiments and of [32] measured by the same method in adiabatic compression experiments. The results of [31] agree with the present ones. The data of [32] seem to be unreasonably high, confirming the opinion of the authors of [32] according to which their results in figure 2.13a) are corrected with broken lines.

2.7.1. The Plasma Composition

Further discussion requires information about the plasma composition. Up to now, such an information has been obtained only from theoretical calculations. In the present work we use the results, obtained by the method described in [33], for nonideal plasma equilibrium composition estimation. The effect of pressure ionization of Cs

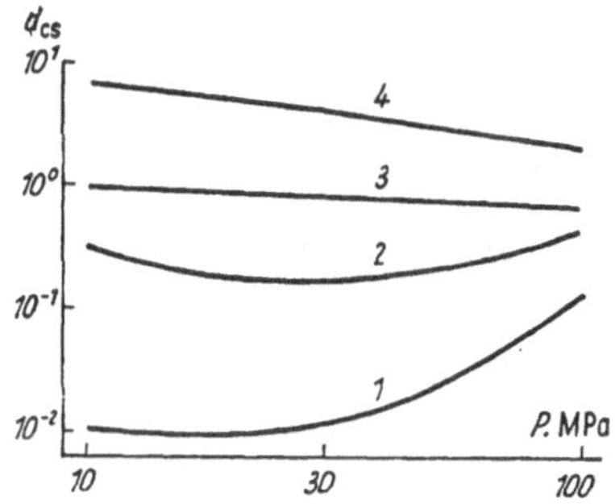

Fig. 2.14. The dependence of the calculated Cs plasma ionization degree $\alpha = \frac{n_e}{n_{a\Sigma}}$ on pressure for different temperatures 1 — T = 4000 K, 2 — T = 6000 K, 3 — T = 8000 K, 4 — T = 12000 K

can be seen in table 1 and figure 2.14 in which the calculated isotherms $\alpha(P)_T = \frac{n_e}{n_{a\Sigma}}(P)_T$ (where $n_{a\Sigma}$ is the total number of neutral particles) are plotted vs. pressure. At $T = 4000$ K the pressure ionization mechanism is predominating in the whole range of pressures, and thus, $\alpha(P)_T$ is a raising function of pressure. At $T = 6000$ K, pressure ionization causes the apparition of a minimum in the pressure dependence of $\alpha(P)_T$. At higher temperatures the minimum disappears, though the values of α are sensibly affected by non-ideality. From the plasma composition data it can be seen that, in the present experiments, the non-ideality parameter $\gamma = e^2 n_e^{1/3} k_B T$ varies in the range $0.3 \leqq \gamma \leqq 2$.

It should be emphasized that, in the whole considered range of temperatures and pressures, the positive, negative and neutral clusters concentrations have been calculated to be small and should not be taken into account to explain the present results for σ and λ. Therefore, practically in the whole considered range of P and T, $n_{a\Sigma} = n_a$.

2.7.2. Coulomb Electrical Conductivity Data

To extract σ_C and σ_C^* from the measured values σ for Cs let us assume that the experimental values of the conductivity can be expressed by the so-called "additive" formula

$$\sigma = (\sigma_{ea}^{-1} + \sigma_c^{-1})^{-1}$$

where

$$\sigma_{ea} = \frac{3}{4} \frac{e^2 n_e}{m_e \bar{v}_e n_a Q_{la}^{(1,1)}} .$$

Here $\bar{v}_e = \sqrt{8k_B T/\pi m_e}$ is the average electron velocity, $Q_{ea}^{(1,1)}$ is the effective transport electron-neutral cross-section for the first-order approximation to an exact value of σ. As it has been shown in [34], these rather simple relations are satisfactory in the case of cesium. Taking $Q_{ea}^{(1,1)}$ for Cs from [34] and using the data of table 2.1, it is easy to

Tab. 2.1. Cesium Dense Plasma Composition

P, MPa	T, K	n_e, cm^{-3}	$n_{a\Sigma}$, cm^{-3}	$\alpha = n_e/n_{a\Sigma}$	γ
10	4000	2.0/18	1.8/20	1.1/$\bar{2}$*)	0.52
	6000	1.8/19	5.6/19	3.2/$\bar{1}$	0.73
	8000	3.2/19	3.3/19	9.6/$\bar{1}$	0.66
	12000	3.2/19	4.7/18	6.7/0	0.44
	16000	2.5/19	9.3/17	2.7/1	0.31
	20000	2.1/19	2.9/17	7.2/1	0.29
30	4000	7.8/18	6.6/20	1.2/$\bar{2}$	0.83
	6000	4.8/19	2.8/20	1.7/$\bar{1}$	1.01
	8000	9.8/19	1.2/20	8.2/$\bar{1}$	0.96
	12000	1.0/20	2.5/19	4.0/0	0.65
	16000	8.0/19	6.3/18	1.3/1	0.45
	20000	6.3/19	2.0/18	3.1/1	0.33
100	4000	1.2/19	7.4/20	1.6/$\bar{1}$	2.06
	6000	2.5/20	5.9/20	4.2/$\bar{1}$	1.75
	8000	2.6/20	4.0/20	6.5/$\bar{1}$	1.33
	12000	2.4/20	1.3/20	1.9/0	0.87
	16000	2.2/20	4.0/19	5.5/0	0.63
	20000	1.9/20	1.6/19	1.2/1	0.48

*) 1.1/$\bar{2}$ should be read as $1.1 \cdot 10^{-2}$

get the coulombic conductivities σ_c from the present experimental results. This operation was performed for the temperature range $T = 5000$ up to 14000 K. The resulting non-dimensional dependences $\sigma_c^*(\gamma)_T$ for Cs at $T = 5000$, 10000 and 14000 K are presented on figure 2.15. At lower non-ideality degrees the three isotherms tend to coincide indicating the position of the universal function $\sigma_c^*(\gamma)$ for classical plasma. At higher values of γ every isotherm goes its separate way: the higher T is the steeper $\sigma_c^*(\gamma)_T$ raises. These data for Cs nonideal plasma at $T \approx 10^4$ K are compared here with the shock-wave experimental results of [16] for Ne, Ar and Xe plasma at $T = 25000$ and 70000 K. Note that at the first temperature $\sigma_c^*(\gamma)_T$ is situated higher and at $T \approx 70000$ K it lowers. This peculiarity can be seen clearlier in the form of $\sigma_c^*(T)_\gamma$ shown on figure 2.16 for $\gamma = 0.9$. The raising part of this curve qualitatively corresponds to the theoretical predictions of [11—15]; the falling one seems to be due to the ionic core quantum effects described in [16]. So the curve on figure 2.16 can be considered as a quantitative confirmation of the qualitative dependence of figure 2.1.

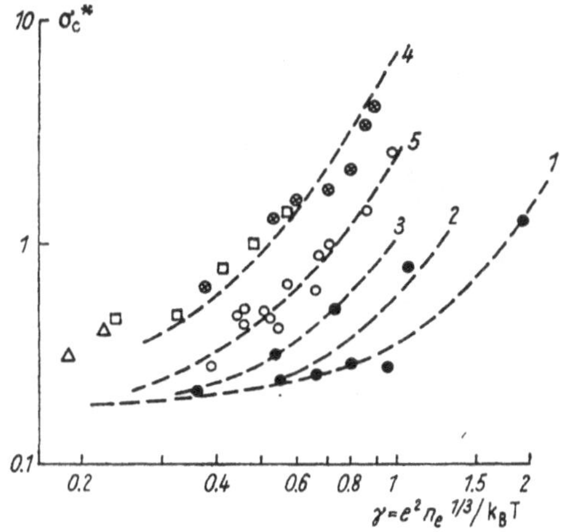

Fig. 2.15. The dependence of the non-dimensional coulombic electrical conductivity on non-ideality degree (the averaged error is ± 40 up to 50%). ● — Cs (present work), $T \approxeq 5000$ up to 14000 K, △ — Ne [16], $T \approxeq 25000$ K, □ — Ar [16], $T = 25000$ K, ⊗ — Xe [16], $T = 25000$ K, ○ — Xe [16], $T \approxeq 70000$ K; isotherms: 1 — 5000 K, 2 — 10000 K, 3 — 14000 K, 4 — 25000 K, 5 — 70000 K

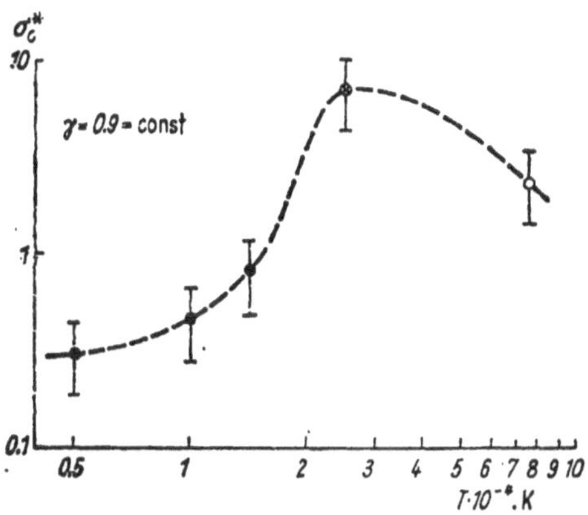

Fig. 2.16. The temperature dependence of σ_c^* for $\gamma = 0.9 =$ const. ● — Cs (present work), ⊗ — Xe [16], ○ — Xe [16]

2.7.3. Pressure Ionization Effect

The isotherms $\sigma(P)$ are shown for $T = 6000$ and 10000 K on figure 2.17. At $T = 6000$ K (fig. 2.17a)) the curves for all the alkali metals show a characteristic minimum corresponding to pressure ionization as predicted in 2.1. (fig. 2.2b)). At $T = 10000$ K (fig. 2.17b)) the monotonous character of the isoterms proves that here both temperature and pressure mechanisms compete in increasing σ. At $P = 100$ MPa the electrical conductivity of all the alkali metals is essentially the same, indicating that plasma is ionized enough (both thermally and by pressure) that coulombic interactions are predominating and the influence on σ of the individual properties of the particles disappears.

2.7.4. The Radiative Thermal Conductivity

The resulting isobars $\lambda(T)_P$ for Li, Na, K and Cs are presented on figure 2.18. They are compared to the quasiideal plasma calculation data of [28]. The lower the temperature is the more the experimental curves $\lambda_{exp}(T)$ exceed theoretical ones. This

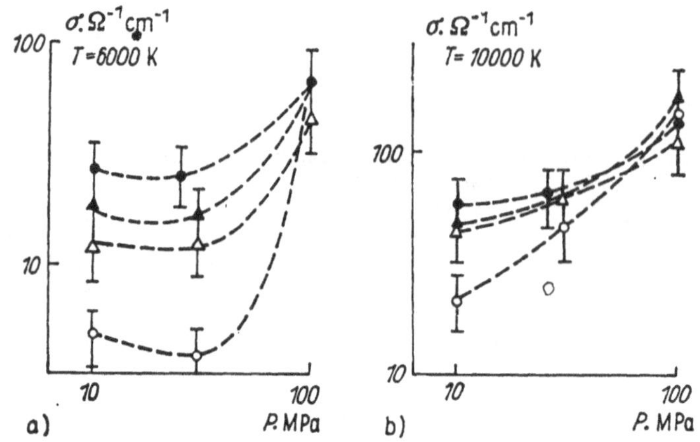

Fig. 2.17. The pressure dependences of the alkali plasma electrical conductivity (○ – Li, △ – Na, ▲ – K, ● – Cs). a) $T = 6\,000$ K = const, b) $T = 10\,000$ K = const

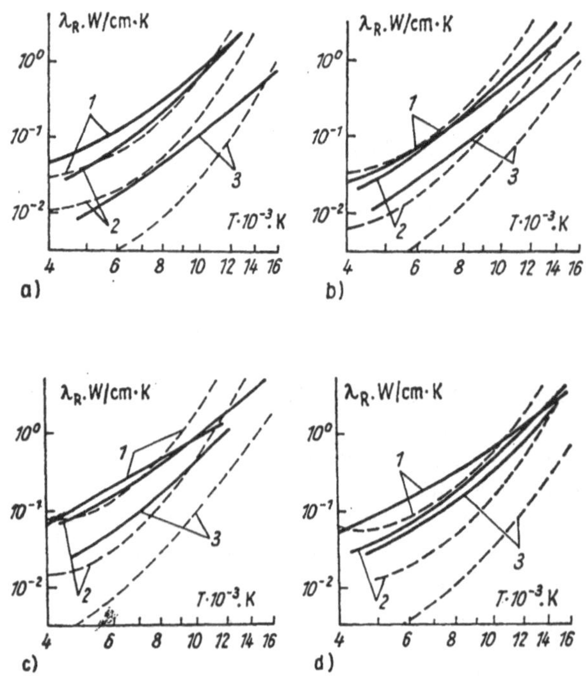

Fig. 2.18. The averaged (with an error of a factor two) experimental dependences (solid lines) of the alkali plasma thermal conductivity on temperature (dotted lines – theory [28]). a) Cs: 1 – $P = 10$ MPa, 2 – $P = 25$ MPa, 3 – $P = 100$ MPa; b) K: 1 – $P = 10$ MPa, 2 – $P = 30$ MPa, 3 – $P = 100$ MPa; c) Na: 1 – $P = 10$ MPa, 2 – $P = 30$ MPa, 3 – $P = 100$ MPa; d) Li: 1 – $P = 10$ MPa, 2 – $P = 30$ MPa, 3 – $P = 100$ MPa

discrepancy grows with pressure and reaches one order of magnitude at $P = 100$ MPa and $T \approx 5000$ K. Such a behaviour of $\lambda(T)$ seems to confirm the above mentioned idea of the enhanced nonideal plasma transparency corresponding to increased photon mean free path. This pecularity of the radiative thermal conductivity of dense plasma can be seen more clearly in the case of the isotherms $\lambda(P)_T$ shown on figure 2.19. Here only data for Cs and Li are presented at two temperatures. The growth of the discrepancy between the theoretical predictions and the results of measurements with the raise of pressure is expressed graphically. One can notice that for a given pressure the lower the temperature is, i.e. the more nonideal the plasma is, the greater becomes this descrepancy. It is interesting to follow this trend in the nondimensional coordinates $\lambda_{exp}/\lambda_{th}$ vs. γ for fixed temperatures (see fig. 2.20). At $T = 6000$ K nearly everywhere $(\lambda_{exp}/\lambda_{th}) > 1$, the dependence going upward with γ. For $T = 10000$ K at the lowest values of γ this ratio is less than unity. If γ raises, it grows up. This is true for all the alkali metals.

These observations seem to confirm the existence of "spectrum windows" increasing the photon mean free path with non-ideality, as discussed in 2.1.

In conclusion the authors thank Dr. B. M. Kovaliov for calculations of the dense Cs composition, Dr. N. V. Yermokhin, Dr. A. V. Pallo and V. V. Vorobiov for their help in ICID-experiments and all of them for helpful and fruitful discussions.

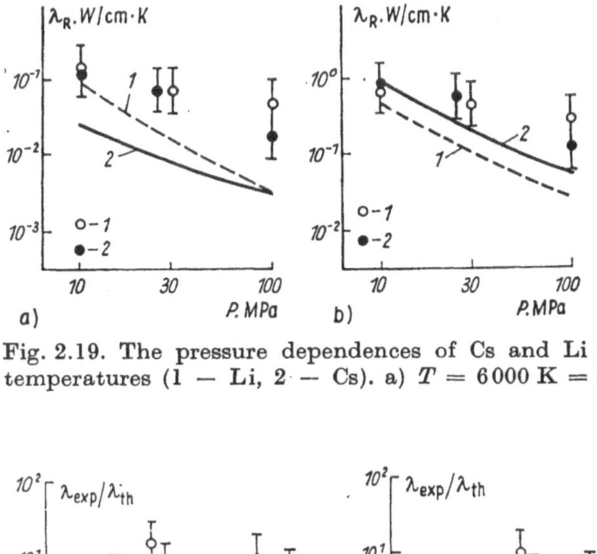

Fig. 2.19. The pressure dependences of Cs and Li thermal conductivities for different temperatures (1 — Li, 2 — Cs). a) $T = 6000$ K = const, b) $T = 10000$ K = const

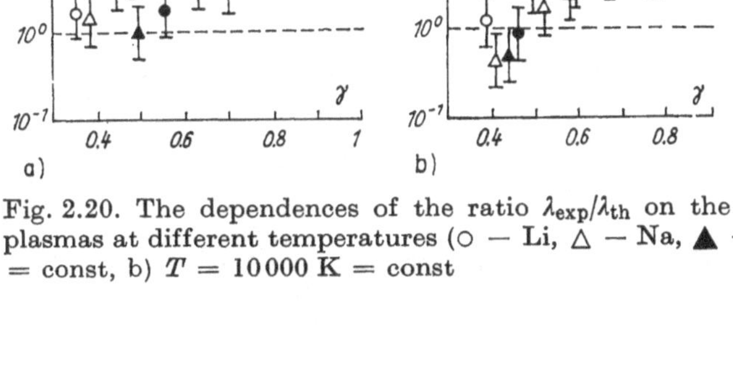

Fig. 2.20. The dependences of the ratio $\lambda_{exp}/\lambda_{th}$ on the non-ideality degree γ for alkali plasmas at different temperatures (○ — Li, △ — Na, ▲ — K, ● — Cs). a) $T = 6000$ K = = const, b) $T = 10000$ K = const

3. Kinetic Equations and Linear Response Theory for Dense Coulomb Fluids

W. Ebeling, Yu. L. Klimontovich, W. D. Kraeft, G. Röpke

3.1. Introduction

As demonstrated in the foregoing chapters the increasing interest to dense nonideal plasmas is due to its importance for many problems of practical interest as thermonuclear fusion, MHD-generators, extreme pressure devices, highly doped or excited semiconductors, liquid metals etc. [1, 2, 3]. On the basis of the new experimental facts and the new physical models which were presented in part in chapters 1 and 2 the theoretical study of nonideal fluids with Coulombic interactions was the subject of many investigations [1—4]. The simplest model of a plasma is the one-component plasma (OCP) model [4]. The OCP-model is clearly insufficient for the study of transport properties which are mainly due to the quantum-mechanical scattering processes between electrons and ions. Therefore our investigation is based on the many component quantum plasma (MCQP) model considering the electrons as light quantum particles subject to Fermi statistics and the ions as heavy nearly classical particles. The plasma is characterized by the following Hamiltonian

$$H = \sum_a \sum_i \frac{\boldsymbol{p}_i^2}{2m_a} + \frac{1}{2} \sum_{ab} \sum_{ij} \Phi_{ab}(|\boldsymbol{r}_i - \boldsymbol{r}_j|) \,, \tag{3.1}$$

where the first term represents the kinetic energy while the second one is the potential part with Φ_{ab} being the two particle interaction energy between species a and b. The summations over a, b, ... run over the species and the summations over i, j, ... run over the number of particles of the corresponding species.

We restrict ourselves to the nonrelativistic region, i.e. to temperatures below 10^7 K. With respect to the densities we try to cover the region up to liquid metals.

The aim of this investigation is the application of recent concepts and results of kinetic theory as well as of the linear response theory for the transport coefficients. Besides the general concepts of the transport theory of the MCQP-model we are interested here especially in the following effects of nonideality:

(i) *Short-range non-Coulombic forces.* While the interaction between the electrons is the pure Coulombic one

$$\Phi_{ee}(r) = \frac{e^2}{r} \,, \tag{3.1 a}$$

for the ions the internal structure leads to deviations from the Coulomb potential.

For the electron-ion interaction we assume pseudopotentials as introduced e.g. 1935 by HELLMANN and GASPAR. We use the special expression

$$\Phi_{ie}(r) = e_i\, e_e \left[\frac{1}{r} - A\,\frac{e^{-Br}}{r} - C\,e^{-Br}\right], \tag{3.1b}$$

which contains the Hellmann potential ($C = 0$) and the Krasko-Gurski ($A = 1$) expression as special cases. For the ion-ion interaction we shall sometimes use hard-core potentials

$$\Phi_{ii}(r) = \begin{cases} e_i^2/r & \text{if} \quad r \geqq 2R_i\,, \\ \infty & \text{if} \quad r < 2R_i\,, \end{cases} \tag{3.1c}$$

R_i being the effective ion radius.

(ii) *Quantum effects in the scattering process.* The consequent quantum-statistical treatment avoids divergences which must, in classical theories, be overcome by arbitrary cut-off procedures at small distances. The essential length in this connection is the thermal de Broglie wave length of the electrons

$$\Lambda_e = h[2\pi m_e k_B T]^{-1/2}\,.$$

(iii) *Dynamical screening effects.* A peculiarity of Coulomb systems is the occurrence of long range divergencies, which are overcome by taking into account screening. The simplest manner to do this is to replace the bare interaction potential by a statically screened one. This means in Fourier space

$$V(q) \to V(q)/\varepsilon(q, 0)\,,$$

$$\varepsilon(q, 0) \approx (1 + \varkappa^2/q^2)\,,$$

$$\varkappa^2 = 4\pi \sum_a n_a\, e_a^2/k_B T\,;$$

$\varkappa$ — reciprocal Debye length ($\varkappa = r_D^{-1}$).
A more refined treatment has to take into account dynamical screening.

(iv) *Debye-Onsager relaxation effect.* Around electrons moving in an electrical field an asymmetrical charge cloud is formed which gives rise to a field in opposite direction. Therefore the effective field is smaller than the external one. This effect was first studied in electrolyte theory.

(v) *Strong interaction effects.* In dense systems it is necessary to go beyond first order perturbations with respect to the potentials in order to describe in a correct manner large angle scattering, formation of bound states etc. Multiple scattering may be accounted for by the introduction of scattering matrices.

(vi) *Equilibrium correlations.* It is known that the equilibrium correlations influence strongly the transport properties. For example the structure factor is of primary importance in liquid metal theory. For this reason the equilibrium correlation are treated here as good as possible.

Finally let us consider briefly a quantitative definition of non-ideality in a plasma. We introduce the following set of dimensionless parameters (n-total density of the charged particles)

$$\begin{aligned}
\Gamma &= (\varepsilon_{\text{pot}}/\varepsilon_{\text{kin}}) = \left(\frac{e^2}{2d}\right)\bigg/\left(\frac{\theta}{2}\right), \\
\xi &= (\varepsilon_{\text{bound}}/\varepsilon_{\text{kin}}) = \left[|E_{10}|\bigg/\frac{\theta}{2}\right]^{1/2}, \\
\mu &= (8\pi n_e e^6)^{1/2}\,\theta^{-3/2}, \\
\delta &= n_e \Lambda_e^3, \qquad d = (3/4\pi n)^{1/3}, \\
\zeta &= \delta \exp\left[-E_{10}/\theta\right],
\end{aligned} \tag{3.2}$$

where $(\theta/2)$ denotes the mean kinetic energy per degree of freedom of the electrons (α-ideal chemical potential)

$$\begin{aligned}
\theta/2 &= 2(n_e\Lambda_e^3)^{-1}\, I_{3/2}(\alpha), \\
\Lambda_e &= h(2\pi m_e\, k_B T)^{-1/2}.
\end{aligned}$$

The region of nonideality is given by $\Gamma \geqq 1$. In our considerations we have to restrict ourselves to cases of weak or moderate nonideality i.e. Γ is not to large. Further we shall exclude here the region $\zeta \geqq 1$ where bound states play the role of new particles (see chap. 4.). The region which is of interest here is shown in figure 3.1.

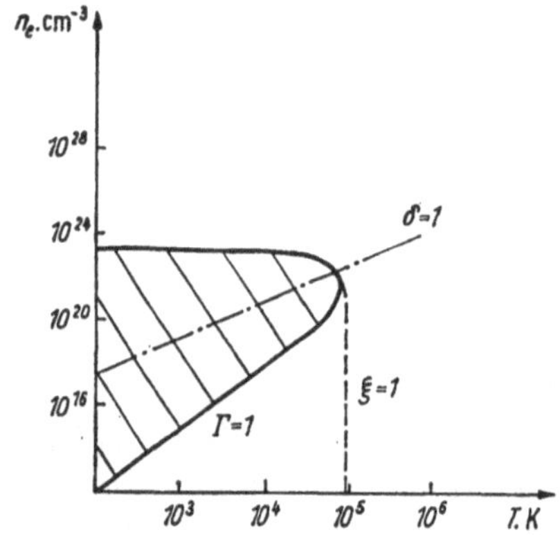

Fig. 3.1. The region of non-ideality in the density-temperature-plane in thermodynamical equilibrium

(vii) *Non-equilibrium correlations*. In thermodynamic equilibrium the correlation length is given by the Debye radius

$$r_{\text{corr}} \approx r_D = [4\pi \sum_a n_a e_a^2/k_B T]^{-1/2}.$$

Far from equilibrium large-scale fluctuations may occur with correlation length much exceeding the Debye radius,

$$r_{\text{corr}} \gg r_D.$$

These large-scale fluctuations may be connected e.g. with ion sound instabilities, two-stream instabilities and turbulent regimes. The nonideality parameter is in such situations given by

$$\Gamma = \varepsilon_{\text{pot}}/\varepsilon_{\text{kin}} = [\int \mathrm{d}\boldsymbol{k} (\delta \boldsymbol{E}\, \delta \boldsymbol{E})_k / 8\pi] : [\sum_a \int \mathrm{d}\boldsymbol{p}\, \boldsymbol{p}^2 f_a / 2m_a] \,, \tag{3.3}$$

where $(\delta \boldsymbol{E}\, \delta \boldsymbol{E})_k$ denotes the correlation function of the electrical field. If the system is pumped by an external source of free energy e.g. by a battery driving the current, the potential energy may be much higher than in equilibrium at the same values of the temperature and density. This may give rise to strong effects of nonideality even at rather low densities and high temperatures (fig. 3.2).

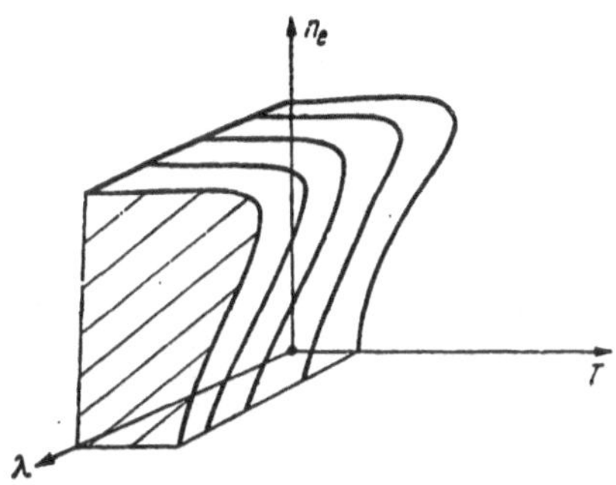

Fig. 3.2. The region of non-ideality under non-equilibrium situations (λ — distance from equilibrium)

3.2. Kinetic Theory

3.2.1. Classical Kinetic Equations

The classical kinetic equation for the distribution function of charges is assumed to be of the form

$$\frac{\partial f}{\partial t} + \frac{\boldsymbol{p}}{m_a} \cdot \frac{\partial f}{\partial \boldsymbol{r}} + e_a \left[\boldsymbol{E} + \frac{1}{c} \boldsymbol{u} \times \mathbf{H} \right] \cdot \frac{\partial f_a}{\partial \boldsymbol{p}} = I_n(\boldsymbol{r}, \boldsymbol{p}, t) + \tilde{I}_a(\boldsymbol{r}, \boldsymbol{p}, t) + \sum_n \nu_{an}(\boldsymbol{r}, \boldsymbol{p}) [f_a^0 - f_a] \,. \tag{3.4}$$

Here I_a is the usual collision term for charged particle collisions which are determined by small scale fluctuations $r_{\text{corr}} \leqq r_{\text{D}}$. On the other hand $\tilde{I}_a$ describes the correlations with $r_{\text{corr}} \geqq r_{\text{D}}$ which are determined by large-scale fluctuations.

In general this contribution may be omitted but there are special situations e.g. in turbulent plasmas where just this will be the dominant one. Further f_a^0 is the Maxwell-distribution and ν_{an} the frequency of collisions of the charged particles of species a with neutrals of species n which is given by

$$\nu_{an} = n_n \frac{|p|}{m_a} q_{an}(p) \,,$$

$$q_{an}(p) = \int \mathrm{d}\Omega (1 - \cos \theta)\, q_{an}(\Omega, p) \,.$$

The transport cross sections q_{an} may be expressed also by the scattering phase shifts

$$q_{an}(p) = \frac{4\pi}{k^2} \sum_{l=0}^{\infty} (l+1) \sin^2(\delta_l - \delta_{l+1}), \quad p = \hbar k .$$

The distribution functions f_a are normalized in such a way that

$$n_a(\boldsymbol{r}, t) = \int f_a(\boldsymbol{r}, \boldsymbol{p}, t) \, \mathrm{d}\boldsymbol{p}$$

corresponds to the mean local density of the species a at the point $\boldsymbol{r}$. Besides we shall use also the distribution function

$$F_a(\boldsymbol{r}, \boldsymbol{p}, t) = V f_a(\boldsymbol{r}, \boldsymbol{p}, t)/N_a ,$$

which is normalized to unity and the binary distribution function of the pair a, b denoted by F_{ab} [5]. Using these quantities the collision term may be written as

$$I_a(\boldsymbol{r}, \boldsymbol{p}, t) = \frac{\partial}{\partial \boldsymbol{p}} \sum_b n_a n_b \int \mathrm{d}\boldsymbol{r}' \, \mathrm{d}\boldsymbol{p}' \frac{\partial \Phi_{ab}}{\partial \boldsymbol{r}} g_{ab}(\boldsymbol{r}, \boldsymbol{r}', \boldsymbol{p}, \boldsymbol{p}', t) ,$$

$$g_{ab}(\boldsymbol{r}, \boldsymbol{r}', \boldsymbol{p}, \boldsymbol{p}', t) = F_{ab}(\boldsymbol{r}, \boldsymbol{r}', \boldsymbol{p}, \boldsymbol{p}', t) - F_a(\boldsymbol{r}, \boldsymbol{p}, t) F_b(\boldsymbol{r}', \boldsymbol{p}', t) . \tag{3.5}$$

Let us introduce now relative and barycentric coordinates

$$\boldsymbol{x} = \boldsymbol{r} - \boldsymbol{r}' , \qquad \boldsymbol{y} = (\boldsymbol{r} + \boldsymbol{r}')/2 ,$$

$$g_{ab}(\boldsymbol{r}, \boldsymbol{r}', \boldsymbol{p}, \boldsymbol{p}', t) \to g_{ab}(\boldsymbol{x}, \boldsymbol{y}, \boldsymbol{p}, \boldsymbol{p}', t) .$$

Assuming that the dependence on $\boldsymbol{y}$ is relatively weak we get by Taylor expansion

$$g_{ab}(\boldsymbol{x}, \boldsymbol{y}, \boldsymbol{p}, \boldsymbol{p}') = g_{ab}(\boldsymbol{x}, \boldsymbol{r}, \boldsymbol{p}, \boldsymbol{p}') - \frac{\boldsymbol{x}}{2} \cdot \frac{\partial}{\partial \boldsymbol{r}} g_{ab}(\boldsymbol{x}, \boldsymbol{r}, \boldsymbol{p}, \boldsymbol{p}') ,$$

$$I_a(\boldsymbol{r}, \boldsymbol{p}, t) = \frac{\partial}{\partial \boldsymbol{p}} \cdot \sum_b n_a n_b \int \mathrm{d}\boldsymbol{x} \, \mathrm{d}\boldsymbol{p}' \frac{\partial \Phi_{ab}}{\partial \boldsymbol{x}} g_{ab}(\boldsymbol{x}, \boldsymbol{r}, \boldsymbol{p}, \boldsymbol{p}')$$
$$- \frac{1}{2} \frac{\partial}{\partial \boldsymbol{p}} \frac{\partial}{\partial \boldsymbol{r}} \cdot\cdot \sum_b n_a n_b \int \mathrm{d}\boldsymbol{x} \, \mathrm{d}\boldsymbol{p}' \, \boldsymbol{x} \frac{\partial \Phi_{ab}}{\partial \boldsymbol{x}} g_{ab}(\boldsymbol{x}, \boldsymbol{r}, \boldsymbol{p}, \boldsymbol{p}') .$$

Another representation of the collision integral follows by defining even and odd parts of the correlations

$$g_{ab} = g_{ab}^{(+)} + g_{ab}^{(-)}; \qquad g_{ab}^{(+-)} = (g_{ab} \pm g_{ba})/2 . \tag{3.6}$$

Introducing this into the collision integral one finds

$$I_a(\boldsymbol{r}, \boldsymbol{p}, t) = I_a^{(+)} + I_a^{(-)} ,$$

$$I_a^{(+)}(\boldsymbol{r}, \boldsymbol{p}, t) = -\frac{1}{2} \frac{\partial}{\partial \boldsymbol{r}} \frac{\partial}{\partial \boldsymbol{p}} \cdot\cdot \sum_b n_a n_b \int \mathrm{d}\boldsymbol{x} \, \mathrm{d}\boldsymbol{p}' \, \boldsymbol{x} \frac{\partial \Phi_{ab}}{\partial \boldsymbol{x}} g_{ab}^{(+)}(\boldsymbol{x}, \boldsymbol{r}, \boldsymbol{p}, \boldsymbol{p}') ,$$

$$I_a^{(-)}(\boldsymbol{r}, \boldsymbol{p}, t) = \frac{\partial}{\partial \boldsymbol{p}} \cdot \sum_b n_a n_b \int \mathrm{d}\boldsymbol{x} \, \mathrm{d}\boldsymbol{p}' \frac{\partial \Phi_{ab}}{\partial \boldsymbol{x}} g_{ab}^{(-)}(\boldsymbol{x}, \boldsymbol{r}, \boldsymbol{p}, \boldsymbol{p}') . \tag{3.7}$$

By transformation into a Fourier space using

$$g_{ab}(\boldsymbol{k}) = \int \exp(-i\boldsymbol{k}\boldsymbol{x}) \, g_{ab}(\boldsymbol{x}) \, \mathrm{d}\boldsymbol{x} \tag{3.8}$$

we see that $g_{ab}^{(+)}$ yields Re $g_{ab}(\boldsymbol{k})$ and $g_{ab}^{(-)}$ yields Im $g_{ab}(\boldsymbol{k})$ respectively. Therefore we get the equivalent representation in the Fourier space

$$I_a^{(+)} = \frac{1}{(2\pi)^3} \frac{\partial}{\partial \boldsymbol{r}} \frac{\partial}{\partial \boldsymbol{p}} \cdot\cdot \sum_b n_a n_b \int \mathrm{d}\boldsymbol{k}\, \mathrm{d}\boldsymbol{p}' \left(\boldsymbol{\delta} + \boldsymbol{k} \frac{\partial}{\partial \boldsymbol{k}}\right) \cdot \\ \cdot \Phi_{ab}(\boldsymbol{k}) \operatorname{Re} g_{ab}(\boldsymbol{k}, \boldsymbol{r}, \boldsymbol{p}, \boldsymbol{p}') ,$$

$$I_a^{(-)} = \frac{1}{(2\pi)^3} \frac{\partial}{\partial \boldsymbol{p}} \cdot \sum_b n_a n_b \int \mathrm{d}\boldsymbol{k}\, \mathrm{d}\boldsymbol{p}'\, \boldsymbol{k}\, \Phi_{ab}(\boldsymbol{k}) \operatorname{Im} g_{ab}(\boldsymbol{k}, \boldsymbol{r}, \boldsymbol{p}, \boldsymbol{p}') . \tag{3.9}$$

Therefore the problem of the calculation of the collision integrals may be reduced to the problem of finding the real and imaginary part of the Fouriertransform of g_{ab} respectively. We note that the even part of the correlation function $g_{ab}^{(+)}$ or the real part Re $g_{ab}(\boldsymbol{k})$ respectively are given in a rather good approximation by the approximation of local equilibrium correlations

$$g_{ab}^{(+)}(\boldsymbol{x}, \boldsymbol{y}) = F_a(\boldsymbol{r}, \boldsymbol{p})\, F_b(\boldsymbol{r}, \boldsymbol{p}')\, h_{ab}(\boldsymbol{x}) ,$$
$$\operatorname{Re} g_{ab}(\boldsymbol{k}, \boldsymbol{y}) = F_a(\boldsymbol{r}, \boldsymbol{p})\, F_b(\boldsymbol{r}, \boldsymbol{p}')\, h_{ab}(\boldsymbol{k}) ,$$

where h_{ab} denotes the equilibrium correlation function. Therefore we may concentrate ourselves on the calculation of the odd part $g_{ab}^{(-)}$ or Im $g_{ab}(\boldsymbol{k})$ respectively considering the coordinates $\boldsymbol{r}$ simply as parameters.

The equation for the pair distribution function reads neglecting the H-field and neutral particle effects

$$\left(\frac{\partial}{\partial t} + \frac{\boldsymbol{p}}{m_a} \frac{\partial}{\partial \boldsymbol{p}} + \frac{\boldsymbol{p}'}{m_b} \frac{\partial}{\partial \boldsymbol{r}'} + e_a \boldsymbol{E} \frac{\partial}{\partial \boldsymbol{p}} + e_b \boldsymbol{E} \frac{\partial}{\partial \boldsymbol{p}'}\right) F_{ab} =$$
$$= \left\{\frac{\partial \Phi_{ab}}{\partial \boldsymbol{r}} \left(\frac{\partial}{\partial \boldsymbol{p}} - \frac{\partial}{\partial \boldsymbol{p}'}\right) + \sum_c n_c \int \left(\frac{\partial \Phi_{ac}}{\partial \boldsymbol{r}} \frac{\partial}{\partial \boldsymbol{p}} + \frac{\partial \Phi_{bc}}{\partial \boldsymbol{r}'} \frac{\partial}{\partial \boldsymbol{p}'}\right) \frac{F_{abc}}{F_{ab}}\, \mathrm{d}\boldsymbol{r}''\, \mathrm{d}\boldsymbol{p}''\right\} F_{ab} = R_{ab} . \tag{3.10}$$

After replacing the conditional third order correlations by their equilibrium values we get [6, 8]

$$\left(\frac{\partial}{\partial t} + \frac{\boldsymbol{p}}{m_a} \frac{\partial}{\partial \boldsymbol{r}} + \frac{\boldsymbol{p}'}{m_b} \frac{\partial}{\partial \boldsymbol{r}'} + \tilde{\boldsymbol{E}}\left(e_a \frac{\partial}{\partial \boldsymbol{p}} + e_b \frac{\partial}{\partial \boldsymbol{p}'}\right)\right) g_{ab} = R_{ab} ,$$
$$R_{ab} = \frac{\partial \psi_{ab}}{\partial \boldsymbol{r}} \left(\frac{\partial}{\partial \boldsymbol{p}} - \frac{\partial}{\partial \boldsymbol{p}'}\right) F_{ab} . \tag{3.11}$$

Here $\tilde{\boldsymbol{E}}$ is the mean field and ψ_{ab} the potential of average force which is given by the equilibrium correlation function in the following way

$$\psi_{ab} = -k_B T \ln \left(1 + h_{ab}(r)\right) . \tag{3.12}$$

Due to the condition (3.12) the basic equation (3.11) is solved in equilibrium with $\boldsymbol{E} = 0$ by the exact distribution function

$$F_{ab}^0 = F_a^0 F_b^0 (1 + h_{ab}(r))$$

where F_a^0 is the Maxwell distribution and $h_{ab}(r)$ the exact equilibrium correlation function.

Let us solve now (3.11) in several approximations neglecting the explicite dependence on $\boldsymbol{r}$:

(i) *Landau-approximation.* Replace the right hand side of (3.11) by

$$\frac{\partial \tilde{V}_{ab}}{\partial \boldsymbol{r}} \left(\frac{\partial}{\partial \boldsymbol{p}} - \frac{\partial}{\partial \boldsymbol{p}'} \right) F_a F_b \,, \tag{3.13}$$

where $\tilde{V}_{ab}$ is the screened potential

$$\begin{aligned} \tilde{V}_{ab} &= (e_a e_b / r) \, \mathrm{e}^{-\varkappa r} \,, \\ \varkappa^2 &= 4\pi \sum n_a e_a^2 / k_B T \,. \end{aligned} \tag{3.14}$$

Then the Fourier-transform of the solution is given by

$$\begin{aligned} g_{ab}(\boldsymbol{k}, \boldsymbol{r}, \boldsymbol{p}, \boldsymbol{p}', t) = {} & \mathrm{i} \tilde{V}_{ab}(k) \int_0^\infty \mathrm{d}\tau \exp . \\ & \cdot \left[-\Delta\tau - \mathrm{i}\boldsymbol{k}(\boldsymbol{v} - \boldsymbol{v}')\tau + i\left(\frac{e_a}{2m_a} - \frac{e_b}{2m_b} \right) \boldsymbol{k} \cdot \tilde{\boldsymbol{E}}\tau^2 \right] \boldsymbol{k} \cdot \\ & \cdot \left(\frac{\partial}{\partial \boldsymbol{p}} - \frac{\partial}{\partial \boldsymbol{p}'} \right) \cdot \boldsymbol{F}_a(\boldsymbol{r}, \boldsymbol{p} - e_a \tilde{\boldsymbol{E}}\tau, t - \tau) \, F_b(\boldsymbol{r}, \boldsymbol{p}' - e_b \tilde{\boldsymbol{E}}\tau, t - \tau) \,. \end{aligned} \tag{3.15}$$

In equilibrium the Debye correlation function is obtained

$$g_{ab}^0(\boldsymbol{k}, \boldsymbol{r}, \boldsymbol{p}, \boldsymbol{p}') = F_a^0(\boldsymbol{p}) \, F_b^0(\boldsymbol{p}') \, (-\tilde{V}_{ab}(k)/k_B T) \,. \tag{3.16}$$

(ii) *Effective-potential-approximation* (EPA). Replace the right hand side of (3.11) by

$$\frac{\partial \tilde{\varphi}_{ab}}{\partial r} \left(\frac{\partial}{\partial \boldsymbol{p}} - \frac{\partial}{\partial \boldsymbol{p}'} \right) F_a F_b \,, \tag{3.17}$$

where $\tilde{\varphi}_{ab}$ is an appropriate effective potential. One possible variant is [8]

$$\tilde{\varphi}_{ab} = V_{ab} \sum e_a^2 \int f_a^0(p) \, |\varepsilon(\boldsymbol{k} \cdot \boldsymbol{v}, \boldsymbol{k})|^{-2} \, (\sum n_c e_c^2)^{-1} \,, \tag{3.18}$$

where $\varepsilon(\omega, \boldsymbol{k})$ is the dielectric function [10, 11] and f_a^0 is the equilibrium distribution. Another possibility follows if F_{ab} in (3.11) is replaced by $F_a F_b (1 + h_{ab})$. Then the right hand side of (3.5) takes the form

$$R_{ab} = \frac{\partial}{\partial \boldsymbol{r}} \varphi_{ab} \left(\frac{\partial}{\partial \boldsymbol{p}} - \frac{\partial}{\partial \boldsymbol{p}'} \right) F_a F_b (1 + h_{ab}) = \frac{\partial \tilde{\varphi}_{ab}}{\partial \boldsymbol{r}} \left(\frac{\partial}{\partial \boldsymbol{p}} - \frac{\partial}{\partial \boldsymbol{p}'} \right) F_a F_b \,, \tag{3.19}$$

$$\tilde{\varphi}_{ab}(r) = -k_B T h_{ab}(r) = k_B T [1 - \exp(-\varphi_{ab}/k_B T)] \,. \tag{3.20}$$

Introducing this into (3.15) follows

$$\begin{aligned} g_{ab}(\boldsymbol{k}, \boldsymbol{r}, \boldsymbol{p}, \boldsymbol{p}') = {} & -\mathrm{i} k_B T h_{ab}(k) \int_0^\infty \mathrm{d}\tau \exp . \\ & \cdot \left[-\Delta\tau - \mathrm{i}\boldsymbol{k} \cdot (\boldsymbol{v} - \boldsymbol{v}')\tau + i\left(\frac{e_a}{2m_a} - \frac{e_b}{2m_b} \right) \boldsymbol{k} \cdot \tilde{\boldsymbol{E}}\tau^2 \right] \cdot \\ & \cdot \boldsymbol{k} \cdot \left(\frac{\partial}{\partial \boldsymbol{p}} - \frac{\partial}{\partial \boldsymbol{p}'} \right) \cdot \\ & \cdot [F_a(\boldsymbol{r}, \boldsymbol{p} - e_a \boldsymbol{E}\tau, t - \tau) \, F_b(\boldsymbol{r}, \boldsymbol{p}' - e_b \boldsymbol{E}\tau, t - \tau)] \,. \end{aligned} \tag{3.21}$$

In this approximation the exact equilibrium correlations are taken into account but on the other hand the dynamics is the ideal one. In the equilibrium case where $\boldsymbol{F}_a$ is given by the Maxwellian F_a^0 and where $\boldsymbol{E} \equiv 0$ the (3.21) yields the exact local equilibrium correlation function

$$g_{ab}^0(\boldsymbol{k}, \boldsymbol{p}, \boldsymbol{p}') = F_a^0(\boldsymbol{p})\, F_b^0(\boldsymbol{p}')\, h_{ab}(k)\,. \tag{3.22}$$

(iii) *Binary collision approximation* (BCA). This method is based on a formal integration of (3.11). We write (3.11) using a two-particle Liouville operator

$$\frac{\partial}{\partial t} g_{ab} = \boldsymbol{L}_{ab}\, g_{ab} + \frac{\partial \varphi_{ab}}{\partial \boldsymbol{r}} \cdot \left(\frac{\partial}{\partial \boldsymbol{p}} - \frac{\partial}{\partial \boldsymbol{p}'}\right) F_a F_b\,. \tag{3.23}$$

Then the formal integration yields [12]

$$g_{ab}(t) = \Lambda_{ab}(t - t_0)\, g_{ab}(t_0) + \int\limits_0^{t-t_0} \mathrm{d}\tau\, \Lambda_{ab}(\tau) \frac{\partial \varphi_{ab}}{\partial \boldsymbol{r}} \cdot \left(\frac{\partial}{\partial \boldsymbol{p}} - \frac{\partial}{\partial \boldsymbol{p}'}\right) F_a(t - \tau)\, F_b(t - \tau)\,; \qquad \Lambda_{ab}(\tau) = \exp\left(\boldsymbol{L}_{ab}\tau\right)\,. \tag{3.24}$$

We perform the limit $t_0 \to -\infty$ and assume that the effect of initial correlations will be destroyed and further that the integral reaches a plateau value. Retaining only first orders with respect to the retardation we find

$$g_{ab}(t) = \int\limits_0^{\infty} \mathrm{d}\tau\, \Lambda_{ab}(\tau) \frac{\partial \varphi_{ab}}{\partial \boldsymbol{r}} \cdot \left(\frac{\partial}{\partial \boldsymbol{p}} - \frac{\partial}{\partial \boldsymbol{p}'}\right) F_a(t)\, F_b(t) - \frac{\partial}{\partial t} \int\limits_0^{\infty} \mathrm{d}\tau\, \tau \Lambda_{ab}(\tau) \frac{\partial \varphi_{ab}}{\partial \boldsymbol{r}} \cdot \left(\frac{\partial}{\partial \boldsymbol{p}} - \frac{\partial}{\partial \boldsymbol{p}'}\right) F_a(t)\, F_b(t)\,. \tag{3.25}$$

Now the following relation will be used

$$\frac{\mathrm{d}}{\mathrm{d}\tau} \Lambda_{ab}(\tau) F_a(t)\, F_b(t) = \Lambda_{ab} \left\{ -\frac{\boldsymbol{p}}{m_a} \frac{\partial}{\partial \boldsymbol{r}} - \frac{\boldsymbol{p}'}{m_b} \frac{\partial}{\partial \boldsymbol{r}'} - e_a \boldsymbol{E} \frac{\partial}{\partial \boldsymbol{p}} - e_b \boldsymbol{E} \frac{\partial}{\partial \boldsymbol{p}'} + \frac{\partial \varphi_{ab}}{\partial \boldsymbol{r}} \left(\frac{\partial}{\partial \boldsymbol{p}} - \frac{\partial}{\partial \boldsymbol{p}'}\right)\right\} F_a(t)\, F_b(t)\,. \tag{3.26}$$

Introducing this into (3.26) and neglecting again the dependence on $\boldsymbol{r}$ and $\boldsymbol{r}'$ we obtain

$$g_{ab}(t) = \left\{ \int\limits_0^{\infty} \mathrm{d}\tau - \frac{\partial}{\partial t} \int\limits_0^{\infty} \mathrm{d}\tau\, \tau \right\} \cdot \left\{ \frac{\mathrm{d}}{\mathrm{d}\tau} \Lambda_a(\tau)\, F_b F_{ao} \quad \cdot \Lambda_{ab}(\tau) \left(e_a \boldsymbol{E} \frac{\partial}{\partial \boldsymbol{p}} + e_b \boldsymbol{E} \frac{\partial}{\partial \boldsymbol{p}'}\right) F_a F_b \right\}. \tag{3.27}$$

As we see the expression (3.27) contains effects of retardation and external fields. By using (3.5) and neglecting all mixing between the effects of nonlocality, retardation and fields we find

$$g_{ab}(\boldsymbol{x}, \boldsymbol{r}, \boldsymbol{p}, \boldsymbol{p}', t) = F_a(\boldsymbol{r}, \boldsymbol{p}(-\infty), t)\, F_b(\boldsymbol{r}, \boldsymbol{p}'(-\infty) -, t) -$$

$$- F_a(\boldsymbol{r}, \boldsymbol{p}, t)\, F_b\boldsymbol{r}, \boldsymbol{p}', (t) + \int_0^\infty \mathrm{d}\tau \left(e_a \boldsymbol{E} \frac{\partial}{\partial \boldsymbol{p}} + e_b \boldsymbol{E} \frac{\partial}{\partial \boldsymbol{p}'}\right) \cdot$$

$$\cdot F_a\left(\boldsymbol{r}, \boldsymbol{p}(-\tau), t\right) F_b(\boldsymbol{r}, \boldsymbol{p}'(-\tau), t) - \frac{\partial}{\partial t} \int_0^\infty \mathrm{d}\tau\, \tau \frac{\mathrm{d}}{\mathrm{d}\tau} F_a(\boldsymbol{r}, \boldsymbol{p}(-\tau), t) \cdot$$

$$\cdot F_b(\boldsymbol{r}, \boldsymbol{p}'(-\tau), t) - \frac{\boldsymbol{x}}{2} \cdot \frac{\partial}{\partial \boldsymbol{r}} \left(F_a(\boldsymbol{r}, \boldsymbol{p}(-\infty), t)\, F_b(\boldsymbol{r}, \boldsymbol{p}'(-\infty), t)\right). \tag{3.28}$$

The various contribution to the collision integral may be represented in the following way

$I_a = I_a^{(0)} + I_a^{(1)} + I_a^{(2)} + I_a^{(3)}$;

$I_a^{(0)} \approx 0(1)$ — usual collision term,

$I_a^{(1)} \approx 0(\boldsymbol{E})$ — effects of the external field,

$I_a^{(2)} \approx 0(\partial/\partial t)$ — effects of retardation in time,

$I_a^{(3)} \approx 0(\partial/\partial \boldsymbol{r})$ — effects of nonlocality in space.

By using the solution (3.21) for the correlation function one obtains

$$I_a^{(0)} + I_a^{(1)} = -k_B T (22\pi)^{-3} \sum_b \frac{\partial}{\partial \boldsymbol{p}} \cdot \int \mathrm{d}\boldsymbol{k}\, \mathrm{d}\boldsymbol{p}'\, \boldsymbol{k} \Phi_{ab}(k)\, h_{ab}(k) \left\{ \mathrm{Re} \int_0^\infty \mathrm{d}\tau \cdot \right.$$

$$\cdot \exp\left[-\Delta\tau - i\boldsymbol{k} \cdot (\boldsymbol{v} - \boldsymbol{v}')\tau + i\left(\frac{e_a}{2m_a} - \frac{e_b}{2m_b}\right) \boldsymbol{k} \cdot \tilde{\boldsymbol{E}} \tau^2\right] \cdot$$

$$\cdot \boldsymbol{k} \cdot \left(\frac{\partial}{\partial \boldsymbol{p}} - \frac{\partial}{\partial \boldsymbol{p}'}\right) f_a(\boldsymbol{r}, \boldsymbol{p} - e_a \tilde{\boldsymbol{E}} \tau, t)\, f_b(\boldsymbol{r}, \boldsymbol{p}' - e_b \boldsymbol{E} \tau, t) +$$

$$+ \mathrm{Im} \int_0^\infty \mathrm{d}\tau\, \tau^2 \exp\left[-\Delta\tau - i\boldsymbol{k}(\boldsymbol{v} - \boldsymbol{v}')\tau\right] \boldsymbol{k} \cdot$$

$$\left. \cdot \left(\frac{\partial}{\partial \boldsymbol{p}} - \frac{\partial}{\partial \boldsymbol{p}'}\right) f_a(\boldsymbol{r}, \boldsymbol{p}, t)\, f_b(\boldsymbol{r}, \boldsymbol{p}', t) \right\}, \tag{3.29}$$

$$I_a^{(2)} = k_B T (2\pi)^{-3} \frac{\partial}{\partial t} \sum_b \frac{\partial}{\partial \boldsymbol{p}} \cdot \int \mathrm{d}\boldsymbol{k}\, \mathrm{d}\boldsymbol{p}'\, \boldsymbol{k} \Phi_{ab}(k)\, h_{ab}(k) \cdot$$

$$\cdot \mathrm{Im} \int_0^\infty \mathrm{d}\tau \cdot \tau \exp\left[-\Delta\tau - i\boldsymbol{k} \cdot (\boldsymbol{v} - \boldsymbol{v}')\tau\right] \boldsymbol{k} \cdot \left(\frac{\partial}{\partial \boldsymbol{p}} - \frac{\partial}{\partial \boldsymbol{p}'}\right),$$

$$\cdot f_a(\boldsymbol{r}, \boldsymbol{p}, t) f_b(\boldsymbol{r}, \boldsymbol{p}', t), \tag{3.30}$$

$$I_a^{(3)} = k_B T (2\pi)^{-3} \frac{\partial}{\partial \boldsymbol{r}} \frac{\partial}{\partial \boldsymbol{p}} : \sum_b \int dk\, dp' \left(\boldsymbol{\delta} + \boldsymbol{k} \frac{\partial}{\partial \boldsymbol{k}}\right) \Phi_{ab}(k)\, h_{ab}(k) \cdot$$

$$\cdot \operatorname{Im} \int_0^\infty d\tau \exp\left[-\Delta\tau - i\boldsymbol{k}(\boldsymbol{v} - \boldsymbol{v}')\tau\right] \boldsymbol{k} \cdot \left(\frac{\partial}{\partial \boldsymbol{p}} - \frac{\partial}{\partial \boldsymbol{p}'}\right) \cdot$$

$$\cdot f_a(\boldsymbol{r}, \boldsymbol{p}, t)\, f_b(\boldsymbol{r}, \boldsymbol{p}', t) . \tag{3.31}$$

We underline again that in this approximation the dynamics of collisions is replaced by the ideal one. More complicated are the collision integrals on the basis of the binary collision approximation (3.25)

$$I_a^{(0)} = \sum_b \int d\boldsymbol{x}\, d\boldsymbol{p}' \frac{\partial \Phi_{ab}}{\partial \boldsymbol{r}} \cdot \frac{\partial}{\partial \boldsymbol{p}} f_a(\boldsymbol{r}, \boldsymbol{p}(-\infty), t)\, f_b(\boldsymbol{r}, \boldsymbol{p}'(-\infty), t) ,$$

$$I_a^{(1)} = \boldsymbol{E} \cdot \sum_b \int d\boldsymbol{x}\, d\boldsymbol{p}' \frac{\partial \Phi_{ab}}{\partial \boldsymbol{r}} \cdot \frac{\partial}{\partial \boldsymbol{p}} \int_0^\infty d\tau \left(e_a \frac{\partial}{\partial \boldsymbol{p}} + e_b \frac{\partial}{\partial \boldsymbol{p}'}\right) \cdot$$

$$\cdot f_a(\boldsymbol{r}, \boldsymbol{p}(-\tau), t)\, f_b(\boldsymbol{r}, \boldsymbol{p}'(-\tau), t) ,$$

$$I_a^{(2)} = -\frac{\partial}{\partial t} \sum_b \int d\boldsymbol{x}\, d\boldsymbol{p}' \frac{\partial \Phi_{ab}}{\partial \boldsymbol{r}} \cdot \frac{\partial}{\partial \boldsymbol{p}} \int_0^\infty d\tau\, \tau \frac{d}{d\tau} \cdot$$

$$\cdot f_a(\boldsymbol{r}, \boldsymbol{p}(-\tau), t)\, f(\boldsymbol{r}, \boldsymbol{p}'(-\tau), t) ,$$

$$I_a^{(3)} = -\frac{1}{2} \frac{d}{d\boldsymbol{r}} \frac{d}{d\boldsymbol{p}} : \sum_b \int d\boldsymbol{x}\ dp'\, \boldsymbol{x}\ \frac{\partial \Phi_{ab}}{\partial \boldsymbol{x}} f_a(\boldsymbol{r}, \boldsymbol{p}(-\infty), t)\, f_b(\boldsymbol{r}, \boldsymbol{p}'(-\infty), t) . \tag{3.32}$$

In the first approximation of perturbation expansion the expresion (3.32) and (3.29) up to (3.31) coincide. In difference to the situation in dilute plasmas the contributions $I_a^{(1)}$, $I_a^{(2)}$, $I_a^{(3)}$ may be very essential in the case of nonideal plasmas. Therefore we shall study these contributions here carefully.

3.2.2. Transport Equations

Multiplying the kinetic equation with certain function $\psi(\boldsymbol{r}, \boldsymbol{p}, t)$ and integrating with respect to the momenta one obtains equations for the density of transport quantities. This method (momentum method) will be deomnstrated in detail in the next section 3.2.3. The statistical transport equations obtained in this way should correspond to the phenomenological transport equations which we shall derive now first on the basis of phenomenological considerations [16, 17]. Let us define the total mass density and the mass velocity by

$$\varrho(\boldsymbol{r}, t) = \sum m_a n_a(\boldsymbol{r}, t); \qquad \boldsymbol{u}(\boldsymbol{r}, t) = \sum m_a n_a \boldsymbol{u}_a(\boldsymbol{r}, t)/\varrho(\boldsymbol{r}, t) . \tag{3.33}$$

Further we define a diffusion flow by

$$\boldsymbol{I}_a = n_a(\boldsymbol{u}_a - \boldsymbol{u}) . \tag{3.34}$$

Then the following equations hold

$$\begin{gathered}\frac{\partial \varrho}{\partial t} + \nabla \cdot (\varrho \boldsymbol{u}) = 0; \qquad \frac{\partial n}{\partial t} + n_a \nabla \cdot \boldsymbol{u} + \nabla \cdot \boldsymbol{J}_a = \sum_b \nu_{ab} w_b ; \\ \varrho\left[\frac{\partial \boldsymbol{u}}{\partial t} + (\boldsymbol{u} \cdot \nabla)\boldsymbol{u}\right] = -\nabla \cdot \boldsymbol{P} + \frac{1}{c} \boldsymbol{I} \times \mathbf{H} + \varrho^* \left(\boldsymbol{E} + \frac{1}{c} \boldsymbol{u} \cdot \boldsymbol{H}\right)\end{gathered} \tag{3.35}$$

where $\boldsymbol{P}$ is the pressure tensor, w_b are the chemical reaction rates and ν_{ab} the stoechiometrical coefficients. Next we write an equation for the density of the internal energy of the plasma e (translational energy and field energy are not included)

$$\frac{\partial}{\partial t} (\varrho e) + \nabla(\varrho e u) = -\nabla \cdot \boldsymbol{J}_q - \boldsymbol{P} \cdot\cdot \nabla \boldsymbol{u} + \boldsymbol{I} \cdot \boldsymbol{E} \tag{3.36}$$

where

$$\varrho^*(\boldsymbol{r}, t) = \sum n_a e_a \boldsymbol{u}_a; \qquad \boldsymbol{I}(\boldsymbol{r}, t) = \sum e_a \boldsymbol{J}_a \tag{3.37}$$

correspond to the electrical charge density and the current density in the medium. We note the continuity equation

$$\frac{d\varrho^*}{\partial t} + \nabla \cdot (\varrho^* \boldsymbol{u}) + \nabla \cdot \boldsymbol{I} = 0 . \tag{3.38}$$

Another important balance equation is that for the entropy density $s(\boldsymbol{r}, t)$ which reads

$$\frac{\partial}{\partial t} (\varrho s) + \nabla \cdot (\varrho s \boldsymbol{u}) + \nabla \cdot \boldsymbol{J}_s = q_s , \tag{3.39}$$

where the entropy flow $\boldsymbol{J}_s$ and the entropy production q_s are given by

$$\begin{gathered}\boldsymbol{J}_s = \frac{1}{T}\left(\boldsymbol{J}_q - \sum_a \mu_a \boldsymbol{J}_a\right), \\ T q_s = -\boldsymbol{J}_q \cdot \frac{\nabla T}{T} - \sum_a \boldsymbol{J}_a \left\{T \nabla \left(\frac{\mu_a}{T}\right) - e_a \left(\boldsymbol{E} + \frac{1}{c} \boldsymbol{u} \times \boldsymbol{H}\right)\right\} - \boldsymbol{\pi} : \mathrm{Grad}\, \boldsymbol{u} .\end{gathered} \tag{3.40}$$

Further $\boldsymbol{J}_q$ denotes the heat flow, μ_a the chemical potential of the species a and $\boldsymbol{\pi}$ the non-static part of the pressure tensor

$$\boldsymbol{P} = p\delta + \boldsymbol{\pi} . \tag{3.41}$$

We should note that the chemical potential is counted here per particle corresponding to our definition of the flows $\boldsymbol{J}_a$ which are defined as particle flows in difference to the mass flows used by many authors.

Now let us take into account that the diffusion flows are not independent which is due to the relation

$$\sum_{a=1}^{n} m_a \boldsymbol{J}_a = 0 . \tag{3.42}$$

Using this relation we may eliminate one of the flows e.g. $\boldsymbol{J}_n$ which possibly corresponds to the heaviest species of particles. Due to

$$\boldsymbol{J}_n = -\sum_{a=1}^{n-1} m_a \boldsymbol{J}_a \tag{3.43}$$

we get finally for the entropy production or dissipation function respectively

$$Tq_s = -\boldsymbol{J}_q \cdot \frac{\nabla T}{T} + \sum_{a=1}^{n-1} \boldsymbol{J}_a \cdot \boldsymbol{X}_a' - \boldsymbol{\pi} : \nabla \boldsymbol{u} ; \tag{3.44}$$

$$\boldsymbol{X}_a' = \boldsymbol{X}_a - \frac{m_a}{m_n} \boldsymbol{X}_n ; \qquad \boldsymbol{E}^* = \boldsymbol{E} + \frac{1}{c} \boldsymbol{u} \times \boldsymbol{H} ; \tag{3.45}$$

$$\boldsymbol{X}_a = e_a\left(\boldsymbol{E}^* - T \operatorname{grad} (\mu_a/T)\right) .$$

Let us consider now the limiting case of a nearly homogeneous plasma assuming that in the first approximation the electrons only contribute to the transport processes and that viscous flows may be neglected. Then one finds approximately (the index 1 denotes the electrons)

$$Tq_s = -\boldsymbol{J}_q(\nabla T/T) + \boldsymbol{I} \cdot \left(\boldsymbol{E} - (T/e_1) \operatorname{grad} (\mu_1/T)\right) , \tag{3.46}$$

where $\boldsymbol{I} = e_1 \boldsymbol{J}_1$ is the electric current. In the framework of Onsager's linear phenomenological relations we assume

$$\begin{aligned} \boldsymbol{J}_q &= L_{qq}(-\nabla T/T) + L_{pe}\left(\boldsymbol{E} - (T/e_1) \nabla(\mu_1/T)\right) , \\ \boldsymbol{I} &= L_{eq}(-\nabla T/T) + L_{ee}\left(\boldsymbol{E} - (T/e_1) \nabla(\mu_1/T)\right) , \end{aligned} \tag{3.47}$$

where microscopic reversibility requires

$$L_{qe} = L_{eq} .$$

Another more familiar writing is

$$\begin{aligned} \boldsymbol{J}_q &= \lambda(-\nabla T) + \pi \boldsymbol{I} , \\ \boldsymbol{I} &= \alpha \nabla T + \frac{1}{R} \left(\boldsymbol{E} - T\nabla(\mu_1/e_1)\right) , \end{aligned} \tag{3.49}$$

or

$$\left(\boldsymbol{E} - T\nabla(\mu_1/e_1 T)\right) = R\boldsymbol{I} - \eta \nabla T , \tag{3.50}$$

where

$$R^{-1} = L_{ee} ; \qquad \alpha = \frac{\eta}{R} = -\frac{L_{eq}}{T} ; \qquad \lambda = \left\{L_{qq} - \frac{L^2_{qe}}{L_{ee}}\right\} : T .$$

Here λ is the coefficient of heat conduction, R the isothermal resistivity, η the thermoelectrical coefficient and π the Peltier coefficient. Due to the symmetry of the coefficients we have $\pi = -T\eta$. Therefore we have only 3 independent transport coefficients: R, λ and η. The first question to each kinetical theory is the calculation of these coefficients from first principles. One possible way to solve this question is based on Grad's momentum method which will be explained in the next section 3.2.3.

Another interesting limiting case is that of an isothermal nonuniform plasma of electrons (index 1) and ions (index 2) imbedded in a background of uniform neutrals (index 3). Then the entropy production reduces to

$$Tq_s = \boldsymbol{J}_1 \cdot \boldsymbol{X}_1 + \boldsymbol{J}_2 \cdot \boldsymbol{X}_2 \,, \tag{3.51}$$

$$\boldsymbol{X}_1 = e_1\boldsymbol{E} - \quad \mu_4 \,, \qquad \boldsymbol{X}_2 = e_2\boldsymbol{E} < \nabla\mu_2 \qquad \boldsymbol{X}_3 = 0 \,. \tag{3.52}$$

The linear relations read

$$\begin{aligned} \boldsymbol{J}_1 &= M_{11}\left(\boldsymbol{E} - \frac{1}{e_1}\nabla\mu_1\right) + M_{12}\left(\boldsymbol{E} - \frac{1}{e_2}\nabla\mu_2\right), \\ \boldsymbol{J}_2 &= M_{21}\left(\boldsymbol{E} - \frac{1}{e_1}\nabla\mu_1\right) + M_{22}\left(\boldsymbol{E} - \frac{1}{e_2}\nabla\mu_2\right). \end{aligned} \tag{3.53}$$

Using these relations we can express the field $\boldsymbol{E}$ by the current $\boldsymbol{I}$

$$\begin{aligned} \boldsymbol{E} = & \frac{\boldsymbol{I}}{e_1M_{11} + e_2M_{22} + e_1M_{12} + e_2M_{21}} + \\ & + \frac{\left(M_{11} + \frac{e_2}{e_1}M_{21}\right)\nabla\mu_1 + \left(M_{22} + \frac{e_1}{e_2}M_{12}\right)\nabla\mu_2}{e_1M_{11} + e_2M_{22} + e_1M_{12} + e^2{}_2M_{21}} \,. \end{aligned} \tag{3.54}$$

Here the first contribution expresses the Ohm field and the second one the Planck field. In the limiting case of ambipolar diffusion the Ohm field is small in comparison to the Planck field. Then $\boldsymbol{E}$, $\boldsymbol{J}_1$ and $\boldsymbol{J}_2$ are functions of (grad μ_i) only. We find e.g.

$$e_2\boldsymbol{J}_2 \approx \frac{(M_{12}M_{21} - M_{11}M_{22})\,(e_1\nabla\mu_2 - e_2\nabla\mu_1)}{e_1M_{11} + e_2M_{22} + e_1M_{12} + e_2M_{21}} \approx -e_1\boldsymbol{J}_1 \,. \tag{3.55}$$

In other words the electrons and the ions are moving nearly with the same velocity. Since the electrons are forced to follow the ion diffusion we get a closed diffusion equation for the ions only

$$\frac{\partial}{\partial t} n_2(\boldsymbol{r}, t) = D_A \Delta n_2(\boldsymbol{r}, t) \,, \tag{3.56}$$

$$D_A = \frac{(M_{11}M_{22} - M_{12}M_{21})\,(e_1\mu_2' - e_2\mu_1')}{e_1M_{11} + e_2M_{22} + e_1M_{12} + e_2M_{21}} \,, \tag{3.57}$$

where the dashes denote the derivatives with respect to the ion densities n_2. We note that due to electroneutrality holds

$$n_1(\boldsymbol{r}, t) \approx -\,(e_2/e_1)\, n_2(\boldsymbol{r}, t) \,. \tag{3.58}$$

3.2.3. The Momentum Method

By multiplying the kinetic equation (3.4) with certain functions $\psi(\boldsymbol{r}, \boldsymbol{p}, t)$ and integrating with respect to $\boldsymbol{p}$ the transport equations are obtained

$$\frac{\partial}{\partial t}(n_a\langle\psi\rangle_a) + \frac{\partial}{\partial r_\alpha}(n_a\langle v_\alpha\psi\rangle_a) - \frac{e_a}{m_a}E_\alpha n_a\left\langle\frac{\partial\psi}{\partial v_\alpha}\right\rangle_a -$$

$$- n_a\left\langle v_\alpha\frac{\partial\psi}{\partial r_\alpha}\right\rangle_a - n_a\left\langle\frac{\partial\psi}{\partial t}\right\rangle_a = I_a\{\psi\}\,, \tag{3.59}$$

with

$$\begin{aligned} n_a\langle\psi\rangle_a &= \int\psi(\boldsymbol{r}, \boldsymbol{p}, t)\, f_a(r, p, t)\,\mathrm{d}\boldsymbol{p}\,,\\ I_a\langle\psi\rangle &= \int\psi(\boldsymbol{r}, \boldsymbol{p}, t)\,[I_a(\boldsymbol{p}) + \sum_n \nu_{an}(\boldsymbol{p})\,(f_a^0 - f_a)]\,\mathrm{d}\boldsymbol{p}\,. \end{aligned} \tag{3.60}$$

The following momenta are of special physical importance, density:

$$n_a(\boldsymbol{r}, t) = \int 1\cdot f_a(\boldsymbol{r}, \boldsymbol{p}, t)\,\mathrm{d}\boldsymbol{p};$$

mean velocity:

$$u_{a\alpha}(\boldsymbol{r}, t) = \int v_\alpha f_a(\boldsymbol{r}, \boldsymbol{p}, t)\,\mathrm{d}\boldsymbol{p}/n_a(\boldsymbol{r}, t);$$

relative velocity:

$$c_\alpha = (p_\alpha/m_a) - u_{a\alpha};$$

temperature:

$$\theta_a(\boldsymbol{r}, t) = (m_a/3n_a)\int c_\alpha c_\alpha f_a(\boldsymbol{r}, \boldsymbol{p}, t)\,\mathrm{d}\boldsymbol{p};$$

kinetic energy density:

$$\varepsilon_a(\boldsymbol{r}, t) = (3/2)\,n_a(\boldsymbol{r}, t)\,\theta_a(\boldsymbol{r}, t)\,;$$

pressure tensor:

$$P_{a\alpha\beta}(\boldsymbol{r}, t) = m_a\int(c_\alpha c_\beta - \delta_{\alpha\beta}\theta_a/m_a)\,f_a\,\mathrm{d}\boldsymbol{p};$$

heat flow tensor:

$$S_{a\alpha\beta\gamma} = m_a\int c_\alpha c_\beta c_\gamma f_a\,\mathrm{d}\boldsymbol{p};$$

heat flow vector:

$$q_{a\alpha}(\boldsymbol{r}, t) = S_{a\alpha\beta\beta}/2 = (m_a/2)\int c_\alpha c_\beta c_\beta f_a\,\mathrm{d}\boldsymbol{p}\,. \tag{3.61}$$

The statistical transport equations (3.59) correspond to the phenomenological transport equations given in the previous section. Following GRAD [18] we develope now the distribution function with respect to Hermite polynomials

$$\begin{aligned} f_a(\boldsymbol{r}, \boldsymbol{p}, t) &= (2\pi m_a\theta_a)^{-3/2}\exp(-m_a\mathbf{c}_a^2/2\theta_a)\\ &\quad\cdot\{1 + \sum(m_a^{\nu-1}/\theta_a^\nu \nu l)\,\boldsymbol{M}_a^{(\nu)}\overline{\mathrm{H}}(\boldsymbol{c}_a)\}\,,\\ \overline{H}^{(0)} &= 1\,,\quad \overline{H}_\alpha^{(1)} = c_{a\alpha}\,,\quad \overline{H}_{\alpha\beta}^{(2)} = c_\alpha c_\beta - \delta_{\alpha\beta}\theta_a/m_a\,,\\ \overline{H}_{\alpha\beta\gamma}^{(3)} &= c_\alpha c_\beta c_\gamma - (c_\alpha\delta_{\beta\gamma} + c_\beta\delta_{\alpha\gamma} + c_\gamma\delta_{\alpha\beta})\,\theta_a/m_a \qquad \text{etc.} \end{aligned} \tag{3.62}$$

The coefficients are given by

$$M_a^{(\mu)} = m_a \int \bar{H}^{(\mu)}(\mathbf{c})\, f_a(\mathbf{r}, \mathbf{p}, t)\, d\mathbf{p} ,$$
$$M_a^{(1)} = 0 , \qquad M_{a\alpha\beta}^{(2)} = m_a \int (c_\alpha c_\beta - \delta_{\alpha\beta}\theta_a/m_a)\, f_a\, d\mathbf{p} = p_{a\alpha\beta} , \tag{3.63}$$
$$M_{a\alpha\beta\gamma}^{(3)} = m_a \int c_\alpha c_\beta c_\gamma f_a\, d\mathbf{p} = S_{a\alpha\beta\gamma} .$$

We see that the first coefficients are connected with important physical momenta.

Let us consider now a plasma with axial symmetry characterized by one specified direction $\mathbf{e}$ parallel to the z-axis (external field). The vectors $\mathbf{E}$, $\mathbf{u}_a$, $\mathbf{q}_a$ etc. are all parallel to $\mathbf{e}$ and all the tensors must be dyadic products. In other words we expect

$$p_{a\alpha\beta} = A\delta_{\alpha\beta} + Be_\alpha e_\beta ,$$
$$S_{a\alpha\beta\gamma} = C(e_\alpha\delta_{\beta\gamma} + e_\beta\delta_{\alpha\gamma} + e_\gamma\delta_{\alpha\beta}) + De_\alpha e_\beta e_\gamma . \tag{3.64}$$

Due to $p_{a\alpha\alpha} = 0$ we find

$$3A + B = 0 .$$

Due to their physical origin we expect $B = O(\mathbf{E}^2)$, $D = O(\mathbf{E}^3)$ or $B = O(\mathbf{u}^2)$, $D = O(\mathbf{u}^3)$. Neglecting such nonlinear terms we have $A = B = D = 0$. Using the relation

$$q_{a\alpha} = S_{a\alpha\beta\beta}/2 = C(3e_\alpha + e_\alpha + e_\alpha)/2 = 5Ce_\alpha/2$$

we get

$$S_{a\alpha\beta\gamma} = (q_{a\alpha}\delta_{\beta\gamma} + q_{a\beta}\delta_{\alpha\gamma} + q_{a\gamma}\delta_{\alpha\beta})\, 2/5 . \tag{3.65}$$

Therefore we obtain in a first approximation

$$f_a(\mathbf{r}, \mathbf{p}, t) = n_a(r, t)\, [2\pi m_a\theta_a]^{-3/2} \exp[-m_a c^2/2\theta_a] \cdot$$
$$\cdot \{1 + q_{a\alpha}(m_a c_\alpha/n_a\theta_a)\, ((m_a c^2/5\theta_a) - 1)\} . \tag{3.66}$$

In order to calculate the transport coefficients let us consider first an ideal stationary plasma. In other words we take into account the usual collision term $I_a^{(0)}$ only. Further we assume that the ions and the neutrals are at rest and all species have the same temperature. Then Grad's method yields the following equations for the electron component [7, 14]

$$-\frac{e_e}{m_e} n_e\mathbf{E} + \frac{1}{m_e}\frac{\partial(n_i\theta)}{\partial\mathbf{r}} = -\mathbf{u}_e\frac{n_e}{t_e} + \mathbf{q}_e\frac{3}{5\theta t_e} ,$$
$$\frac{5n_e\theta}{2m_e}\frac{\partial\theta}{\partial\mathbf{r}} = \frac{3}{2}\mathbf{u}_e n_e\theta\frac{1}{t_e} - \frac{5}{2}\mathbf{q}_e\left[\frac{8}{25\sqrt{2}\, t_{ee}} + \frac{13}{25t_e}\right]. \tag{3.67}$$

Introducing

$$\mathbf{I} \approx n_e e_e(\mathbf{u}_e - \mathbf{u}) , \qquad \mathbf{J}_q \approx \mathbf{q}_e$$

we get

$$\left[\mathbf{E} - \frac{1}{n_e e_e}\frac{\partial}{\partial\mathbf{r}}(n_e\theta)\right] = \mathbf{I}\cdot\frac{m_e}{n_e e_e^2 t_e}\cdot\left(\frac{8t_e + 4\sqrt{2}\, t_{ee}}{8t_e + 13\sqrt{2}\, t_{ee}}\right) - \frac{\partial\theta}{\partial\mathbf{r}}\left(\frac{15\sqrt{2}\, t_{ee}}{|e_e|\, (8t_e + 13\sqrt{2}\, t_{ee})}\right),$$
$$\mathbf{J}_q = -\frac{\partial\theta}{\partial\mathbf{r}}\cdot\frac{n_e\theta}{m_e}\cdot\left(\frac{10\sqrt{2}\, t_e t_{ee}}{2t_e + \sqrt{2}\, t_{ee}}\right) + \mathbf{I}\frac{\theta}{e_e}\left(\frac{15\sqrt{2}\, t_{ee}}{8t_i + 13\sqrt{2}\, t_{ee}}\right). \tag{3.68}$$

Comparing this with the phenomenological equations one finds the following phenomenological coefficients

$$R = \frac{m_e}{n_e e_e^2 t_e} \cdot \left(\frac{8t_e + 4\sqrt{2}\,t_{ee}}{8t_e + 13\sqrt{2}\,t_{ee}}\right), \qquad \lambda = \frac{k_B^2 T n_e}{m_e}\left(\frac{10\sqrt{2}\,t_e t_{ee}}{2t_e + \sqrt{2}\,t_{ee}}\right),$$

$$\eta = \frac{k_B}{|e_e|}\left(\frac{15\sqrt{2}\,t_{ee}}{8t_e + 13\sqrt{2}\,t_{ee}}\right), \qquad \pi = -T\eta\,. \tag{3.69}$$

The relaxation times which appear in our formulae are defined by

$$t_e = \frac{t_{ei}t_{en}}{t_{ei} + t_{en}} \text{ or } \frac{1}{t_e} = \frac{1}{t_{ei}} + \frac{1}{t_{en}},$$

$$t_{en} = 3m_e\theta[\int \mathrm{d}p\, p^2 \nu_{en}(p)\, f_e^0]^{-1}\,,$$

$$t_{ab} = \frac{3m_a^{3/2}(k_B T)^{3/2}}{4(2\pi)^{1/2}\, n_b e_b^2 e_a^2 L_{ab}}\,. \tag{3.70}$$

Here t_{en} corresponds to a relaxation due to electron-neutral interaction, t_{ei} due to electron-ion interaction and t_{ee} due to electron-electron interaction. Further L_{ab} denotes the Coulomb Logarithmus which is in the Landau Approximation (3.13) given by

$$L_{ab} = (4\pi e_a e_b)^{-2} \int_0^{\alpha_{ab}} \mathrm{d}k\, k^3 \Phi_{ab}(k)\, \tilde{V}_{ab}(k)\,. \tag{3.71}$$

Here α_{ab} is a short-wave cut-off parameter which is usually defined by

$$\alpha_{ab} = 3\theta(-e_a e_b)^{-1}\,. \tag{3.72}$$

Another approximation corresponding to the effective potential approximation (3.21) is given by

$$L_{ab} = -\theta(4\pi e_a e_b)^{-2} \int_0^{\alpha_{ab}} \mathrm{d}k\, k^3 \Phi_{ab}(k)\, h_{ab}(k)\,, \tag{3.73}$$

where $h_{ab}(k)$ is the Fourier transform of the exact equilibrium correlation function.

Instead of (3.69) we may write also

$$\begin{aligned} R &= R_L(1 + (\sqrt{2}\,t_e/t_{ee})/(1 + (4\sqrt{2}\,t_e/t_{ee})\,, \\ \alpha &= \alpha_L[1 + (4\sqrt{2}\,t_e/13t_{ee})]^{-1}\,, \\ \lambda &= \lambda_L[1 + (\sqrt{2}\,t_e/t_{ee})]^{-1}\,, \end{aligned} \tag{3.74}$$

where R_L, α_L and λ_L are the transport coefficients for the corresponding Lorentz gas ($\Phi_{ee} \equiv 0$, $t_{ee} \to \infty$). So far all the calculations have been done in the framework of the 13-moment-approximation. For the Lorentz gas much better methods are available. Of special importance is the relaxation time approximation [19]. Assuming

$$f_a(p) = f_a^0(p) + f_a^1(\boldsymbol{p}); \qquad f_a^1(p) = \boldsymbol{p} \cdot \boldsymbol{E}\varphi(|\boldsymbol{p}|)\,,$$

which is in agreement also with the Grad expansion (3.66), the electron-ion collision term for a Lorentz plasma may be written as

$$I_e^{(0)} = \sum_i \nu_{ei}(\boldsymbol{r}, \boldsymbol{p})\, [f_e^0 - f_e]\,,$$

where the sum is to be extended over all the species. The collision frequencies may be represented by the transport cross sections.

$$\nu_e(\boldsymbol{r}, \boldsymbol{p}) = n_i \frac{p}{m_e} q_{ei}(p) ,$$

where

$$q_{ei}(p) = \int (1 - \cos\theta)\, q_{ei}(\Omega, p)\, \mathrm{d}\Omega = (4\pi\hbar^2/p^2) \cdot \\ \cdot \sum_{l=0}^{\infty} (2l+1) \sin^2 [\delta_l(p/\hbar) - \delta_{l+1}(p/\hbar)] .$$

In the effective potential approximation we find e.g.

$$\nu_{ei} = n_i(r)\, 4\pi e^2 e_i^2 L_{ei} m_e/p^3$$

where the Spitzer logarithm is given by (3.71). In the relaxation time approximation the kinetic equation of ideal plasmas is exactly solvable and yields the electron distribution

$$f_e(\boldsymbol{r}, \boldsymbol{p}) = f_e^0(\boldsymbol{r}, \boldsymbol{p}) - \frac{e\boldsymbol{E}}{\nu_e(\boldsymbol{p})} \cdot \frac{\partial f_e^0(\boldsymbol{r}, \boldsymbol{p})}{\partial \boldsymbol{p}} , \\ \nu_e(\boldsymbol{p}) = \sum_i \nu_{ei}(\boldsymbol{p}) + \sum_n \nu_{en}(\boldsymbol{p}) . \tag{3.75}$$

Using this solution one obtains the transport coefficients

$$R_{\mathrm{L}}^{-1} = -\frac{4\pi}{3m_e^4} n_e e^2 \int_0^\infty \frac{\partial f_e^0}{\partial p} \frac{p^3\, \mathrm{d}p}{\nu_e(\boldsymbol{p})} ,$$

$$\alpha_{\mathrm{L}} = \frac{4\pi}{3m_e^4} \cdot \frac{e n_e}{k_{\mathrm{B}} T} \cdot \int_0^\infty f_e^0(p) \left(\frac{p^2}{2m_e k_{\mathrm{B}} T} - \frac{3}{2} \right) \frac{p^4\, \mathrm{d}p}{\nu_e(\boldsymbol{p})} ,$$

$$\lambda_{\mathrm{L}} = \frac{2\pi}{3m_e^7} \cdot \frac{m_e n_e}{k_{\mathrm{B}} T} \cdot \int_0^\infty f_e^0(p) \left(\frac{p^2}{2m_e k_{\mathrm{B}} T} - \frac{3}{2} \right) \frac{p^6\, \mathrm{d}p}{\nu_e(\boldsymbol{p})} . \tag{3.76}$$

Now let us consider the effect of nonideality on the transportprocesses in nonideal plasmas. These effects are described by the additional collision terms $I_a^{(1)}$, $I_a^{(2)}$ and $I_a^{(3)}$. The equations (3.31) and (3.32) show that these terms are proportional to $\boldsymbol{E}$, $(\partial/\partial t)$ and $(\partial/\partial \boldsymbol{r})$ respectively. Therefore one may combine the moments of these terms with corresponding expressions on the left hand side of the transport equations. The result is

$$\frac{\partial}{\partial t} n_a(r, t) + n_a \frac{\partial}{\partial r_\alpha} u_{a\alpha} + \frac{\partial}{\partial r_\alpha} J_{a\alpha} = 0 ,$$

$$\frac{\partial}{\partial t} (m_a n_a u_{a\alpha}) + \frac{\partial}{\partial r_\beta} (m_a n_a u_{a\alpha} u_{a\beta}) + \frac{\partial}{\partial r_\beta} \overline{P}_{a\alpha\beta} - \bar{e}_a \overline{E}_\alpha^* = I_a^{(0)}\{p_\alpha\} ,$$

$$\frac{\partial}{\partial t} \bar{\varepsilon}_a (r, t) + \frac{\partial}{\partial r_\alpha} (\bar{\varepsilon}_a u_{a\alpha}) + \frac{\partial}{\partial r_\alpha} \bar{q}_{a\alpha} + \overline{P}_{a\alpha\beta}\, \partial u_{a\beta}/\partial r_\alpha = (m_a/2)\, I_a^{(0)}\{\boldsymbol{c}^2\} ,$$

$$\begin{aligned}
&\frac{\partial}{\partial t}\boldsymbol{M}_a^{(\nu)} + \frac{\partial}{\partial r_\alpha}(\boldsymbol{M}_a^{(\nu)}u_{a\alpha}) + \frac{\partial}{\partial r_\alpha}\left(\overline{\boldsymbol{M}}_{a\alpha}^{(\nu+1)} + \delta_\alpha \overline{\boldsymbol{M}}_a^{(\nu-1)}\frac{\theta_a}{m_a}\right) + \\
&+ \frac{\partial}{\partial r_\alpha}u_{a\alpha}\delta_\beta\left(\overline{\boldsymbol{M}}_{a\alpha}^{(\nu)} + \delta_\alpha\overline{\boldsymbol{M}}_a^{(\nu-2)}\frac{\theta_a}{m_a}\right) + \frac{\partial}{\partial r_\alpha}\left(\frac{\theta_a}{m_a}\right)\boldsymbol{\delta}\left(\overline{\boldsymbol{M}}_{a\alpha}^{(\nu-1)} + \delta_\alpha\overline{\boldsymbol{M}}_a^{(\nu-3)}\frac{\theta_a}{m_a}\right) - \\
&- \frac{1}{m_a n_a}\frac{\partial}{r\partial_\beta}P_{a\alpha\beta}\,\delta_\alpha\boldsymbol{M}_a^{(\nu-1)} - \left(\overline{P}_{a\alpha\beta}\frac{\partial}{\partial r_\alpha}u_{a\beta} + \frac{\partial}{\partial r_\alpha}\bar{q}_{a\alpha}\right)\left(\frac{2}{3m_a n_a}\boldsymbol{\delta}\boldsymbol{M}_a^{(\nu-2)}\right) - \\
&- \frac{e_a}{m_a c}\varepsilon_{\alpha\beta\gamma}\mathrm{H}_\gamma\delta_\alpha\left(\boldsymbol{M}_{a\beta}^{(\nu)} + \delta_\beta\boldsymbol{M}_a^{(\nu-2)}\frac{\theta_a}{m_a}\right) = \\
&= I_a\{\overline{\mathrm{H}}^{(\nu)}\}m_a - I_a\{v_\alpha\}\,(\delta_\alpha\boldsymbol{M}_a^{(\nu-1)}/n_a) - I_a\,\{\mathbf{c}^2\}\,(\boldsymbol{\delta}\boldsymbol{M}_a^{(\nu-2)}/3n_a)\,. \qquad (3.77)
\end{aligned}$$

Here the Einstein convention with respect to double indices and the Grad conventions with respect to tensor multiplications have been used [18].

The renormalized "nonideal" transport quantities are in the effective field approximation (3.19) given by

$$\overline{\boldsymbol{E}}_a^* = \tilde{\boldsymbol{E}}^* + \frac{1}{e_a}\sum_b \frac{n_b}{3\theta(2\pi)^3}\left(\frac{e_a}{m_a} - \frac{e_b}{m_b}\right)\left(\frac{1}{m_a} + \frac{1}{m_b}\right)^{-1}\int \mathrm{d}\boldsymbol{k}\Phi_{ab}(k)\,h_{ab}(k)\,\tilde{\boldsymbol{E}}\,, \quad (3.78)$$

$$\bar{\varepsilon}_a = \frac{3}{2}n_a\theta_a + \frac{1}{2(2\pi)^3}\sum_b n_b\int \mathrm{d}\boldsymbol{k}\,\Phi_{ab}(k)\,h_{ab}(k)\,, \quad (3.79)$$

$$\overline{P}_{a\alpha\beta} = n_a\theta_a + \frac{1}{2(2\pi)^3}n_a\sum_b n_b\int \mathrm{d}\boldsymbol{k}\cdot(\delta_{\alpha\beta}\Phi_{ab}(k) + k_\alpha\frac{\partial}{\partial k_\beta}\Phi_{ab}(k)\,h_{ab}(k)\,, \quad (3.80)$$

$$\begin{aligned}
\bar{q}_{a\alpha} &= q_{a\alpha} + (u_{a\alpha} - u_\alpha) + \frac{1}{2(2\pi)^3}\frac{n_a}{\theta_a}\sum_b n_b\int \mathrm{d}\boldsymbol{k}\, - \\
&- \left(\delta_{\alpha\beta} + k_\alpha\frac{\partial}{\partial k_\beta}\right)\Phi_{ab}(k)\,h_{ab}(k)\,; \qquad (3.81)
\end{aligned}$$

$$\overline{M}_{a\alpha}^{(1)} = n_a m_a(\,u_{a\alpha} - u_\alpha)\,\theta_a^{-1}\,, \quad (3.82)$$

$$\overline{M}_{a\alpha\beta}^{(2)} = \overline{P}_{a\alpha\beta} - \delta_{\alpha\beta}\overline{P}_{a\gamma\gamma}/3\,, \quad (3.83)$$

$$\begin{aligned}
\overline{M}_{a\alpha}^{(\nu)} &= M_{a\alpha}^{(\nu)} + \frac{1}{2(2\pi)^3}n_a\sum_b n_b\int \mathrm{d}k\left(\delta_{\alpha\beta} + k_\alpha\frac{\partial}{\partial k_\beta}\right)\cdot \\
&\cdot\Phi_{ab}(k)\,h_{ab}(k)\,\delta_\beta M^{(\nu-2)} - \delta_\alpha(\overline{M}_a^{(\nu-2)} - M_a^{(\nu-2)})\,\theta_a/m_a\,. \qquad (3.84)
\end{aligned}$$

3.2.4. Nonideal Transport Coefficients

Let us consider now the standard transport situation where heat and electric current is transported by the electrons only. By specialization of (3.77) we obtain

$$\begin{aligned}
\boldsymbol{I} &= \frac{1-Q}{R}\boldsymbol{E} - \frac{1}{e_e R n_e}\nabla((1-G)\,n_e\theta) + \frac{\eta(1-G)}{R}\nabla T\,, \\
\boldsymbol{J}_q &= -\lambda(1-G)\,\nabla T + \pi(1-G)\,\boldsymbol{I}\,, \qquad (3.85)
\end{aligned}$$

where the coeffiicents of nonideality are given by

$$Q = \frac{\theta}{3(2\pi)^3} \int_0^\infty \mathrm{d}k\, k^2 \Phi_{ei}(k)\, h_{ei}(k)\,,$$
$$G = 1 - \bar{P}_e / n_e \theta_e\,, \tag{3.86}$$
$$\bar{P}_e = n_e \theta_e - \frac{1}{6} n_e \sum_b \int r \frac{\partial}{\partial r} \Phi_{eb}(r)\, F_{eb}(r)\, 4\pi r^2\, \mathrm{d}r\,.$$

In the approximation of the limiting law one finds

$$Q = \mathrm{e}^2 \varkappa / 6\theta\,,$$
$$G = \mathrm{e}^2 \varkappa / 6\theta\,, \tag{3.87}$$
$$\bar{P}_e = n_e \theta (1 - \mathrm{e}^2 \varkappa / 6\theta)\,.$$

We see that the nonideality effects are of the order of magnitude of the plasma parameter

$$\mu = \mathrm{e}^2 \varkappa / \theta\,.$$

For the effective nonideal conductivity we get e.g. [20, 21]

$$\sigma = \frac{1}{R}(1 - Q)\,. \tag{3.88}$$

The problem of the calculation of Q in higher order approximations is not yet solved. Therefore we shall give here now a practicable formula which reduces Q to a thermodynamic quantity. Considering one ion species only we have

$$G = G_{ee} + G_{ei}\,,$$
$$G_{ee} = -\frac{1}{6} n_e \int \mathrm{d}r\, r \Phi'_{ee}(r)\, h_{ee}(r)\,,$$
$$G_{ei} = -\frac{1}{6} n_i \int \mathrm{d}r\, r \Phi'_{ei}(r)\, h_{ei}(r)\,.$$

For pure Coulombic interactions we find

$$r \Phi'_{ei}(r) = -e_e e_i / r = -\Phi_{ei}(r)\,,$$
$$G_{ei} = \frac{1}{6} n_i \int \mathrm{d}r\, \Phi_{ei}(r)\, h_{ei}(r) = \frac{1}{2} Q\,.$$

Therefore we may consider also in the general case of small deviations of the Coulomb law the formula

$$Q = 2 G_{ei}\,,$$

as a rather good approximation. Then the calculation of Q is reduced to a pure thermodynamic problem.

Let us consider now the second limiting situation of a nonuniform isothermal plasma of electrons (index e) and ions (index i) imbedded in a background of quiescent neutrals (index n). Then we obtain from the momentum equations the flow relations

$$\begin{aligned} \boldsymbol{J}_e &= M_{ee}\left(\boldsymbol{E} - \frac{1}{n_e e_e}\nabla\overline{P}_e\right) + M_{ei}\left(\boldsymbol{E} - \frac{1}{n_i e_i}\nabla\overline{P}_i\right), \\ \boldsymbol{J}_i &= M_{ie}\left(\boldsymbol{E} - \frac{1}{n_e e_e}\nabla\overline{P}_e\right) + M_{ii}\left(\boldsymbol{E} - \frac{1}{n_i e_i}\nabla\overline{P}_i\right) \end{aligned} \tag{3.89}$$

with

$$\begin{aligned} M_{ee} &= \frac{n_e e_e}{m_e}\left[\frac{1}{t_{ei}} + \frac{1}{t_{en}} - \frac{m_e}{m_e t_{ei} + m_i t_{ei}^2 t_{in}^{-1}}\right]^{-1}, \\ \frac{M_{ei}}{e_i} &= \frac{M_{ie}}{e_e} = \frac{n_i}{m_i}\left[\left(1 + \frac{t_{ei}}{t_{en}}\right)\left(\frac{1}{t_{in}} + \frac{m_e}{m_i t_{ei}}\right) - \frac{m_e}{m_i t_{ei}}\right]^{-1}, \\ M_{ii} &= \frac{n_i e_i}{m_i}\left[\frac{1}{t_{in}} + \frac{m_e}{m_i t_{ei}} - \frac{m_e}{m_e t_{ei} + m_i t_{ei}^2 t_{in}^{-1}}\right]^{-1}. \end{aligned} \tag{3.90}$$

Correspondingly the coefficient of ambipolar diffusion is given by

$$D_{\text{A}} = \frac{M_{ee}M_{ii} - M_{ei}M_{ie}}{e_e M_{ee} + e_i M_{ii} + e_e M_{ei} + e_i M_{ii}}\left(\frac{e_e}{n_i}\overline{P}_i' - \frac{e_i}{n_e}\overline{P}_e'\right). \tag{3.91}$$

A remarkable result seems to be that in the statistical theory the gradients of partial pressures appear instead of the gradients of chemical potentials. Both approaches are equivalent if and only if a Gibbs-Duhem equation is valid for each species separately i.e.

$$\nabla\overline{P}_e = n_e\nabla\mu_e\,, \qquad \nabla\overline{P}_i = n_i\nabla\mu_i\,. \tag{3.92}$$

Let us mention that Walter Nernst [23] already considered the gradient of the partial osmotic pressure as the driving force of diffusion instead of the gradient of the chemical potential. Note that in the Debye approximation the condition (3.92) is always fulfilled since

$$\begin{aligned} \nabla\overline{P}_a &= \theta\nabla n_a - \frac{1}{4}e_a^2\varkappa\nabla n_a\,, \\ n_a\nabla\mu_a &= \theta\nabla n_a - \frac{1}{4}e_a^2\varkappa\nabla n_a\,. \end{aligned} \tag{3.93}$$

Let us consider finally the case of strong nonideality where the coefficients Q and G are of the order of the unity

$$Q \gtrapprox 1\,, \qquad G \gtrapprox 1\,.$$

This would lead to negative values of $(1 - Q)$ and $(1 - G)$ and therefore possibly to unrealistic negative transport coefficients. Let us show here very briefly how this difficulty can be avoided in the case of conductivity. In the kinetic theory explained so far the binary correlation function g_{ab} is a function of the external field

$$g_{ab} = g_{ab}^{(0)} + E g_{ab}^{(1)}\,.$$

This leads to two contributions to the collision integral and we find in the case of uniform isothermal plasmas

$$E = RI + QE \, .$$

where R is determined by $g_{ab}^{(0)}$ and Q by $g_{ab}^{(1)}$.

Another writing is

$$\bar{E} = RI \, ,$$

where $\bar{E}$ is the effective field. Now let us take into account correlation effects on the three-particle-level which will lead to a replacement of E by $\bar{E}$ on the two-particle level i.e.

$$g_{ab} = g_{ab}^{(0)} + \bar{E} g_{ab}^{(1)} \, .$$

Let us assume that we have in the new theory

$$\bar{E} = \gamma E \, ,$$

where γ is still unknown. Using the new formula for g_{ab} we get

$$E = RI + \bar{E}Q \, ,$$

$$\bar{E} = (1 - \gamma Q) \, E = RI \, .$$

This yields finally

$$\gamma = 1 - \gamma Q \, , \qquad \gamma = 1/(1 + Q) \, ,$$

$$E = RI(1 + Q) \, ,$$

$$\sigma = R/(1 + Q) \, . \tag{3.94}$$

In the new conductance formula there appears $(1 + Q)$ in the denominator instead of $(1 - Q)$ in the numerator. A similar procedure has been developed in the theory of electrolytic conductance [24, 25] where the effect has been named feedbackeffect.

3.2.5. Quantum Effects and Large Scale Fluctuations

The quantum kinetic theory may be developed formally in the same way as the classical theory if the one-particle Wigner function is introduced by [5, 8, 15]

$$f_a(\boldsymbol{r}, \boldsymbol{p}, t) = \frac{N_a}{V(2\pi)^3} \int \varrho_a \left(\boldsymbol{r} + \frac{1}{2} \hbar \boldsymbol{y}, \boldsymbol{r} - \frac{1}{2} \hbar \boldsymbol{y} \right) \exp\left(-i \boldsymbol{y} \boldsymbol{p}\right) \mathrm{d}\boldsymbol{y} \, , \tag{3.95}$$

ϱ_a — being the one-particle density matrix.

Then the kinetic equations remain the same structure as in the classical case, except changes of the integral kernels. We do not want to go into detail and give just the necessary changes in the classical formulae restricting ourselves to the effective potential approximation (EPA). The new expressions which have to be used instead of equations (3.73) and (3.86) are [21, 22]

$$L_{ab} = -\theta(4\pi e_a e_b)^{-2} \int_0^\infty \mathrm{d}k\, k^3 \Phi_{ab}(k)\, h_{ab}(k) \exp\left[-\frac{\hbar^2 k^2}{8\theta}\left(\frac{1}{m_a} + \frac{1}{m_l}\right)\right], \tag{3.96}$$

$$Q = \frac{\varkappa^2 e^2 \theta^2}{6(2\pi)^3} \int_0^\infty \mathrm{d}k\, |\Phi_{ei}(k)|\, h_{ei}(k) \cdot$$

$$\cdot \int_0^\infty \mathrm{d}\tau \exp\left[-\frac{1}{2}\hbar^2\tau^2\theta\left(\frac{1}{m_e} + \frac{1}{m_i}\right)\right] \sin\left[\frac{1}{2}\hbar^2 k^2 \tau\left(\frac{1}{m_e} + \frac{1}{m_i}\right)\right], \tag{3.97}$$

$$P_a = n_a\theta - \frac{\partial}{\partial V} \int_0^1 \frac{\mathrm{d}\lambda}{\lambda} U_a^{\mathrm{pot}}(\lambda), \tag{3.98}$$

$$U_a^{\mathrm{pot}}(\lambda) = \frac{1}{2(2\pi)^3} n_a \sum_b n_b \int \lambda \Phi_{ab}(k)\, h_{ab}(k, \lambda)\, \mathrm{d}\boldsymbol{k},$$

where λ is a charging parameter and h_{ab} the Fourier transform of the quantum correlation function.

More elaborated quantum transport theories based on the linear response theory and the Greens function theory will be developed in the next two paragraphs and in the last chapter of this book.

Finally let us consider the question of nonideal behaviour at large distances from equilibrium. Now the correlation parameter is given by (3.3). If the system is pumped by an external source of free energy the potential energy may be much higher than near equilibrium where $\Gamma = (e^2/k_B T d)$. We expect that at critical distances from the equilibrium nonequilibrium transitions occur which are connected with the appearance of zeroes of the dielectric function. As an example of such dissipative structures in the sense of Prigogine we present measurements of the conductance of H-plasmas due to Demidov and Hamberger which clearly show a staircase structure (fig. 3.3). With in-

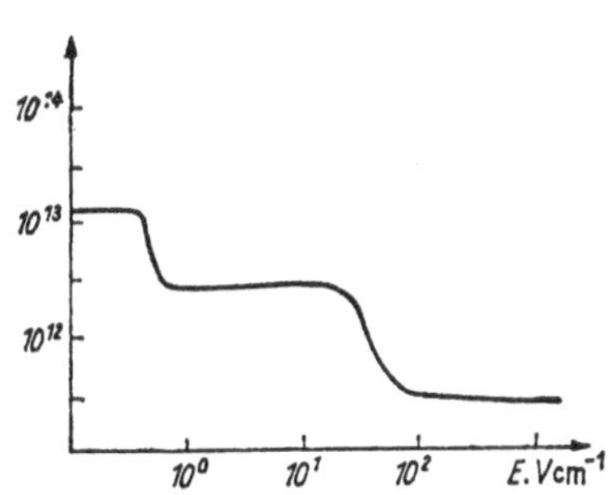

Fig. 3.3. Conductivity of low-density hydrogen plasmasin electrostatic untis ($n_i = 10^{12}$ cm^{-3}) in dependence on the external electrical field

creasing field the conductivity decreases two times — first decrease is due to ion-sound instabilities and the second one to two-stream instabilities. These kinds of explanations of the anomalous resistivity have been worked out by Bunemann, Sagdeev, Kadomtsev and others. There is a quite new statistical theory of the anomalous resistivity due to

Belyi, Klimontovich and Nalivaiko [25]. In this theory the additional contribution I_a to the collision integral is taken into account. We remember that I_a is due to large-scale fluctuations with a radius of correlations greater than the Debye radius. The additional collision integral has the form

$$\tilde{I}_e = \frac{n_e e^2}{(2\pi)^4} \frac{\partial}{\partial \boldsymbol{p}} \int \mathrm{d}\boldsymbol{k}\, \mathrm{d}\omega \frac{\boldsymbol{k}}{k^2} \operatorname{Re}\left[\frac{(i/\pi)}{\omega - \boldsymbol{k}\boldsymbol{v} + i\nu_e}\right] \cdot$$
$$\cdot \left[(\delta \boldsymbol{E}\, \delta \boldsymbol{E})_{\omega,\boldsymbol{k}}\, \boldsymbol{k} \cdot \frac{\partial f_e}{\partial \boldsymbol{p}} + \frac{8\pi \operatorname{Im} \tilde{\varepsilon}(\omega, \boldsymbol{k})}{|\varepsilon(\omega, \boldsymbol{k})|^2} f_e\right], \tag{3.99}$$

where the wavy line means smoothing out over small scales. The large-scale fluctuations lead in a first approximation to the replacement of the relaxation time t_{ei} by an effective one,

$$t_{ei}^{\text{eff}} = t_{ei} \Big/ \left[1 + N_\mathrm{D} \frac{e^2 \langle E^2 \rangle r_\mathrm{D}^2}{(k_\mathrm{B} T)^2}\right], \tag{3.100}$$

which may be much smaller than the relaxation time near equilibrium and is responsible for the decrease of the conductivity.

3.3. Linear Response Theory

Alternatively to the kinetic theory, the linear response theory has been developed to describe transport phenomena in many particle systems [26, 27]. Within linear reponse theory the transport coefficients like electrical conductivity are related to equilibrium correlation functions. On the other hand, it was possible recently [28] to express the collision integral in kinetic theory by correlation functions of fluctuating quantities so that the equivalence of both approaches may be shown. It is one of the advantages of linear response theory that the quantum mechanical many body theory immediately can be applied to evaluate the transport coefficients. Therefore the linear response theory may serve to find systematic approximations for the evaluation of transport coefficients valid in the entire density range between normal plasmas and liquid metals.

3.3.1. Statistical Operator of Non-Equilibrium and Correlation Function Method

Let us consider a system (system Hamiltonian H_S) under the influence of an external electrical field $\boldsymbol{E}$. The interaction is described by

$$H_\mathrm{E} = -\sum_a e_a \boldsymbol{R}_a \cdot \boldsymbol{E}\,, \tag{3.101}$$

where $\boldsymbol{R}_a = \sum_i \boldsymbol{r}_{a,i}$ is the center of mass of species a, charge e_a. Under the influence of the field $\boldsymbol{E}$ an electrical current arises in the system. In order to evaluate the mean value of the electrical current density

$$\boldsymbol{j}(t) = \frac{1}{V} \operatorname{Sp} \{\varrho(t) \sum_{a,i} e_a \boldsymbol{v}_{a,i}\} \tag{3.102}$$

(V – normalization volume, $\boldsymbol{v}_{a,i}$ – particle velocity), we have to determine the non-equilibrium statistical operator $\varrho(t)$.

Linear response theory starts from the solution of the equation of motion of the statistical operator of the system

$$\frac{\partial}{\partial t}\varrho(t) + \frac{\mathrm{i}}{\hbar}[H, \varrho(t)] = -\varepsilon\big(\varrho(t) - \varrho_q(t)\big) \tag{3.103}$$

in first order with respect to the external field $\boldsymbol{E}$; $H = H_\mathrm{S} + H_\mathrm{E}$. We are interested in the stationary case where H_E, ϱ and ϱ_q are not explicitely time dependent. In contrast to the von Neumann equation, in (3.103) on the right hand side a source term has been introduced [27] in order to select out the retarded solutions $\varrho(t)$ obeying the correct boundary conditions (weakening of correlations [29], see also chap. 4); the limit $\varepsilon \to +0$ should be taken after the thermodynamic limit. The solution of (3.103) reads (stationary case)

$$\varrho = \varrho_q - \int\limits_{-\infty}^{0} \mathrm{d}t\, \mathrm{e}^{\varepsilon t}\, \mathrm{e}^{\mathrm{i}Ht/\hbar} \frac{\mathrm{i}}{\hbar}[H, \varrho_q]\, \mathrm{e}^{-\mathrm{i}Ht/\hbar} . \tag{3.105}$$

The Kubo theory will be obtained from (3.104) if we take $\varrho_q = \varrho_0$, where the equilibrium statistical operator ϱ_0 is given by

$$\varrho_0 = \exp(-\beta H_\mathrm{S} + \beta\mu N)/\mathrm{Sp}\exp(-\beta H_\mathrm{S} + \beta\mu N) . \tag{3.105}$$

According to Zubarev [27], this special choice of ϱ_q may be generalized using the quasi-equilibrium statistical operator obtained from the condition of maximum of information entropy if, besides energy and total particle number, the mean values of further observables B_m relevant for the non-equilibrium state are fixed:

$$\begin{aligned} \varrho_q &= \exp\Big(-\beta\big(H_\mathrm{S} - \mu N - \sum_m F_m B_m\big)\Big)/Z_q ; \\ Z_q &= \mathrm{Sp}\exp\Big(-\beta\big(H_\mathrm{S} - \mu N - \sum_m F_m B_m\big)\Big) . \end{aligned} \tag{3.106}$$

It was shown [30–32] that different choices of ϱ_q lead to equivalent results if the correct partial summations are performed within the perturbative treatment of the scattering process which is responsible for the transport process. The advantage of the Zubarev formalism is that good results for transport coefficients can be obtained avoiding partial summations, if the ϱ_q, i.e. the set of observables B_m, is suitably choosen. As shown below, the freedom in the choice of a set of observables B_m which determines ϱ_q is equivalent to the freedom to choose a class of trial functions in a variational problem [32, 39].

In first order with respect to $\boldsymbol{E}$ and assuming $F_m \propto \boldsymbol{E}$ (linear response) we have from (3.104) (see [33])

$$\varrho = \varrho_q - \varrho_0 \int\limits_{-\infty}^{0} \mathrm{d}t\, \mathrm{e}^{\varepsilon t} \int\limits_{0}^{\beta} \mathrm{d}\tau \left\{ - \sum_a \frac{e_a}{m_a} \boldsymbol{E} \cdot \boldsymbol{P}_a (t - \mathrm{i}\hbar\tau) + \sum_m F_m \dot{P}_m (t - \mathrm{i}\hbar\tau) \right\} \tag{3.107}$$

with $\dot{B} = \frac{\mathrm{i}}{\hbar}[H_\mathrm{S}, B]$; $\boldsymbol{P}_a = m_a \dot{\boldsymbol{R}}_a$; $B(t) = \exp(\mathrm{i}H_\mathrm{S}t/\hbar)\, B \exp(-\mathrm{i}H_\mathrm{S}t/\hbar)$.

The thermodynamic parameters F_m are determined by the condition of stationarity for the mean values of the observables B_m

$$\frac{\mathrm{d}}{\mathrm{d}t} \mathrm{Sp}\,\{\varrho B_\mathrm{m}\} = \mathrm{Sp}\left\{\varrho \frac{\mathrm{i}}{\hbar} [H, B_\mathrm{m}]\right\} = 0 \tag{3.108}$$

which leads with expression (3.107) to the foollwing system of coupled equations

$$\sum_a e_a \boldsymbol{E} \cdot [(\boldsymbol{P}_a; \dot{B}_m) + \langle \boldsymbol{P}_a; \dot{B}_m \rangle] = \sum_{m'} F_{m'} \langle \dot{B}_{m'}; \dot{B}_m \rangle, \tag{3.109}$$

where the scalar product is defined by

$$(A; B) = \int_0^\beta \mathrm{d}\tau\, \mathrm{Sp}\,\{\varrho_0 A(-\mathrm{i}\hbar\tau)\, B\}$$

and

$$\langle A; B \rangle = \int_{-\infty}^0 \mathrm{d}t\, \mathrm{e}^{\varepsilon t}\, (A(t); B)$$

is the Laplace transform of the correlation function $(A(t);\ B)$. The solution of the system of linear equations (3.109) for given $\boldsymbol{E}$ means that the parameters F_m (and by this the statistical operator ϱ and the mean values of observables in non-equilibrium) are expressed by correlation functions in equilibrium. On the other hand, equilibrium correlation functions can be evaluated systematically using the methods of many body theory (Green functions, see sect. 3.3.5.)

An interesting property of the correlation functions is the validity of the Onsager symmetry relations which follow from

$$\mathrm{Sp}\,\{\varrho_0 A(t)\, B(t')\} = \mathrm{Sp}\,\{\varrho_0 B^+(t)\, A^+(t')\}$$

(see [27], § 16) so that for Hermitean observables B_m (zero magnetic field)

$$\langle \dot{B}_{m'}; \dot{B}_m \rangle = \langle \dot{B}_m; \dot{B}_{m'} \rangle\,. \tag{3.110}$$

We mention that similar to the electrical conductivity also the other transport coefficients may be treated by correlation functions. The generalization from the stationary case to arbitrary frequencies is straightforward, see [33].

3.3.2. Expression for the Electrical Conductivity of a Fully Ionized Gas in the Adiabatic Approximation

In ion plasmas, because of the large mass ratio m_i/m_e (m_i — ion mass, $m_e = m$ — electron mass) we can employ the adiabatic approximation i.e. the electrons are moving in the given (time dependent or fixed) potential $V_{ei}(\boldsymbol{r}, t) = \sum_n \Phi_{ei}(\boldsymbol{r} - \boldsymbol{r}_{i,n}(t))$, where $\boldsymbol{r}_{i,n}(t)$ is the position of the nth ion at time t, so that

$$\begin{aligned} H_\mathrm{S}^\mathrm{ad} &= \sum_\mathrm{k} E_k c_k c_k + 1/2 \sum_{kk'q} \frac{4\pi e^2}{q^2} c_{k+q}^+ c_{k'-q}^+ c_{k'} c_k + \sum_{kq} \mathrm{V}_q(t)\, c_{k+q}^+ c_k \\ &= H_0 + H_{ee} + H_{ei}; \end{aligned} \tag{3.111}$$

$$V_q(t) = \sum_h \exp(\mathrm{i}\boldsymbol{q} \cdot \boldsymbol{r}_{i,n}(t))\, v_q;$$

$$v_q = \int \mathrm{d}^3 r \exp(\mathrm{i}\boldsymbol{q} \cdot \boldsymbol{r})\, \Phi_{ei}(\boldsymbol{r})\,.$$

The averaging procedure will be performed in two steps: First over the electron coordinates at given ion distribution, then over the ion dynamics (dynamical ion structure factor). The adiabatic approximation is widely used in liquid metal physics, see chapter 5., but also in semiconductor and ion plasmas to evaluate transport coefficients.

Furthermore, we neglect the occurrence of bound states (bound states are considered in sect. 3.3.7.). This means that an appropriate choice of the set of observables B_m occuring in ϱ_q ("weakening of initial correlations") is to take the occupation numbers $n_k = c_k^+ c_k$ of free one-particle states k, and the corresponding parameters F_m will be denoted by $\delta\mu_k$.

The parameters $\delta\mu_k$ are determined in the stationary case by the set of equations (3.109):

$$e\boldsymbol{E} \cdot [(\boldsymbol{P}; n_k) + \langle \boldsymbol{P}; \dot{n}_k \rangle] = \sum_{k'} \delta\mu_{k'} \langle \dot{n}_{k'}; \dot{n}_k \rangle \tag{3.112}$$

which have the meaning of generalized Boltzmann equations for the mean occupation numbers

$$f(k) = \mathrm{Sp}\,\{\varrho n_k\} = \mathrm{Sp}\,\{\varrho_0 n_k\} + \sum_{k'} \delta\mu_{k'} [(n_{k'}; n_k) - \mathrm{Sp}\,\{\varrho_0 n_{k'}\}\,\mathrm{Sp}\,\{\varrho_0 n_k\}]\,. \tag{3.113}$$

The right hand side of equation (3.112) may be considered as a generalized collision integral. By virtue of the Onsager symmetry relations (3.110) this collision operator is symmetric; furthermore it is bounded from below so that the Kohler variational principle [34] may be used to solve the generalized Boltzmann equation. The Chapman-Enskog method as well as the Grad method described in section 3.2.3. may be considered as realizations of this variational principle expanding $\delta\mu_k$ with respect to a suitable set of basis functions $\varphi(k)$.

In order to demonstrate the equivalence to the 13 momenta approach by Grad of section 3.2.3. we use the variational ansatz

$$\delta\mu_k = \hbar\boldsymbol{k} \cdot \left(\gamma_0 + \gamma_2 \frac{\hbar^2 k^2}{m k_\mathrm{B} T}\right)$$

which corresponds to the following form of the quasi-equilibrium statistical operator

$$\varrho_q = Z_q^{-1} \exp\left(-\beta(H_S - \mu N - \gamma_0 \cdot \boldsymbol{P}_0 - \gamma_2 \cdot \boldsymbol{P}_2)\right) \tag{3.114}$$

where $\boldsymbol{P}_n = \sum_k \hbar\boldsymbol{k}(\hbar^2 k^2/m k_\mathrm{B} T)^n c_k^+ c_k$ are related to the particle current and the thermal current. The parameters γ_0, γ_2 are determined from the corresponding stationarity conditions (3.108) and are explicitely given by correlation functions solving the corresponding two linear equations (3.109) by Cramer's rule.

The mean value of the electrical current density (3.102) is given by the parameters γ_n according to

$$\boldsymbol{j} = \frac{e}{mV}\,\mathrm{Sp}\,\{\varrho_q \boldsymbol{P}_0\} = \frac{ie}{\hbar V}\,\mathrm{Sp}\,\{\varrho_0[(\gamma_0 \cdot \boldsymbol{P}_0 + \gamma_2 \cdot \boldsymbol{P}_2), \boldsymbol{R}]\} = \sigma\boldsymbol{E}\,. \tag{3.115}$$

For the isotropic case, $\boldsymbol{E}, \gamma_n \parallel \boldsymbol{e}^x$, finally we arrive at the following expression for the conductivity [32]:

$$\sigma = \frac{e^2}{V}\,\frac{Q_0(N_0\langle F_2; F_2\rangle - N_2\langle F_0; F_2\rangle) - Q_2(N_0\langle F_2; F_0\rangle - N_2\langle F_0; F_0\rangle)}{\langle F_0; F_0\rangle\,\langle F_2; F_2\rangle - \langle F_0; F_2\rangle\,\langle F_2; F_0\rangle} \tag{3.116}$$

where

$$F_n = (i/\hbar)\,[H_S, P_n^x],$$
$$N_n = (i/\hbar)\,\mathrm{Sp}\,\{\varrho_0[P_n^x, R^x]\};$$
$$Q_n = N_n + 1/m \cdot \langle P^x; F_n\rangle\,. \tag{3.117}$$

Expression (3.116) relates the conductivity to correlation functions of the form $\langle A; B\rangle$ and may be considered as a generalization of the force-force correlation function expression for the resistivity obtained from (3.114) if $\gamma_2 = 0$ [33]:

$$\sigma_0 = \frac{e^2}{V}\frac{Q_0 N_0}{\langle F_0; F_0\rangle}\,. \tag{3.118}$$

The extension of the variational ansatz (3.114) (inclusion of γ_4) was considered for semiconductors in [35] and gives only small corrections. Furtheron, expressions similar to (3.116) can be derived for other transport processes as thermal conductivity and thermopower, see [35].

3.3.3. Born Approximation and Structure Factor Effects

To evaluate the correlation functions occuring in the expression (3.116) for the conductivity we discuss the first Born approximation. A Green function technique will be outlined in section 3.3.5.

With the Hamiltonian (3.111) we find for the F_n, (e.117)

$$F_0 = -i \sum_{kq} V_q q^x c_{k+q}^+ c_k \equiv F_{0,ei}\,, \tag{3.119}$$

$$F_2 = -i\frac{\hbar^2}{mk_BT}\sum_{kq} V_q[(k^x + q^x)\,(\boldsymbol{k} + \boldsymbol{q})^2 - k^x k^2]\,c_{k+q}^+ c_k -$$
$$-i\frac{\hbar^2}{mk_BT}\sum_{kk'q}\frac{4\pi e^2}{q^2}\,[(k^x + q^x)\,(\boldsymbol{k} + \boldsymbol{q})^2 - k^x k^2]\,c_{k-q}^+ c_{k'-q}^+ c_{k'} c_k \equiv$$
$$\equiv F_{2,ei} + F_{2,ee}\,. \tag{3.120}$$

Because F_n is of first order in the interaction, in first Born approximation H_S is replaced by H_0 in evaluating the correlation functions, and we have by using Wick's theorem and performing spin summations

$$\langle F_{n,ei}; F_{n',ei}\rangle = \frac{2\pi\hbar}{3}\left(\frac{\hbar^2}{mk_BT}\right)^{(n+n')/2}\sum_{kq}\int d(\hbar\omega)\left(-\frac{df}{dE_k}\right)\cdot$$
$$\cdot\,\delta(E_{\boldsymbol{k}+\boldsymbol{q}} - E_{\boldsymbol{k}} - \hbar\omega)\,k^{(n+n')}q^2 N S(\boldsymbol{q}, \omega)\,|v_q^s|^2;$$

$$\langle F_{2,ee}; F_{2,ee}\rangle = \frac{2\pi}{3}\frac{\hbar^5}{m^2(k_BT)^3}\sum_{kk'q} f_{\boldsymbol{k}+\boldsymbol{q}} f_{\boldsymbol{k}'-\boldsymbol{q}}(1 - f_{\boldsymbol{k}'})\,(1 - f_{\boldsymbol{k}})\,\cdot$$
$$\cdot\,\delta(E_{\boldsymbol{k}+\boldsymbol{q}} + E_{\boldsymbol{k}'-\boldsymbol{q}} - E_{\boldsymbol{k}'} - E_{\boldsymbol{k}})\left(\frac{4\pi e^2}{q^2 + \varkappa^2}\right)^2\cdot$$
$$\cdot\,[(\boldsymbol{k} + \boldsymbol{q})\,(\boldsymbol{k} + \boldsymbol{q})^2 + (\boldsymbol{k}' - \boldsymbol{q})\,(\boldsymbol{k}' - \boldsymbol{q})^2 - \boldsymbol{k}k^2 - \boldsymbol{k}'k'^2]^2; \tag{3.121}$$

$$Q_n = N_n = (1 + 2n) \left(\frac{mk_B T}{2\pi\hbar^2}\right)^{3/2} \frac{2V}{\Gamma\left(\frac{n+3}{2}\right)} \int_0^\infty \frac{y^{(n+1)/2}}{\exp(y - \mu_e/k_B T) + 1} \, dy \tag{3.122}$$

are expressed by Fermi integrals. Exchange terms and higher order of density contributions are neglected in (3.121); the correlation functions $\langle F_{n,ei}; F_{2,ee}\rangle$ vanish in the order V_q^2 (first Born approximation).

In order to avoid divergencies for Coulomb like potentials (where $\lim_{q\to 0} V_q \propto q^{-2}$), screening effects must be taken into account in expressions (3.121). Within this Section we use the approximation where the potentials V_q are replaced by the electronically screened static potentials

$$V_b^s = V_q/\varepsilon(q) \approx V_q/(1 + \varkappa_e^2/q^2) , \tag{3.123}$$

$$\varkappa_e^2 = \lim_{q\to 0} 4\pi e^2 \frac{1}{V} \sum_k \frac{f_{k+q} - f_k}{E_{k+q} - E_k} = 4\pi e^2 \frac{\partial}{\partial\mu} \frac{1}{V} \sum_k f_k;$$

has the meaning of the inverse Debye length and is given in the non-degenerate limit by $\varkappa_e = (4\pi n_e e^2/k_B T)^{1/2}$. In section 3.3.5. we show that the screened potential can be introduced in a consequent manner by using many particle theory techniques.

The dynamical structure factor $S(\boldsymbol{q}, \omega)$ defined by

$$S(\boldsymbol{q}, \omega) = N^{-1} \int dt \, e^{-i\omega t} \langle \sum_{nn'} \exp\left(i\boldsymbol{q} \cdot (\boldsymbol{r}_{i,n}(0) - \boldsymbol{r}_{i,n'}(t)\right)\rangle$$

(N — number of ions) is obtained, if $V_q(t)^2$ which depends on the ion position, see (3.111), is averaged over the ion dynamics. In the static case where the ion positions are fixed we have $S(\boldsymbol{q}, \omega) = S(\boldsymbol{q})\,\delta(\omega)$, and if furthermore the ions are assumed to be randomly distributed the static structure factor is given by $S(\boldsymbol{q}) = 1$. For $S(q) = 1$, we find from (3.121) in first Born approximation

$$\langle F_{n,ei}; F_{n',ei}\rangle = \frac{4e^4 Nm}{3\pi\hbar} \left(\frac{\hbar^2}{mk_B T}\right)^{(n+n')/2} \int_0^\infty dk \left(-\frac{df_k}{dE_k}\right) k^{n+n'+1} \Phi\left(\frac{4k^2}{\varkappa_e^2}\right),$$

$$\Phi(b) = \frac{1}{2}\left(\ln(1 + b) - \frac{b}{1 + b}\right). \tag{3.124}$$

Neglecting electron-electron interaction, in the non-degenerate case ($n_e \Lambda_e^3 \ll 1$) results from (3.116) (with $N_2 = 5N$; $\langle F_0; F_{2,ei}\rangle = 2\langle F_0; F_0\rangle$; $\langle F_{2,ei}; F_{2,ei}\rangle = 8\langle F_0; F_0\rangle$)

$$\sigma_{ei}(n_e, T) = \frac{13}{4} \sigma_0(n_e, T) , \tag{3.124a}$$

whereas including electron-electron interaction ($\langle F_{2,ee}; F_{2,ee}\rangle = \sqrt{2}\, 4\langle F_0; F_0\rangle$) (see for comparison eqs. (3.69) and (3.74)):

$$\sigma_{ei,ee}(n_e, T) = \frac{13 + 4\sqrt{2}}{4 + 4\sqrt{2}} \sigma_0(n_e, T) \tag{3.124b}$$

follows. In the strongly degenerate case ($n_e \Lambda_e^3 \ll 1$) we have $\sigma_{ei,ee}(n_e, T) = \sigma_{ei}(n_e, T) = \sigma_0(n_e, T)$. Numerical evaluations of equations (3.116), (3.121), (3.122) for $S(\boldsymbol{q}) = 1$ are given in section 3.5., see figure 3.10.

In general, the static structure factor defined by

$$S(\boldsymbol{q}) = 1 + n_i \int g_{ii}(\boldsymbol{r}) \exp(\mathrm{i}\boldsymbol{q} \cdot \boldsymbol{r}) \, \mathrm{d}^3 r \tag{3.125}$$

where $g_{ii}(\boldsymbol{r})$ is the pair distribution function of ions, which is determined by the ion-ion interaction [36]. In the case of a pure Coulomb interaction between the ions we may use (for $\varkappa e^2/k_B T < 1$) the well-known nonlinear Debye expression

$$g_{ii}(r) = \exp\left(-\frac{e^2}{r k_B T} \mathrm{e}^{-\varkappa r}\right). \tag{3.126}$$

In the case that there is additionally a hard core contribution to the ion-ion interaction (hard-core radius R_i), an expression for $g_{ii}(r)$ can be found in the mean spherical approximation. According to [37] where were considered charged hard spheres immersed in a uniform neutralizing background the following expression for the Laplace transform

$$G_{ii}(s) = \int_0^\infty x g_{ii}(x) \, \mathrm{e}^{-sx} \, \mathrm{d}x \,, \qquad x = r/2R_i \,,$$

of the radial distribution function is used

$$G_{ii}(s) = \frac{sL(s)}{12\eta[L(s) + S(s) \, \mathrm{e}]} \,, \tag{3.127}$$

$$L(s) = 12\eta(-Ms^2 + Ps + \tau); \qquad \tau = 2R_i \varkappa_i;$$

$$S(s) = s^4 + Rs^3 + R^2 s^2/2 + 12\eta(\tau - P) s - 12\eta\tau;$$

$$R = \frac{6\eta}{1-\eta} - Q; \qquad P = \frac{1 + 2\eta}{(1-\eta)^2} - \frac{Q}{2(1-\eta)} + \frac{\tau}{2};$$

$$M = \frac{Q^2}{24\eta} - \frac{1 + \eta/2}{(1-\eta)^2}; \qquad Q = \frac{1 + 2\eta}{1-\eta}\left[1 - \left(1 + \frac{2\tau(1-\eta)^3}{(1+2\eta)^2}\right)^{1/2}\right];$$

$\eta = 4\pi n_i R_i^3/3$ is the packing parameter. From $G_{ii}(s)$ the structure factor $S(q)$ is obtained by the relation

$$S(\boldsymbol{q}) = 1 + \frac{8\pi n_i R_i^2}{\mathrm{i}q} [G_{ii}(-2\mathrm{i}\boldsymbol{q}\boldsymbol{R}_i) - G_{ii}(2\mathrm{i}\boldsymbol{q}\boldsymbol{R}_i)] \,.$$

This approximation coincides in the low density limit with the linearized Debye expression. In the case that the charges of the hard spheres may be neglected ($\tau = 0$) it coincides with the Percus-Yevick solution for uncharged hard spheres.

The structure factor $S(\boldsymbol{q})$ according to equations (3.126), (3.127) is presented for different parameter values in figures 3.4 [36]. Numerical evaluations of the conductivity including structure factor effects are given in section 3.5. The inclusion of structure effects is necessary if the theory of plasma conductivity is extrapolated to the high density region up to the densities of liquid state, see chapter 1. The inclusion of the dynamical structure factor, see also chapter 5. and [38], enables us to improve the adiabatic approximation by including dynamical degrees of freedom of the ion system, as, for instance, the scattering by phonons as well known in solid state physics.

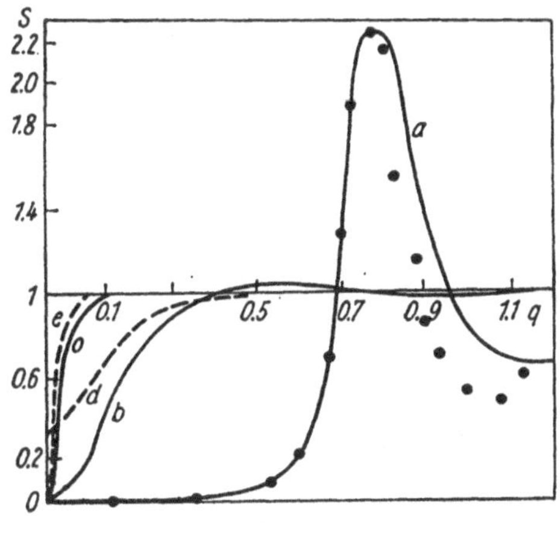

Fig. 3.4. Ion structure factor $S(q)$ in dependence on the wave vector modulus q. Full lines: MSA theory, eq. (3.127); $R_i = 1.69 \cdot 10^{-10}$ m (crystallografic radius of Cs); a — T_1, n_{i3}; b — T_2, n_{i2}; c — T_2, n_{i1}. Dashed lines: Debye theory, eq. (3.126); d — T_2, n_{i2}; e — T_2, n_{i1}. Experimental values for Cs at the melting point T_1 [48]. $T_1 = 303.15$ K, $T_2 = 10^4$ K; $n_{i1} = 1.196 \cdot 10^{25}$ m^{-3}, $n_{i2} = 6.245 \cdot 10^{26}$ m^{-3}, $n_{i3} = 8.292 \cdot 10^{27}$ m^{-3} [36]

3.3.4. Statically Screened *t*-Matrix Approximation

In order to improve the Born approximation, higher order scattering effects must be taken into account in evaluating the correlation functions (3.121). In this section we give a more intuitive approach; for a systematic approach see section 3.3.5.

Multiple scattering effects are taken into account by introducing the t-matrix. This can be done by using the method of Kubo-Greenwood [39] in evaluating the correlation functions. We propose different approximations in evaluating the electron-ion collision and the electron-electron collision correlation functions, (3.121), [32, 36].

To describe multiple scattering of electrons by the statically screened electron-ion potential V_q^s we replace the system Hamiltonian H_S in the time evolution operator $\exp(\mathrm{i}H_S t/\hbar)$ by $H_0 + H_{ei}^s$, see (3.111), instead of H_0 (Born approximation). We obtain

$$\langle F_{n,ei}; F_{n,ei}\rangle = \frac{2\pi\hbar}{3}\left(\frac{\hbar^2}{mk_BT}\right)^{(n+n')/2} \sum_{kk'}\left(-\frac{\partial f}{\partial E_k}\right)\delta(E_k - E^{k'}) \cdot k^{(n+n')}(\boldsymbol{k} - \boldsymbol{k}')^2\,|T^S(\boldsymbol{k}, \boldsymbol{k}')|^2 \tag{3.128}$$

where the electron-ion t-matrix $T^S(\boldsymbol{k}, \boldsymbol{k}')$ may be taken in the single site approximation as

$$|T^S(\boldsymbol{k}, \boldsymbol{k}')|^2 = NS(\boldsymbol{k} - \boldsymbol{k}')\,|t^S(\boldsymbol{k} - \boldsymbol{k}')|^2\,,$$

and the single ion t-matrix is given by the scattering phase shifts $\delta_1(E)$:

$$t^S(\boldsymbol{k} - \boldsymbol{k}') = -\frac{2\pi\hbar^2}{m}\sum_{l=0}^{\infty}(2l + 1)\,P_l(\cos\vartheta)\,\frac{1}{k}\,e^{i\delta_l(E)}\sin\delta_1(E) \tag{3.129}$$

where $E = \hbar^2k^2/2m$, ϑ — scattering angle $\sphericalangle\,(\boldsymbol{k}, \boldsymbol{k}')$. We give the result for randomly distributed scatterers, $S(q) = 1$ [36, 40, 41]:

$$\langle F_{n,ei}; F_{n,ei}\rangle = \frac{4N}{3\pi}\frac{mk_BT}{\hbar}\int_0^\infty dE\left(\frac{2E}{k_BT}\right)^{(n+n)/2+1}\left(-\frac{\partial f}{\partial E}\right)\sum_{l=0}^{\infty}(l + 1) \cdot \sin^2[\delta_l(E) - \delta_{l+1}(E)] \tag{3.130}$$

which is generalized for arbitrary structure factor $S(q)$:

$$\langle F_{n,ei}; F_{n',ei}\rangle = \frac{2N}{3\pi}\frac{mk_BT}{\hbar}\int_0^\infty dE\left(\frac{2E}{k_BT}\right)^{(n+n')/2+1}\left(-\frac{\partial f}{\partial E}\right)\sum_{l=0}^\infty\sum_{l'=0}^\infty (2l+1)\cdot$$
$$\cdot(2l'+1)\, e^{i\delta_l(E)}\sin\delta_l(E)\, e^{-i\delta_{l'}(E)}\sin\delta_l(E)$$
$$\cdot\int_{-1}^{1} du\,(1-u)\,P_l(u)\,P_{l'}(u)\,S\left(k\sqrt{2(1-u)}\right). \qquad (3.131)$$

On the other hand, multiple electron-electron scattering is described by the following approximation in the time evolution operator $\exp(iH_s t/\hbar)$: H_S is replaced by $H_0 + H_{ee}$ (eq. (3.111)), instead of H_0 (Born approximation) [32]. Within the screened ladder t-matrix approximation ($\varepsilon \to +0$)

$$t^S_{ee}(kk', k+qk'-q) = \frac{4\pi e^2}{\varkappa^2+q^2} + \sum_{q'}\frac{4\pi e^2}{\varkappa^2+q'^2}\left(E - \frac{(k+q')^2}{2m} - \frac{(k'-q')^2}{2m} + i\varepsilon\right)^{-1}\cdot$$
$$\cdot\, t^S_{ee}(k+q', k'-q', k+qk'-q)$$

the expression

$$\langle F_{2,ee}; F_{2,ee}\rangle = \frac{2\pi}{3}\frac{\hbar^5}{m^2(k_BT)^3}\sum_{kk'q} f_{k+q}f_{k'-q}(1-f_{k'})(1-f_k)\cdot$$
$$\cdot\,\delta(E_{k+q}+E_{k'-q}-E_{k'}-E_k)\,|t^S(kk', k+q\,k'-q)|^2\cdot$$
$$\cdot\,[(k+q)(k+q)^2 + (k'-q)(k'-q)^2 - kk^2 - k'k'^2]^2 \qquad (3.132)$$

is derived, and similar to (3.129) the t-matrix can be expressed by the scattering phase shifts of the screened electron-electron (Debye) potential. In the non-degenerate limit we have

$$\langle F_{2,ee}; F_{2,ee}\rangle = \frac{128}{3}\sqrt{\pi}\,\hbar n_e N\left(\frac{\hbar^2}{mk_BT}\right)^{7/2}\int_0^\infty dP\,P^5\exp(-\hbar^2P^2/mk_BT)\sum_{l=0}^\infty$$
$$\cdot\frac{(l+1)(l+2)}{2l+3}\sin^2[\delta_l(\hbar^2P^2/m) - \delta_{l+2}(\hbar^2P^2/m)]. \qquad (3.133)$$

Scattering phase shifts are obtained from the numerical solution of the Schrödinger equation for the screened electron-ion and electron-electron potential, respectively [36, 42]. Results and discussions are presented in section 3.5. Expressions for the resistivity in terms of phase shifts are well known in liquid metal physics [40, 41].

3.3.5. Systematic Perturbative Expansion of Correlation Functions

After having derived expressions for the correlation functions in Born approximation (sect. 3.3.) and in the statically screened t-matrix approximation (sect. 3.4.) we outline the systematic expansion of correlation functions with the aid of quantum statistical many body theory [21]. The application of Green's function technique in order to work out a quantum kinetic transport theory in plasmas, solids and liquid metals has been put forward in some recent papers [43—45], see also sect. 3.4.

The correlation functions occurring in the expression for the conductivity (3.116) contain quantum statistical averages of operator products like

$$\langle c^+_{k_1}(t)\, c_{k_2}(t)\, c^+_{k_3} c_{k_4}\rangle = \int_{-\infty}^{\infty} \frac{d\omega}{\pi} \frac{e^{i\omega t}}{e^{\beta\omega}-1} A(\boldsymbol{k}_1\boldsymbol{k}_2\boldsymbol{k}_3\boldsymbol{k}_4, \omega) \tag{3.134}$$

where the spectral density $A(\boldsymbol{k}_1\boldsymbol{k}_2\boldsymbol{k}_3\boldsymbol{k}_4, \omega)$ is connected with the Green's function $G(\boldsymbol{k}_1\boldsymbol{k}_2\boldsymbol{k}_3\boldsymbol{k}_4, z)$ according to [46]

$$A(\boldsymbol{k}_1 \dots \boldsymbol{k}_4, \omega) = i[G(\boldsymbol{k}_1 \dots \boldsymbol{k}_4, \omega + i\varepsilon) - G(\boldsymbol{k}_1 \dots \boldsymbol{k}_4, \omega - i\varepsilon)]\,. \tag{3.135}$$

The Green's function $G(\boldsymbol{k}_1\boldsymbol{k}_2\boldsymbol{k}_3\boldsymbol{k}_4, z)$ can be evaluated by using perturbative expansions and partial summations which are compactly represented by Feynman diagrams. Within the usual notation [46] we have for the free electron propagator

$$G_0(\boldsymbol{p}, z_\nu) = \overset{\boldsymbol{p},\, z_\nu}{\longrightarrow} = (z_\nu - E_p)^{-1}; \qquad z_\nu = \frac{\pi\nu}{-i\beta} + \mu\,, \qquad \nu \text{ odd}\,,$$

for the Coulomb propagator

$$D(\boldsymbol{q}, \omega_\lambda) = \overset{\boldsymbol{q},\, \omega_\lambda}{- - - - -} = i\frac{4\pi}{q^2}\,; \qquad \omega_\nu = \frac{\pi\lambda}{-i\beta}\,, \qquad \lambda \text{ even};$$

energy and momentum are conserved at each vertex e. The electron-ion interaction is denoted by

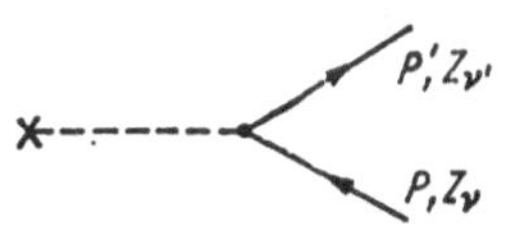

$$= (-i\beta)\, \delta_{\nu\nu'} i V_{p-p'}\,.$$

Furtheron, we introduce the Green's function $G_T(\boldsymbol{p}, \boldsymbol{p}', z)$ describing the multiple scattering by the ion potential

$$G_T(\boldsymbol{p}, \boldsymbol{p}', z) =$$

$$= G_0(\boldsymbol{p}, z_\nu)\, \delta_{pp'} + \sum_{\bar{p}} G_0(\boldsymbol{p}, z_\nu)\, V(\boldsymbol{p} - \bar{\boldsymbol{p}})\, G(\bar{\boldsymbol{p}}, \boldsymbol{p}', z_\nu) \tag{3.136}$$

and the electronically screened electron-electron interaction potential

$$V^s_{ee}(q, \omega_\lambda) =$$

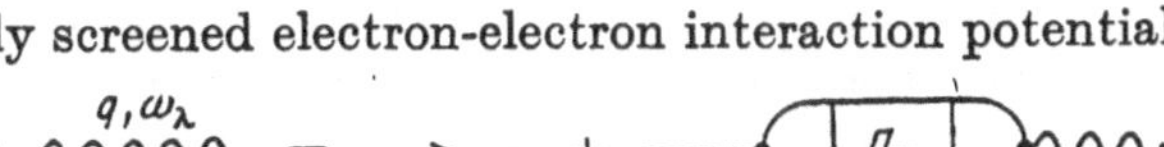

$$= \frac{4\pi e^2/q^2}{1 - (4\pi e^2/q^2)\, \Pi_{ee}(q, \omega_\lambda)} = \frac{4\pi e^2/q^2}{\varepsilon_{ee}(q, \omega_\lambda)}\,, \tag{3.137}$$

where $\Pi_{ee}(q, \omega_\lambda)$ is the irreducible polarization fragment with only electron-electron interactions.

For example, we discuss different approximations for the evaluation of the force-force correlation function $\langle \boldsymbol{F}_{0,ei}; \boldsymbol{F}_{0,ei}\rangle$:

$$\langle \boldsymbol{F}_0; \boldsymbol{F}_0\rangle = -\sum_{qq'} V_q V_{q'} \boldsymbol{q} \cdot \boldsymbol{q}' \sum_{kk'} \int \frac{d\omega}{2\pi} \delta(\omega) \frac{1}{\omega} \cdot$$
$$\cdot \operatorname{Im} G(\boldsymbol{k}, \boldsymbol{k} + \boldsymbol{q}, \boldsymbol{k}', \boldsymbol{k}' + \boldsymbol{q}'; \omega + i\varepsilon)\,. \tag{3.138}$$

In the most simple approximation for $P(q, q', \omega_\lambda) = \sum_{kk'} G(\boldsymbol{k}, \boldsymbol{k} + \boldsymbol{q}, \boldsymbol{k}', \boldsymbol{k}' + \boldsymbol{q}', \omega_\lambda)$ we have

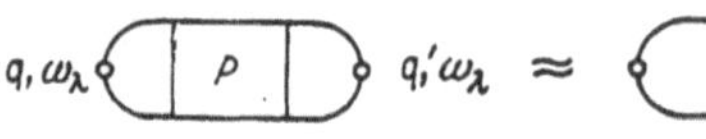

$$= \frac{1}{-i\beta} \sum_{kz_\nu} \frac{1}{\omega_\lambda + z_\nu - E_K} \frac{1}{z_\nu - E_{k+q}} \delta_{q,-q'};$$

$$\operatorname{Im} P(\boldsymbol{q}, \boldsymbol{q}', \omega) = \delta_{\boldsymbol{q},-\boldsymbol{q}'} \sum_k (f_k - f_{k+q}) \, 2\pi\delta\,(\hbar\omega + E_k - E_{k+q}) \qquad (3.139)$$

which corresponds to the Landau approximation for the collision integral and diverges in the case of a Coulomb potential.

The Born approximation discussed in section 3.3.3. corresponds to an approximation where all orders of electron-electron interactions are taken into account:

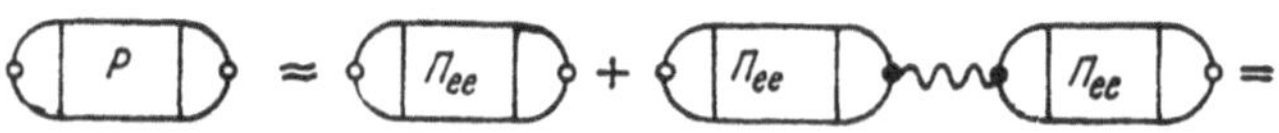

$$= \delta_{q,-q} \frac{\Pi_{ee}(q, \omega_\lambda)}{1 - 4\pi e^2/q^2 \, \Pi_{ee}(q, \Psi\omega_\lambda)};$$

which yields

$$\operatorname{Im} P(q, q', \omega) = \delta_{q,-q} \frac{1}{|\varepsilon_{ee}(q, \omega)|^2} \operatorname{Im} \Pi_{ee}(q, \omega) \qquad (3.140)$$

and corresponds to the Lenard-Balescu approximation of the collision intgral.— and corresponds to the Lenard-Balescu approximation of the collision integral. If $\Pi_{ee}(q, \omega)$ is taken in random phase approximation, (3.139), the result (3.121) follows for $\langle F_0; F_0 \rangle$. We point out that the factor $|\varepsilon_{ee}(q, 0)|^{-2}$ will not occur in the other correlation functions. We remark that equation (3.140) leads to improvements of the expression (3.121) if, for instance, instead of the RPA for $\Pi_{ee}(q, \omega)$, exchange terms are taken into account, (see [38]).

The unscreened t-matrix approximation which corresponds to the Boltzmann collision integral is obtained if the following partial sum including all orders of electron-ion interaction is considered

(3.141)

It leads to the result (3.128) containing the t-matrix for the unscreened potential which is diverging for $q \to 0$ in theCoulomb case. The correct description of dynamical screening (which avoids this divergency) is one of the most important problems in plasma conductivity theory. We present two expressions for the correlation function $\langle F_0; F_0 \rangle$ including multiple electron-ion scattering and screening effects which may be considered asupperand lowerbond for the correct treatment of a dynamically screened t-matrix: (i)

(3.142)

where in higher than second order only the unscreened electron-ion interaction is taken into account [49, 50]. In the case of pure Coulomb interaction, because of the pecularity that the full t matrix coincides with the Born approximation, this expression (3.142) is identical with theBorn approximation (3.121) with a statically screened potential. Screening effects are underestimated in this approximation.

(ii)

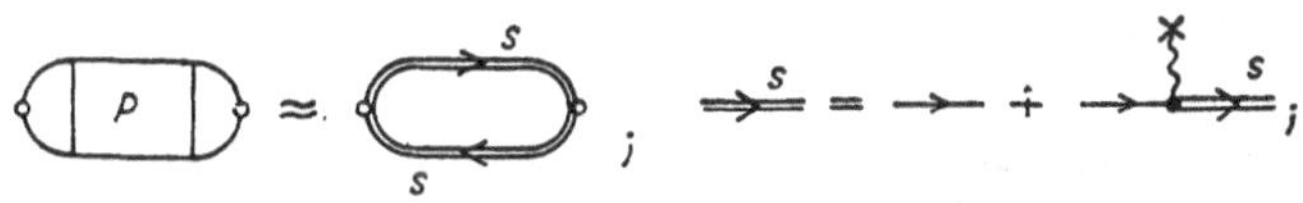

(3.143)

the electron-ion potential is statically screened. This approximation which overestimates screening effects leads to the statically screened t-matrix approximation (3.128), (3.129).

For more details of application of Green's function technique in quantum transport theory see section 3.4.

3.3.6. The Debye-Onsager Relaxation Effect

Degeneration effects, structure factor effects, and the effects of dynamical screening are not the only effects which may become important in the evaluation of conductivity of high dense plasmas. As a further effect the evaluation of the correlation functions Q_n should be considered in equations (3.116), (3.118) which are connected with the Debye-Onsager relaxation effect [20] (see sect. 3.2.4.). For instance, in the non-degenerate case we have

$$\begin{aligned}\langle P^x; F_0\rangle &= \beta \int \mathrm{d}t\, \mathrm{e}^{-\varepsilon t} \int \mathrm{d}\Gamma f(\Gamma)\, P^x F^x(t) \\ &= \beta \frac{1}{3} \int \mathrm{d}\Gamma f(\Gamma)\, [\mathrm{V_D}(r = \infty) - \mathrm{V_D}(r = 0)] \\ &= -\frac{1}{6} N\beta\varkappa e^2 \,. \end{aligned} \tag{3.144}$$

This correction to the electrical conductivity is further discussed in section 3.5.

An other possibility to introduce density effects of this order into the quantum transport theory which stems from correlation functions of the type $\langle P^x; F_n\rangle$ in (3.116), (3.118) or $\langle \boldsymbol{P}; n_k\rangle$ in (3.112) employs the transition from the free particle representation to the quasi-particle representation. From the principle of weakening of initial correlations it is clear that good asymptotic states in the plasma are not free states k, (3.112), but we have to consider in constructing ϱ_q, (3.106), the occupation number of quasiparticle states k, (see also sect. 3.3.7.). Quasiparticle states can be determined in such a way that the correlation functions $\langle \boldsymbol{P}; \dot{n}_k\rangle$ etc. mentioned above become relatively small. The quasiparticle representation can be found with the aid of Green's function technique:

$$\begin{aligned} G^{-1}_{\mathrm{qu}}(\hat{p}, z) &= \hbar z - E_{\mathrm{qu}}(\hat{p}) \,, \\ E_{\mathrm{qu}}(\hat{p}) &= \hat{p}^2/2m + \Delta(\hat{p}); \\ \Delta(\hat{p}) &= \mathrm{Re}\, \Sigma(\check{p}, z)|_{\hbar z = E_{\mathrm{qu}}(\hat{p})} \,, \end{aligned} \tag{3.145}$$

$\Sigma(\hat{p}, z)$ denotes the self energy. From (3.145), the transition to the quasiparticle representation can be performed if the bare electron mass m is replaced by the effective mass m_{eff}, (see sect. 3.5.), with $m_{\mathrm{eff}}^{-1} = m^{-1} + \partial^2\Sigma(\hat{p})/\partial\hat{p}^2$. It should be mentioned that after this transition to the quasiparticle representation also the interaction should be renormalized.

3.3.7. Influence of Bound States on the Conductivity. The Polarization Potential

In general, a plasma is not fully ionized but contains neutral bound states (atoms) as well as partially ionized ions. Therefore, a general quantum transport theory should take into account not only the interaction between charged particles but also the interaction with neutral bound states on the same footing.

As was pointed out in the previous Section, in linear response theory it is of importance to formulate the correct boundary conditions, i.e. the principle of weakening of initial correlations. If bound states occur, the quasi-equilibrium statistical operator ϱ_q, (3.106), should be constructed with the occupation numbers n_ν of the set of base states $|\nu\rangle = \{|k\rangle, |\alpha m\rangle\}$ (see also chap. 4.)

$$1 = \sum_\nu |\nu\rangle\langle\nu| = \sum_k |k\rangle\langle k| + \sum_{\alpha,m} |\alpha m\rangle\langle\alpha m| , \tag{3.146}$$

where α denotes the internal quantum number of bound states at localization centre $\boldsymbol{r}_{i,m}$ (ion position), k are extended states (orthogonalized plane waves), see figure 3.5. With that choice of ϱ_q, a generalized Boltzmann equation similar to (3.112) is derived [47]:

$$e\boldsymbol{E} \cdot [(\boldsymbol{P}; n_\nu) + \langle \boldsymbol{P}; n_\nu\rangle] = \sum_{\nu'} \delta\mu_{\nu'}\langle \dot{n}_{\nu'}; \dot{n}_\nu\rangle . \tag{3.147}$$

As before (sect. 3.3.2.), the solution $\delta\mu_\nu$ of this generalized Boltzmann equation may be found using the Kohler variational principle.

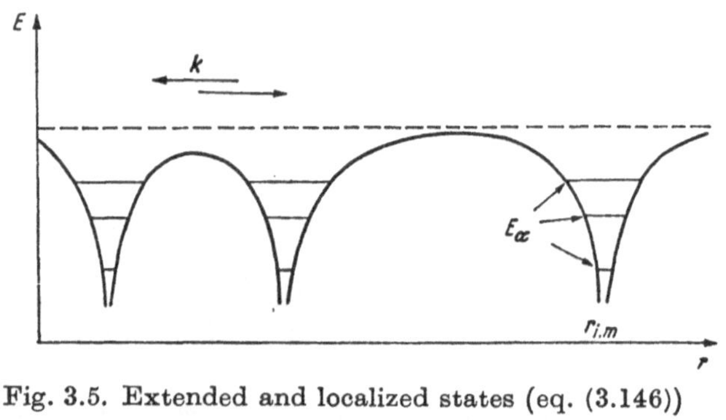

Fig. 3.5. Extended and localized states (eq. (3.146))

In order to evaluate the correlation functions occuring in (3.147) we perform a canonical transformation to represent the system Hamiltonian H_S^{ad}, (3.111), with respect to the new base states $|\nu\rangle$ (3.146):

$$H_S^{ad} = \sum_{\nu\nu'} E_{\nu'\nu} a_\nu^+ a_{\nu'} + \sum_{\nu_1\nu_2\nu_3\nu_4} M_{\nu_1\nu_2\nu_3\nu_4} a_{\nu_1}^+ a_{\nu_2}^+ a_{\nu_3} a_{\nu_4} . \tag{3.148}$$

Approximating the extended states $|k\rangle$ by plane waves $\exp(i\boldsymbol{k}\cdot\boldsymbol{r})$ (free particles instead of quasiparticles) and the localized states $|\alpha m\rangle$ by the bound states of the isolated ion $\psi_\alpha(\boldsymbol{r} - \boldsymbol{r}_{i,m})$, we obtain for the matrix elements in (3.148)

$$\begin{aligned}
&E_{kk} = \hbar^2 k^2/2m; \qquad E_{k,k+q} = V_q;\\
&E_{\alpha m\alpha' m} = E_\alpha \delta_{\alpha\alpha'}; \qquad E_{\alpha m\alpha m'} = J_{mm'\alpha};\\
&M_{kk'k+q\,k'-q} = 4\pi e^2/q^2 = M_{kk'q};\\
&M_{kk'k+q\alpha m} = (4\pi e^2/q^2)\exp(\mathrm{i}\cdot(\boldsymbol{k}'-\boldsymbol{q})\cdot\boldsymbol{r}_{i,m})\,\psi_2(\boldsymbol{k}'-\boldsymbol{q}) = M_{\alpha k'q}.\\
&M_{k\alpha m k+q\alpha' m} = (4\pi e^2/q^2)\exp(-\mathrm{i}\boldsymbol{q}\cdot\boldsymbol{r}_{i,m})\sum \psi_\alpha^*(\boldsymbol{p})\,\psi_{\alpha'}(\boldsymbol{p}-\boldsymbol{q}) = M_{\alpha\alpha' q'},
\end{aligned} \tag{3.149}$$

where V_q is the ion (pseudo) potential; $J_{mm'\alpha}$ has the meaning of the hopping integral; the further matrix elements in (3.149) can be connected within the first Born approximation (golden rule) with different scattering processes as electron-electron scattering, recombination and ionization, and free-bound electron scattering. We assume that the Coulomb repulsion term $U_{\alpha\alpha'} = M_{\alpha m\alpha' m'\alpha m\alpha' m'}$ is large so that a localization center will hardly be occupied by two bound state electrons (atomic limit of the Hubbard model); the Coulomb interaction $M_{\alpha m\alpha' m'\alpha m\alpha' m'}$ between different localization centres may be taken into account by stochastic fluctuating energy eigenvalues E_α. Further matrix elements are not considered here.

The correlation functions $\langle \dot{n}_{\nu'}; \dot{n}_\nu \rangle$ have been evaluated in first Born approximation; furtheron, in second Born approximation the terms of the correlation function $\langle \dot{n}_{k'}; \dot{n}_k \rangle$ describing an elastic scattering process of free electrons by a ground state atom $\left(\propto \delta(E_{k'} - E_k) f_k f_0\right)$ have been considered because by these contributions the dynamical response of the atom (polarization potential) is taken into account. We give immediately the final result for the conductivity [47] if the variational ansatz

$$\delta\mu_{\alpha,m} = 0; \qquad \delta\mu_k = \gamma \boldsymbol{E} \cdot \boldsymbol{k} \tag{3.150}$$

is used to solve the generalized Boltzmann equation (3.147) $(f_{\alpha m} \equiv f_\alpha)$

$$\sigma = \frac{2\pi}{\hbar} \frac{e^2}{V} \sum_{mm'\alpha} |J_{mm'\alpha}|^2 \, r_{i,m}^z (r_{i,m}^z - r_{i,m'}^z) \, \delta(E_{\alpha m} - E_{\alpha m'}) \frac{\partial}{\partial E} f_\alpha + $$
$$+ \gamma \frac{e}{V} \frac{1}{\hbar} \sum_k k^z \frac{\partial}{\partial k^z} E_k \frac{\partial}{\partial E} f_k , \tag{3.151}$$

$$\gamma = \frac{e}{\beta\pi} \sum_k k^z \frac{\partial}{\partial k^z} E_k \frac{\partial}{\partial E} f_k (R_{\text{el}} + R_{\text{inel}} + R_{\text{ioniz}} + R_{\text{recomb}})^{-1} , \tag{3.152}$$

$$R_{\text{el}} = \sum_{kqm} (q^z)^2 f_k (1 - f_k) \, \delta(E_k - E_{k-q}) \left[\left(1 - \sum_\alpha f_\alpha\right) |V_q^{\text{S}}|^2 + \right.$$
$$\left. + \sum_\alpha f_\alpha \left| v_q^{\text{s}} + M_{\alpha\alpha' q}^{\text{s}} + W_{00,q} \delta_{\alpha,0} \right|^2 \right] , \tag{3.153}$$

$$R_{\text{inel}} = \sum_{kqm\alpha\alpha'} (q^z)^2 f_\alpha f_k (1 - f_{\alpha'}) \, (1 - f_{k+q}) \, |M_{\alpha\alpha',q}^{\text{s}}|^2 \cdot$$
$$\cdot \delta(E_\alpha - E_{\alpha'} + E_k - E_{k+q}) \, (1 - \delta_{\alpha\alpha'}) , \tag{3.154}$$

$$R_{\text{ioniz}} = \sum_{kqmk'\alpha} |q^z + k'^z|^2 f_\alpha f_{k'} (1 - f_{k'}) \, (1 - f_{k+q}) \, |M_{\alpha k',q}^{\text{s}}|^2 \cdot$$
$$\cdot \delta(E_\alpha + E_k - E_{k'} - E_{k+q}) , \tag{3.155}$$

$$R_{\text{recomb}} = \sum_{kqmk'\alpha} |q^z - k'^z|^2 f_k f_{k'} (1 - f_{k+q}) \, (1 - f_\alpha) \, |M_{k\alpha,q}^{\text{s}}|^2 \cdot$$
$$\cdot \delta(E_k + E_{k'} - E_{k+q} - E_\alpha) . \tag{3.156}$$

The polarization potential occuring in (3.153) is given in local approximation $(E_0 - E_\alpha \gg E_k)$ by

$$W_{00,q} = \int \mathrm{d}^3 r \exp(\mathrm{i}\boldsymbol{q}\boldsymbol{r}) \, W_{00}(\boldsymbol{r}) ,$$

$$W_{00}(\boldsymbol{r}) = \sum_\nu \frac{2e^4}{E_0 - E_\nu} \int \mathrm{d}^3 \boldsymbol{r}_1 \int \mathrm{d}^3 \boldsymbol{r}_2 \frac{1}{|\boldsymbol{r}_1 - \boldsymbol{r}|} \frac{1}{|\boldsymbol{r}_2 - \boldsymbol{r}|} \psi_\nu^*(\boldsymbol{r}_1) \, \psi_0(\boldsymbol{r}_1) \, \psi_0^*(\boldsymbol{r}_2) \, \psi_\nu(\boldsymbol{r}_2) \tag{3.157}$$

and can be approximated by

$$W_{00}(r) = \frac{e^2\alpha_{\text{pol}}}{(r^2 + r_0^2)^2}, \tag{3.158}$$

where α_{pol} is the ground state polarizability.

The expression (3.151) corresponds to the superposition of a hopping current between localized states and an itinerant electron current. Extensions of the set of trial functions (3.150) will give improvements of the conductivity equivalent, for instance, to the solution of the Miller-Abrahams random resistor network in the limit of only localized states or to the Brooks-Herring formula if localized states are absent.

In (3.151), the conductivity is presented as a function of the grand canonical variables β, μ (via f_α, f_k). It is more convenient to express the conductivity as function of the temperture and the total electron (or total ion) density n. The chemical potential is eliminated by the thermodynamic equation of state $\mu = \mu(\beta, n)$. In the simplest approximation we have

$$n = n_{\text{free}} + n_{\text{bound}} = n_{\text{free}}(1 - \sum_\alpha f_\alpha)^{-1}, \tag{3.159}$$

$$\begin{aligned} n_{\text{free}} &= 2(2\pi\hbar^2/mk_{\text{B}}T^{-3/2}) \exp(\beta\mu), \\ (1 - \sum_\alpha f_\alpha)^{-1} &= 1 + \sum_\alpha \exp\left(-\beta(E_\alpha - \mu)\right), \end{aligned} \tag{3.160}$$

if only one electron can be bound to a given localization center. In this approximation follows

$$\sum_k k^z \frac{\partial}{\partial k^z} E_k f_k(1 - f_k) = \frac{V}{\beta} n_{\text{free}}.$$

The influence of bound states on the conductivity is further discussed in section 3.5.

3.4. Green's Function Theory

3.4.1. Basic Equations

In section 3.3.5. was shown, how correlation functions in terms of which the conductivity may be expressed, are derived by means of Green's function techniques. In this connection the spectral function $A(k, \omega)$ of the two particle equilibrium Green's function plays an essential role and contains all information of the relevant microscopic quantities and its dependence on the properties of the many particle system.

In sections 3.3.1. and 3.3.2. the correlation functions occur in expressions for the conductivity which are derived using a quasi equilibrium statistical operator which differs from that of the grand canonical one by additional field dependent quantities.

Besides the conventional method of the application of kinetic equations (sect. 3.2.) and the more recent technique of the quasi equilibrium statistical operator just mentioned (sect. 3.3.) there is another method which relates directly transport quantities to Green's functions which is in a more restricted sense refered to as Kubo's method. Especially the two latter methods are closely connected with each other. As an alternative method we mention moreover the possibility of determining the conductivity from the complex dielectric constant which in turn must be derived in terms of Green's functions [51].

The direct connection of transport properties and Green's functions has a lot of advantages, among which we mention only the possibility of a consequent perturbation technique and the screening. However, there are also drawbacks which we shall discuss later. Let us start with the following expression for the conductivity which may be derived in the framework, e.g. of the technique outlined in [27]

$$\sigma(\boldsymbol{k}, \omega) = -\,\varepsilon(\boldsymbol{k}, \omega) \sum_{ab} \frac{4\pi\, e_a\, e_b}{k^2} \int \frac{\mathrm{d}\boldsymbol{q}}{(2\pi)^3} \frac{\hbar \boldsymbol{k} \cdot \left(\boldsymbol{q} - \dfrac{\boldsymbol{k}}{2}\right)}{4\pi\, i m_a} G_{ab}(\boldsymbol{k}\boldsymbol{q}\omega + \mathrm{i}0)\,. \tag{3.161}$$

Here $\varepsilon(k, \omega)$ is the dielectric constant, which is defined by

$$\varepsilon(\boldsymbol{k}, \omega) = 1 - \sum_{ab} \frac{4\pi e_a e_b}{k^2} \int \frac{\mathrm{d}\boldsymbol{q}}{(2\pi)^3} \Pi_{ab}(\boldsymbol{q}\boldsymbol{k}\omega + \mathrm{i}0)\,. \tag{3.162}$$

In (3.161), (3.162) e_a and m_a are the mass and the charge of species a, and G_{ab} and Π_{ab} are the causal two particle Green's function and the polarisation function, respectively, which are used in quantum statistics [41].

More precise, one should mention that in deriving (3.161) initially occurs a correlation function which may, by extension of the integration interval, be replaced by a retarded commutator Green's function, which, in turn is the limiting value of the Fourier coefficient of the causal Green's function of the thermodynamic equilibrium, if we approach the real frequency axis from above. This latter point is of practical use because we may apply in this case the method of functional derivatives which allows automatically for a consequent perturbation technique. Before going into the details we give some useful relations.

If we apply the identity

$$G_{ab} = G_a G_b + L_{ab} \tag{3.163}$$

we have

$$\varepsilon(\boldsymbol{k}, \omega) \int \frac{\mathrm{d}\boldsymbol{q}}{(2\pi)^3} L_{ab}(\boldsymbol{q}\boldsymbol{k}\omega + \mathrm{i}0) = \int \frac{\mathrm{d}\boldsymbol{q}}{(2\pi)^3} \Pi_{ab}(\boldsymbol{q}\boldsymbol{k}\omega + \mathrm{i}0)\,. \tag{3.164}$$

A more general function $\tilde{\Pi}_{ab}$ depends under translational invariance on three space coordinates (differences) and thus, in the Fourier transforms, on three wave vectors; we have

$$\int \frac{\mathrm{d}\boldsymbol{l}}{(2\pi)^3} \tilde{\Pi}_{ab}(\boldsymbol{q}, \boldsymbol{q} - \boldsymbol{k}, \boldsymbol{l} + \boldsymbol{k}, \boldsymbol{l}; \omega + \mathrm{i}0) = \Pi(\boldsymbol{q}\boldsymbol{k}\omega + \mathrm{i}0)\,.$$

For our purposes it is sufficient to take only one time difference and thus one frequency, ω. The connection between the (longitudinal part of) the dielectric function and the conductivity is given by

$$\sigma(k, \omega) = \frac{\mathrm{i}\omega}{4\pi} \left(\varepsilon(k, \omega) - 1\right)\,. \tag{3.165}$$

Here we mention that, in general, both $\sigma(k, \omega)$ and $\varepsilon(k, \omega)$ are complex quantities.

3.4.2. Green's Functions

Here we give only the most essential relations; for a more detailed representation we refer to the literature [46, 52].

For instance, the two particle causal Green's function describes the motion of two particles a, b which are created, generally spoken, at different space time points, 1′, 2′ and annihilated at other ones, 1, 2 ($1 \equiv \{r_1, t_1, \sigma_1\}$; σ_1 — spin-variable). We need only the following special function

$$G_{ab}(121'2^+)|_{t_1'=t_1^+} = \frac{1}{i^2} \langle T\{\psi_a(1)\,\psi_b(2)\,\psi_b^+(2^+)\,\psi_a^+(rt)\}\rangle\,. \tag{3.166}$$

Here ψ, ψ^+ are the annihilation and creation operators, respectively, $2^+ \equiv \{r_2, t_2 + 0, \sigma_2\}$, T — Wick time — ordering operator and $\langle ... \rangle$ grand canonical average.

In order to get σ from (3.161) we must determine G_{ab} up to the approximation in question.

A powerful technique to do this is the functional derivation technique (e.g. [46]). By means of such a procedure formally a truncation of the hierarchy of the equations of motion is achieved from which G_{ab} must be determined. One of the most essential features of this technique is that screening is taken into account in a natural way if we define an effective external potential:

$$U_a^{\text{eff}}(1) = U_a(1) - i \sum_b G_b(22^+)\, V_{ab}(12)\,. \tag{3.167}$$

Here and in the following equations doubly occuring variables demand integration and spin summation, e.g.

$$\sum_{\sigma_2} \int_0^{-i\beta} dt_2 \int d\boldsymbol{r}_2; \qquad \beta = \frac{1}{k_B T}\,.$$

The external potential in (3.167), $U_a(1)$, tends to zero after carrying out all calculations, as we have, in the framework in which our Kubo formula (3.161) shall be considered, to deal only with the equilibrium Green's functions G_{ab}. G_b is the single particle Green's function.

$$G_b(12) = \frac{1}{i} \langle T\{\psi_b(1)\,\psi_b^+(2)\}\rangle\,. \tag{3.168}$$

While for the function L_{ab} (density fluctuation fctn.) the relation holds

$$G_{ab} - G_a G_b = -\frac{\delta G_a}{\delta U_b} \equiv L_{ab} \tag{3.169}$$

the polarization function follows from the functional derivative with respect to the potential (3.167)

$$\Pi_{ab} = -\frac{\delta G_a}{\delta U_b^{\text{eff}}}\,. \tag{3.170}$$

From (3.169) and (3.170) follows that it is necessary to have an equation for the determination of G_a. This is the Dyson equation, namely

$$G_a^{-1}(11') = G_a^{0-1}(11') - U_a^{\text{eff}}(1)\,\delta(11') - \Sigma_a'(11')\,. \tag{3.171}$$

Here G_a is the Green's function of free particles, G_a^{-1} is the inverse function:

$$G_a^{-1}(11')\, G_a(1'2) = \delta(12)\,.$$

Σ' is the self energy Σ without the Hartree contribution, Σ_H, which was used to construct U^{eff}. Σ' obeys the equation

$$\Sigma'_a(11') = \mathrm{i} V^s_{aa}(11')\, G_a(1^+1') + \mathrm{i} \sum_c V^s_{ac}(12)\, G_a(13)\, \frac{\delta \Sigma'_a(31')}{\delta U^{\text{eff}}_c(2)} \,. \tag{3.172}$$

The screened potential V^s must be determined from

$$V^s_{ab}(12) = V_{ab}(12) + \mathrm{i} \sum_{-cd} V^s_{ca}(13)\, \Pi_{dc}(4343)\, V_{db}(42) \,. \tag{3.173}$$

We want to mention here, that only in the single case in which we have e.g. in a two component system $V_{11} = V_{22} = -V_{12} = -V_{24}$ (relevant for the Coulomb potential) the screening may be the same one for all interactions. But this simplification does not hold if for the electron-ion potential a pseudopotential is used, and for the ion-ion interaction also a short range contribution is added to the Coulomb potential.

The polarization function must be determined from a Bethe-Salpeter-type equation

$$\Pi_{ab}(121'2') = -\delta_{ab} G_a(12')\, G_a(21') + \sum_c G_a(1\bar{1})\, G_a(\bar{2}1')\, \frac{\delta \Sigma'_a(\bar{1}\bar{2})}{\delta\, G_a(34)}\, \Pi_{cb}(3242') \tag{3.174}$$

while G_{ab} is related to Π_{ab} by [53]

$$G_{ab}(121'2') = G_a(11')\, G_b(22') + \Pi_{ab}(121'2') + \mathrm{i} \sum_{cd} \Pi_{ac}(131'3)\, V^s_{cd}(34)\, \Pi_{db} \cdot (4242') \,.$$

Equation (3.174) is the appropriate starting point from which we may, together with (3.171), (3.173), derive any desired approximation [53, 54].

In the theory of the conductivity it is necessary to consider e.g. the following approximation for GL, which we give for a shorthand notation in a graphical version [51, 55]

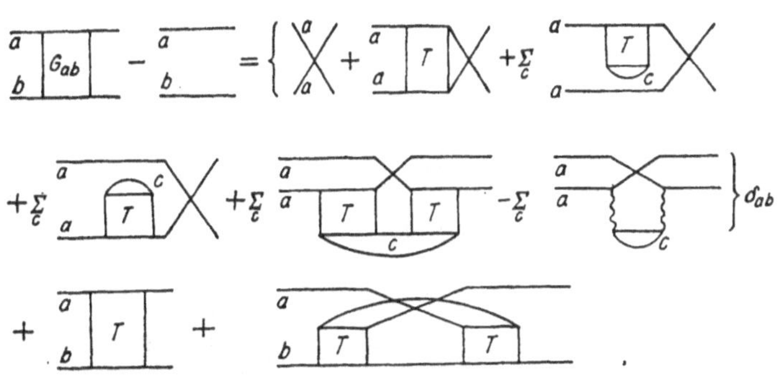

(3.175)

Here T means the ladder sum of the screened potential V^s and must be determined from the integral equation:

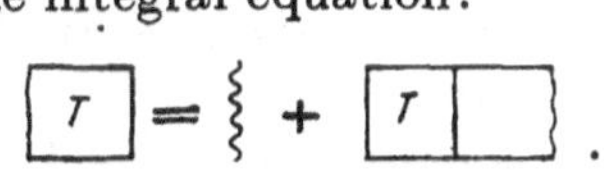

The full line represents the single particle Green's function, and the wavy line means the dynamically screened potential (3.173), which reads in a special approximation ($\Pi_{cd} = \delta_{cd} \Pi_{cc}$)

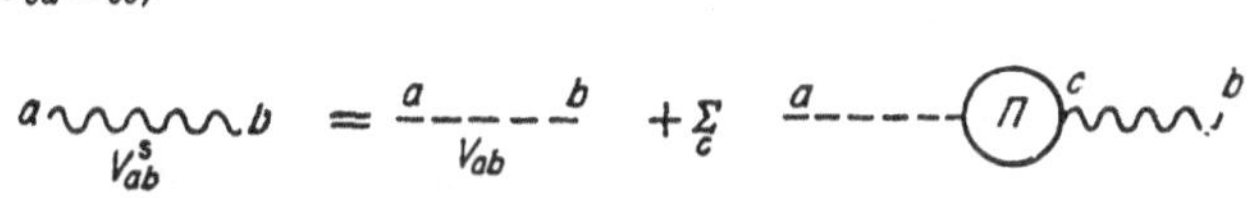

Physical processes which are described by (3.175) were taken into account in section 3.3.5. The influence of bound states which is considered in section 3.3.7. and in chapter 4. requires higher order diagrams than those incorporated in (3.175) (see also section 3.4.4.).

3.4.3. Collision Transport Frequency

In this section we want to deal with some more general questions and difficulties. For the sake of simplicity, we restrict ourselves to some simple approximation of G_{ab}, which corresponds to the Born approximation of (3.175).

In the conductance formula (3.161) G_{ab} occurs with special weight factors, namely

$$\sigma \to \sum_{ab} \frac{e_a e_b}{m_a} G_{ab} \equiv \sum_a \frac{e_a^2}{m} \tilde{G}_a \tag{3.176}$$

where $\tilde{G}_a = \sum_b e_{ea}^{eb} G_{ab}$; $a, b = 1, 2$. We will consider only a system of two species with $e_2 = -e_1$. The formal quantity $\tilde{G}_a$ is for practical purposes subdivided into two parts,

$$\tilde{G}_a = \tilde{G}_a^0 + \tilde{G}_a^{\text{corr}} . \tag{3.177}$$

Here $\tilde{G}_a^0 \equiv G_{aa}^{\text{HF}}$ is the Hartree-Fock approximation of G_{ab},

$$G_{ab}^{\text{HF}} = -\delta_{ab} G_a G_b .$$

We are now faced with the general fact that a series expansion with respect to the interaction parameter leads to $1/\omega$ and higher order divergencies in the $\omega \to 0$ limit which are usually overcome by the application of Drude like formulae [51].

However, it is more useful to perform a partial summation with respect to the interaction parameter, which is achieved, e.g., if we cast G_{ab} into a shape which is known for the single particle Green's function as a solution of Dyson's equation. This means we write instead of (3.177)

$$\tilde{G}_a = \frac{1}{\tilde{G}_a^{0-1} - \Gamma_a} . \tag{3.178}$$

This procedure was applied, e.g., by Plakida [56] and is in agreement with the considerations in [30]. From (3.177) follows with the definition (3.178)

$$\Gamma_a = \tilde{G}_a^{-1} \tilde{G}_a^{\text{corr}} \tilde{G}_a^{0-1} \tag{3.179}$$

If G_a^{corr} would be known exactly, (3.179) would be an exact equation, too. In practical cases, of course, (3.179) represents another type of approximation than (3.177) does. For the special approximation we are interested in it is sufficient to use instead of (3.179)

$$\Gamma_a = \tilde{G}_a^{0-1} \tilde{G}_a^{\text{corr}} \tilde{G}_a^{0-1} , \tag{3.180}$$

which will be correct up to the second order of the interaction potential. For a more detailed discussion of this question see [57].

In the approximation discussed here we take the following expression

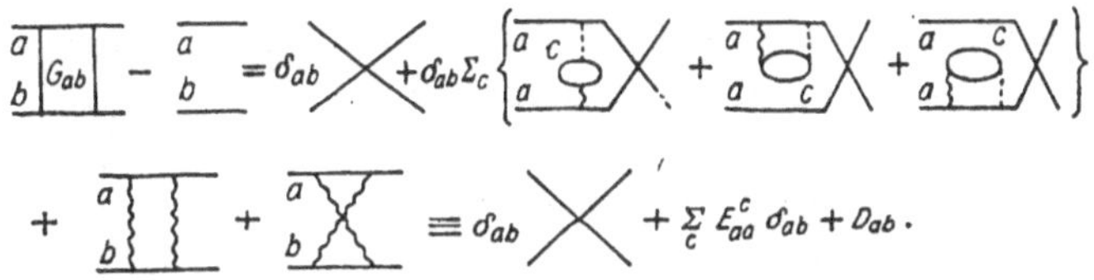

(3.181)

These diagrams were discussed in [58] too. The diagrams are denoted by E (exchange; contribution in the curly bracket) and by D (direct terms). Further exchange terms which do not contain a closed loop of Green's functions may be omitted in our considerations.

In frecency and momentum representation (3.178) reads

$$\tilde{G}_a(\boldsymbol{pk}\omega + \mathrm{i}0\,) = \mathrm{i}\,\frac{f_a(\boldsymbol{p}) - f_a(\boldsymbol{p}+\boldsymbol{k})}{\hbar\omega + \varepsilon^a_{\boldsymbol{p}} - \varepsilon^a_{\boldsymbol{p}+\boldsymbol{k}} + \mathrm{i}\nu_a(pk\omega)}\,.$$

Here we introduced the collision transport frequency ν_a by the relation

$$\Gamma_a(\boldsymbol{pk}\omega) = -\nu_a(\boldsymbol{pk}\omega)\,(f_a(\boldsymbol{p}) - f_a(\boldsymbol{p}+\boldsymbol{k}))\,. \qquad (3.182)$$

f_a is the Fermi function, and $\varepsilon^a_{\boldsymbol{p}} = \hbar^2 p^2/2m_a$. For the Hartree-Fock approximation we have

$$\tilde{G}^0_a(\boldsymbol{pk}\omega + \mathrm{i}0) = \mathrm{i}\,\frac{f_a(\boldsymbol{p}) - f_a(\boldsymbol{p}+\boldsymbol{k})}{\hbar\omega + \mathrm{i}0 + \varepsilon^a_{\boldsymbol{p}} - \varepsilon^a_{\boldsymbol{p}+\boldsymbol{k}}}\,.$$

From (3.180) and (3.182) we get then

$$\nu_a(\boldsymbol{pk}\omega) = -\frac{(\hbar\omega + \varepsilon^a_{\boldsymbol{p}} - \varepsilon^a_{\boldsymbol{p}+\boldsymbol{k}})^2}{f_a(\boldsymbol{p}) - f_a(\boldsymbol{p}+\boldsymbol{k})}\,\tilde{G}^{\mathrm{corr}}_a(\boldsymbol{pk}\omega + \mathrm{i}0) \equiv \sum_c (\nu^E_{ac} + \nu^D_{ac})\,. \qquad (3.183)$$

Here we introduced formal quantities ν^E_{ac} and ν^D_{ac}, which correspond to the different contributions of (3.176), (3.181). The quantity $\tilde{G}_a^{\mathrm{corr}}(\boldsymbol{pk}\omega + \mathrm{i}0)$ follows after Fourier-transformation or Fourier series expansion respectively from the corresponding space-time quantity, which is, e.g., for the expression in the curly brackets of (3.181)

$$\begin{aligned} G^E_{aa}(121'2^+)|_{t_1'=t_1^+} \equiv \tilde{G}^0_a + \sum_c E^c_{aa} = \\ &- [G_a(12^+)\,G_a(21') + \mathrm{i}G_a(13)\,G_a(32^+)\,\cdot \\ &\cdot \{V^{\mathrm{s}}_{aa}(34) - V_{aa}(34)\}\,G_a(24)\,G_a(41') + \\ &+ \mathrm{i}G_a(13)\,\{V^{\mathrm{s}}_{aa}(34) - V_{aa}(34)\}\,G_a(34)\,\cdot \\ &\cdot G_a(42^+)\,G_a(21') + \mathrm{i}G_a(12^+)\,G_a(23)\,\cdot \\ &\cdot \{V^{\mathrm{s}}_{aa}(34) - V_{aa}(34)\}\,G_a(34)\,G_a(41')]_{t'_1=t^+_1}\,. \end{aligned}$$

(3.184)

The first right hand side term of (3.184) is the RPA (random phase approximation), the second one is the contribution of the scattering between two particles, and the last two terms represent the self energy corrections. In our approximation all single particle functions in (3.184) are assumed to be free ones. For the latter we use the Fourier coefficients.

$$G^0_a(\boldsymbol{p}z^a_\nu) = \frac{1}{z^a_\nu - \varepsilon^a_{\boldsymbol{p}}}\,; \qquad z^a_\nu = \frac{\pi\nu}{-\mathrm{i}\beta} + \mu_a\,, \qquad \nu - \text{odd}. \qquad (3.185)$$

z_ν^a are the Matsubara frequencies, μ_a — chemical potential [46]. For the dynamically screened potential the spectral representation is applied,

$$V^{\mathrm{s}}_{aa}(k\Omega_\mu) = V_{aa}(k) + \int_{-\infty}^{+\infty} \frac{\mathrm{d}\omega'}{2\pi} \frac{2 \operatorname{Im} V^{\mathrm{s}}_{aa}(k, \omega' - \mathrm{i}0)}{\Omega_\mu - \omega'} . \tag{3.186}$$

The Matsubara frequency is now $\Omega_\mu = \pi\mu/(-\mathrm{i}\beta)$, μ — even. The spectral function is expressed by the dielectric function in the following manner (in RPA)

$$\operatorname{Im} V^{\mathrm{s}}_{aa}(k, \omega - \mathrm{i}0) = \sum_c \frac{V^2_{aa}(k) \operatorname{Im} \Pi_{cc}(k, \omega - \mathrm{i}0)}{\varepsilon'^2_{\mathrm{RPA}}(k\omega) + \varepsilon''^2_{\mathrm{RPA}}(k\omega)} . \tag{3.187}$$

While in the general situation one must take into account the dynamical screening we shall consider here only the static case. The influence of the dynamics is discussed in [59].

In RPA the real part of the dielectric function reads in the static case

$$\varepsilon'_{\mathrm{RPA}}(k, 0) = 1 + \sum_d \frac{4\pi n_d e_d^2}{k^2 k_{\mathrm{B}} T} {}_1F_1\left(1, \frac{3}{2}; -\frac{\hbar^2 k^2}{8 m_d k_{\mathrm{B}} T}\right) . \tag{3.188}$$

From (3.188) follows in the case $\hbar \to 0$ the Debye-screening

$$V^{\mathrm{s}}_{aa}(k, 0) = \frac{V_{aa}(k)}{\varepsilon'_{\mathrm{RPA}}(k, 0)} = \frac{4\pi e_a^2}{k^2 + \varkappa_D^2} ,$$
$$\varkappa_D^2 = \sum_a 4\pi n_a e_a^2 / k_{\mathrm{B}} T . \tag{3.189}$$

After some algebra, which is given in more detail in [55] and [60] we get in the $k \to 0$, $\omega \to 0$ limit (the $\omega \to 0$ limit must be carried out as the second one)

$$\begin{aligned} \nu_a(p) &\equiv \sum_c \left(\nu^E_{ac}(p) + \nu^D_{ac}(p)\right) \equiv \sum_c \nu_{ac} = 2\pi \sum_c \left(1 - \frac{e_c m_a}{e_a m_c}\right) \cdot \\ &\cdot \int \frac{\mathrm{d}\boldsymbol{q}}{(2\pi)^3} \int \frac{\mathrm{d}\boldsymbol{p}'}{(2\pi)^3} \cdot \left.\frac{\boldsymbol{q} \cdot \boldsymbol{k}}{\boldsymbol{p} \cdot \boldsymbol{k}}\right|_{k=0} \cdot \left|V^{\mathrm{s}}_{aa}(\boldsymbol{q}, \varepsilon^a_p - \varepsilon^a_{\boldsymbol{p}'+\boldsymbol{q}})\right|^2 \cdot \\ &\cdot f_c(\boldsymbol{p}' + \boldsymbol{q})\,(1 - f_c(\boldsymbol{p}')) \cdot \\ &\cdot \delta(\varepsilon^c_{\boldsymbol{p}'+\boldsymbol{q}} - \varepsilon^c_{\boldsymbol{p}'} + \varepsilon^a_{\boldsymbol{p}} - \varepsilon^a_{\boldsymbol{p}+\boldsymbol{q}}) . \end{aligned} \tag{3.190}$$

While the unity in the first brackets of the right hand side corresponds to the E-diagrams of (3.181), the quantity $\frac{e_c m_a}{e_a m_c}$ is characteristic of the D-diagrams.

It is to be seen, that for equal masses and charges there is no contribution to the transport relaxation frequency. This means that e.g. the electron-electron interaction does not influence the conductivity. Because this property follows generally in the considered frame of equilibrium correlation functions [44, 61] it is a necessary condition which should be fulfilled by a special approximation. For $e_1 = -e_2$ and $m_1 \ll m_2$ only the E-diagrams contribute,

$$\nu_{12} = \nu^E_{12} \quad \text{for} \quad \frac{m_1}{m_2} \to 0 ,$$

while for the quantity ν_{21} follows

$$\nu_{21} = \nu_{21}^D \quad \text{for} \quad \frac{m_1}{m_2} \to 0 .$$

Here we used $a \equiv 1$ and $b \equiv 2$.

But, from the final result for the conductivity is to be seen that the contribution of ν_{21} may be neglected with respect to that of ν_{12}.
Moreover from (3.190) is to be seen, that for the situation $m_a/m_c \to 0$ or $m_c \to \infty$ we have $\varepsilon^c_{p'+q} = \varepsilon^c_{p'} = 0$ and from the argument of the δ-distribution follows that in this case we have to take exactly only the statically screened potential in (3.190).

We give now some analytical expressions and some numerical values which follow from (3.190) in the nondegenerate case.

For the electron-ion scattering (Lorentz-gas situation), we get the well known Mott-Brooks-Herring result [62, 63] if the screened potential in (3.190) is the Debye potential:

$$\nu_e^E(p) \to \nu_{ei}(p) = \frac{2\pi n_i e^4 m_e}{\hbar^3 p^3} \left[\ln\left(1 + \frac{4p^2}{\varkappa_D^2}\right) - \frac{4p^2}{\varkappa_D^2 + 4p^2} \right]. \tag{3.191}$$

This quantity is indicated in figure 3.6 [60]. The exact screening in this situation would be the application of (3.188). However, in this case eq. (3.190) may be evaluated only numerically. The result is shown in figure 3.7 [60].

For finite masses m_a, m_c of the two species the evaluation of (3.190) gives, if instead of the dynamically screened potential only the Coulomb potential is used

$$\begin{aligned} \nu_a(p) &= \left(1 + \frac{m_a}{m_c}\right) \frac{n_c \Lambda_c^3 m_a^2 e^4 m_c}{p^3 \beta^2 \hbar^7 \pi} \int\limits_0^\infty \frac{dt}{t^2} \left\{ \ln\left(1 + \frac{4t^2}{\varkappa_D^2}\right) - \frac{4t^2}{\varkappa_D^2 + 4t^2} \right\} \cdot \\ &\quad \cdot \exp\left\{ -\frac{\beta\hbar^2 m_c}{2} \left[\frac{t^2}{m_{ac}^2} + \frac{p^2}{m_a^2} \right] \right\} (v \operatorname{ch} v - \operatorname{sh} v); \\ v &= \frac{\beta\hbar^2 m_c}{m_a m_{ac}} tp; \qquad m_{ac} = \frac{m_a m_c}{m_a + m_c}; \qquad \Lambda_c^2 = \frac{h^2}{2\pi m_c k_B T}. \end{aligned} \tag{3.192}$$

Equation (3.192) was integrated numerically, the result was represented in figure 3.6 for $m_a = m_c$. The integration in (3.192) may be carried out analytically if the first curly bracket is taken out of the integral at that momentum, where the main contribution comes from.

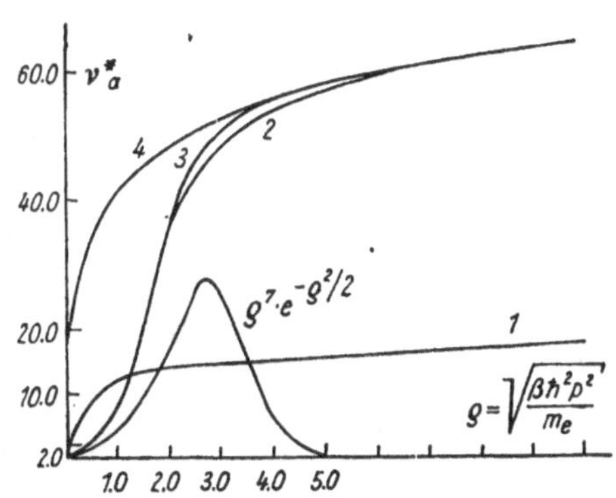

Fig. 3.6. $\nu^*(\varrho) = \nu(\varrho)\, \pi\hbar^3\varrho^3/(n\Lambda/m_e e^4)$. 1 — Lorentz gas, eq. (3.191), 2 — eq. (3.192), $m_a/m_c = 1$, 3 — eq. (3.193), $m_a/m_c = 1$, 4 — eq. (3.193) with first curly bracket equal to unity

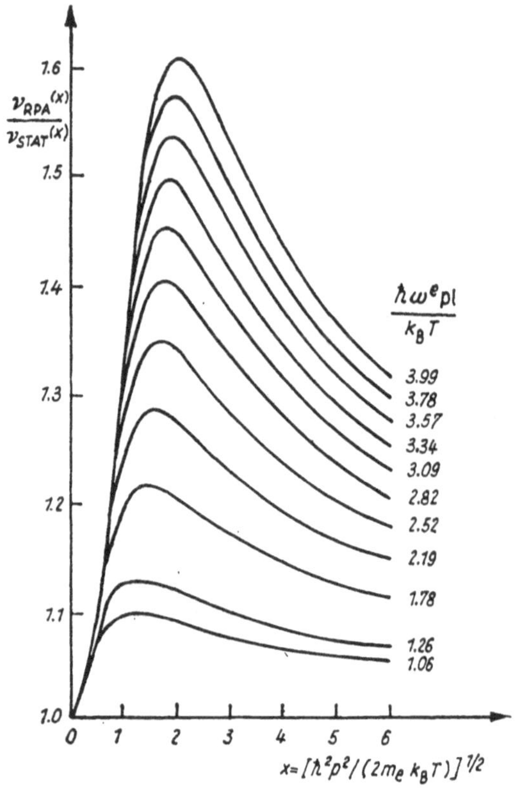

Fig. 3.7. $\nu_{\text{RPA}}(x)$ — values of the expression (3.190) (nondegenerate case) using eq. (3.188) instead of eq. (3.189), compared with $\nu_{\text{stat}}(x)$ according to (3.191) for $m_e/m_i = 0$. $(\omega_{\text{pl}}^e)^2 = 4\pi ne^2/m_e$

The result is

$$\nu_a^{(1)}(\varrho) = \left(1 + \frac{m_a}{m_c}\right) \frac{n_c}{\sqrt{2\pi}} \frac{\Lambda_c^3 m_c^3 e^4}{\hbar^3 m_a m_{ac} \varrho^3} \left\{\Phi\left(\frac{\varrho}{\sqrt{2}}\right) - \sqrt{\frac{2}{\pi}}\, \varrho\, e^{-\frac{\varrho^2}{2}}\right\} \cdot \left\{\ln\left(1 + A\varrho^2\right) - \frac{A\varrho^2}{1 + A\varrho^2}\right\}; \tag{3.193}$$

$$A = \frac{4m_{ac}^2}{\beta\hbar^2 m_c \varkappa_D^2}; \qquad \Phi \text{ — error function}; \qquad \varrho^2 = \frac{\beta\hbar^2 m_c}{m_a^2} p^2 .$$

Expression (3.193) was shown in figure 3.6 together with the further approximation that the first curly brackets of (3.193) are replaced by unity.

3.4.4. Conductivity. Discussion

In the framework of this section we get the conductivity if (3.190) together with (3.176), (3.181) is inserted in (3.161). In the $k \to 0$, $\omega \to 0$ limit we have

$$\sigma = \varepsilon \frac{e^2 n \Lambda_a^3 \hbar^2}{3m_a^2 k_B T} \int \frac{\mathrm{d}\boldsymbol{p}}{(2\pi)^3} p^2 \frac{e^{-\beta\varepsilon_p^a}}{\nu_a(p)} + \varepsilon \frac{e^2 n \Lambda_c^3 \hbar^2}{3m_c^2 k_B T} \int \frac{\mathrm{d}\boldsymbol{p}}{(2\pi)^3} p^2 \frac{e^{-\beta\varepsilon_p^c}}{\nu_c(p)} . \qquad 3.(194)$$

If (3.190) is evaluated with a Debye potential we get

$$\sigma \equiv \sigma_a + \sigma_c = \varepsilon \frac{(k_B T)^{3/2}}{e^2 m_a^{1/2}} \frac{16\sqrt{2}}{3\pi} \cdot \frac{1}{\left(1 + \frac{m_a}{m_c}\right)^2} \cdot \int\limits_0^\infty dx\, x^7\, e^{-x^2} \cdot$$

$$\cdot \left[\psi_1^{-1}(x, A_1) + \left(\frac{m_a}{m_c}\right)^{5/2} \psi_2^{-1}(x, A_2)\right]. \tag{3.195}$$

$$\psi_1(x, A_1) = \int\limits_0^\infty \frac{dt}{t^2} \left\{\ln(1 + A_1 t^2) - \frac{A_1 t^2}{1 + A_1 t^2}\right\} \cdot$$

$$\cdot \left\{4\sqrt{\frac{m_c}{m_a}}\, tx \,\mathrm{ch}\left[2\sqrt{\frac{m_c}{m_a}}\, tx\right] - 2\,\mathrm{sh}\left[2\sqrt{\frac{m_c}{m_a}}\, tx\right]\right\} \cdot$$

$$\exp\left\{-\left[\frac{m_c}{m_a} x^2 + t^2\right]\right\}; \qquad A_1 = \frac{8 m_{ac}^2}{\beta \hbar^2 m_c \varkappa_D^2}. \tag{3.196}$$

$\psi_2(x, A_2)$, A_2 follow from ψ_1, A_1 by $m_a \to m_c$. In figure 3.7 the evaluation of (3.195) is shown for different mass ratios [60].

In the Lorentz gas situation the contribution with ψ_2 vanishes in (3.194). The result is shown in figure 3.8.

An approximate evaluation of (3.195) gives

$$\sigma^L = \varepsilon \frac{(k_B T)^{3/2}}{e^2 m_e^{1/2}} \frac{8\sqrt{2}}{\pi^{3/2}} \left[\ln(1 + 3A_1^L) - \frac{3A_1^L}{1 + 3A_1^L}\right]^{-1},$$
$$A_1^L = \frac{8 m_e}{\beta \hbar^2 \varkappa_D^2}. \tag{3.197}$$

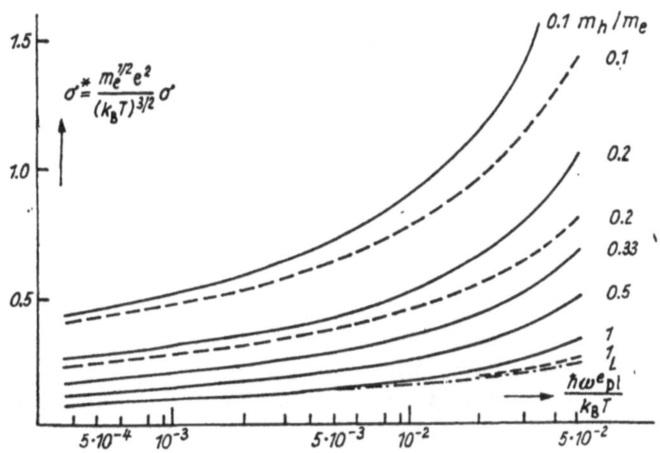

Fig. 3.8. Conductivity according to (3.195), (3.196) for different mass ratios, relevant for electron (e) — hole (h) — scattering in optically excited semiconductors and for scattering between ions. ———— collision frequency according to (3.192) ($m_a = m_e$, $m_c = m_h$), — — — — only the logarithmic term in the curly brackets of (3.192) retained, —·—·—·— Lorentz gas (collision frequency according to (3.191))

If in the case of the Lorentz gas, the exact screening according to (3.188) is taken into account, the evaluation must be carried out numerically. The result is shown in figure 3.9, too [60].
The main drawback of the results included in section 3.4 is the fact, that in the framework of the Kubo formalism used the electron-electron interaction does not come into the play (see also the remarks after (3.190)). This demands, e.g., to take a more general statistical operator or to apply the kinetic theory (see sects. 3.2. and 3.3.).

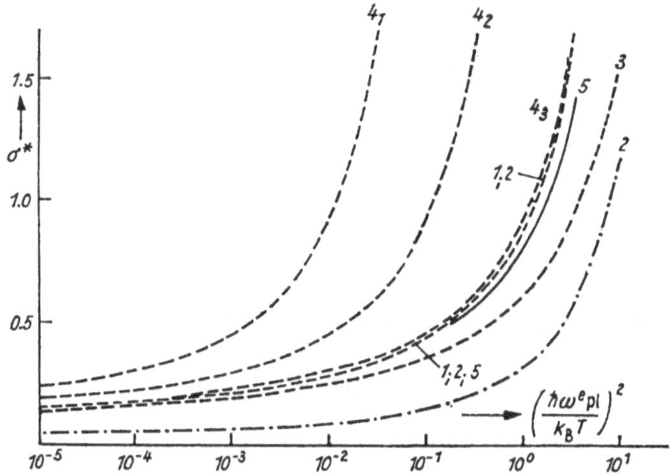

Fig. 3.9. Conductivity σ^* of the Lorentz gas, $\sigma^* = \sigma e^2 m^{1/2} (k_B T)^{-3/2}$, $(\omega_{\mathrm{pl}}^e)^2 = 4\pi n e^2/m_e$. 1 — eq. (3.195) with $m_a/m_e = 0$, 2 — eq. (3.197), 3 — eq. (3.197) with only the logarithmic term in the denominator, 4 — Spitzers result $\sigma^* = 8\sqrt{2}\,\pi^{-3/2} \ln(1 + 9/\mu^2)$, $\mu^2 = 4\pi n e^6 \times \times (k_B T)^{-3}$; 4_1 — 2800 K, 4_2 — 28000 K, 4_3 — 280000 K, 5 — collision frequency used according to fig. 3.7; Z — Ziman formula (nondegenerate case) $\sigma^* = 3\sqrt{2}/(4\sqrt{\pi}) \ln(1 + 3A_1^{\mathrm{L}})$, A_1^{L} — see eq. (3.197)

One of the main advantages, however, of the application of Green's functions is to include the screening in a natural way and to incorporate higher order contributions in a consequent manner.

3.5. Discussion

In this chapter we discussed the transport properties, especially the electrical conductivity, of homogeneous electro-neutral plasmas near the thermodynamic equilibrium. One essential result in the conductance theory which is mostly used as a standard expression, e.g., for the comparision with experimental results is the well-known Spitzer-Härm formula [15]

$$\sigma_{\mathrm{sp}} = \frac{(k_B T)^{3/2}}{e^2 m_e^{1/2}} \sigma_{\mathrm{sp}}^*, \qquad \sigma_{\mathrm{sp}}^* = 0.591 \cdot \frac{1}{\ln \frac{3}{\mu}}; \qquad \mu^2 = \frac{8\pi n e^6}{(k_B T)^3} \tag{3.198}$$

which is valid for the fully ionized plasma.

However, for more dense plasmas the experimental values of the Coulombic part of the conductivity are significantly lower than the Spitzer ones. For this reason, it seemed to be necessary already since longer time to improve the conductance theory. In this chapter we presented several approaches to quantum statistical transport theory, in which the common methods of many particle physics are applied: kinetic theory (sect. 3.2.) and linear response theory (sects. 3.3, 3.4.).

Especially we were interested in the extension of the conductance theory of Coulomb plasmas to strongly nonideal plasmas, i.e. to cover the region up to liquid metals. In detail, we want to point out the influence of different effects which give, e.g., deviations from Spitzer's result. Such effects are:

screening and cut-off procedures,
structure factor effects,
higher order T-matrix contributions,
Debye-Onsager relaxation effects,
bound state effects,
deviations from Coulomb's potential.

The quantitative influence of the effects just mentioned on the conductivity will be explained below.

3.5.1. Lorentz Model

A simple treatment of transport phenomena in a plasma is possible for the Lorentz model, where the ions are fixed and only scattering processes between electrons and ions are taken into account. For the conductivity the following formula is obtained (see sect. 3.2.3., eq. (3.76) and section 3.4.4., eq. (3.194)):

$$\sigma_{\mathrm{L}} = \varepsilon \frac{e^2 n V_e^3 \hbar^2}{3 m_e^3 k_{\mathrm{B}} T} \int \frac{\mathrm{d}\boldsymbol{p}}{(2\pi)^3} p^2 \exp\left(-\beta \varepsilon_{\mathrm{p}}^e\right) \nu_e^{-1}(p) \tag{3.199}$$

This expression is obtained from (3.194) in the limit $m_i/m_e \to \infty$.

In Born approximation the transport collision frequency $\nu_e(p)$ is derived from (3.190) and (3.188) and reads, see also (3.124),

$$\nu_e(\boldsymbol{p}) = 2\pi n \int \frac{\mathrm{d}\boldsymbol{q}}{(2\pi)^3} \frac{q^2}{2p^2} \delta(\varepsilon_{\boldsymbol{p}+\boldsymbol{q}}^e - \varepsilon_{\boldsymbol{p}}^e) \cdot [V_{ei}(q)]^2 \cdot \left[1 + \sum_d \frac{4\pi n e_p^2}{q^2 k_{\mathrm{B}} T} \, {}_1F_1\left(1, \frac{3}{2}; -\frac{\hbar^2 q^2}{8 m_d k_{\mathrm{B}} T}\right)\right] \tag{3.200}$$

From this expression follows the Brooks-Herring formula [63, 64]

$$\sigma_{\mathrm{BH}}^* = \frac{2^{5/2}}{\pi^{3/2}} \varepsilon \Phi_{\mathrm{BH}}^{-1}(A_1^{\mathrm{L}}); \qquad \Phi^{\mathrm{BH}}(A_1^{\mathrm{L}}) = \ln\sqrt{1 + 3A_1^{\mathrm{L}}} - \frac{3}{2} \frac{A_1^{\mathrm{L}}}{1 + 3A_1^{\mathrm{L}}} \tag{3.201}$$

$$A_1^{\mathrm{L}} = \frac{8 m_e k_{\mathrm{B}} T}{\hbar_2 \varkappa_{\mathrm{D}}^2}; \qquad \varkappa_{\mathrm{D}}^2 = \begin{cases} 8\pi n e^2/(k_{\mathrm{B}} T \varepsilon) & \text{for (heavy) moving ions,} \\ 4\pi n e^2/(k_{\mathrm{B}} T \varepsilon) & \text{for fixed ions,} \end{cases}$$

which is usually applied also in semiconductor physics.

We get this formula from (3.199), (3.200) if we replace the screened potential by the Debye one and evaluate the integration in such a way that the logarithmic term is taken out of the integral at the maximum momentum $p^2\hbar^2(2m_ek_BT)^{-1} = 3$.

Numerical values for the reduced conductivity $\sigma = \sigma^*(k_BT)^3\, m_e^{-1/2}e^{-2}$ according to (3.199), (3.200) and (3.201) are presented in figure 3.9. as a function of the quantity $\hbar^2 4\pi ne^2/(m_e k_2^B T^2) = 4/A_1^L$. In this representation the function is universal for any temperature. As also seen from figure 3.9 the approximate evaluation of the integral just mentioned gives only a negligible error as compared with the exact mentioned gives only a negligible error as compared with the exact evaluation. However, the correct static screening according to (3.200) gives a remarkable effect for $4/A_1^L > 0.1$ as compared with the result using the Debye scréening.

If in expression (3.200) $(V^s)^2$ is replaced by VV^s, which means that in formulat (3.201) only the logarithmic term in Φ is retained, we get a remarkable lowering of the conductivity in the high density region as shown also in figure 3.9. This effect is in accordance with the fact that the dynamical screening tends to be more Coulombic than the static screening (see sect. 3.2.). We want to point out that in our quantum statistical theory automatically a convergent result is obtained, namely (3.199), (3.200). Instead of the Landau length which appears in classical theories now the de Broglie wavelength plays the role of a short range "cut-off" length (see Rogov [65], Gould and de Witt [50] and Williams and de Witt [50]. In classical theories usually this (arbitrary) cut-off is chosen in such way that our A_L^1 is replaced by $3/\mu_2 = 3(k^BT)^3/(8\pi ne^6)$. In this way the same Coulomb logarithm occurs as in the Spitzer result (3.198). The effect of the cut-off procedure is also shown in figure 3.9. We point out that with respect to the abcissa the Spitzer result is no universal function but depends additionally on the temperature.

We want to mention that different screening procedures and different cut-off techniques do not influence the contribution ln n in the Coulomb logarithm of the conductivity formula. One advantage of the Lorentz gas model is the fact that the exact solution of the linearized Boltzmann equation may be found using the relaxation time ansatz. Therefore, a comparison with other methods to solve the Boltzmann equation may be given. Such a comparision is of interest because in the case of inclusion of

Tab. 3.1. Conductivity Values in the Nondegenerate Limit

method	f without $e-e$ interaction	including $e-e$ interaction
relaxation time approximation (Brooks-Herring) [63, 64]	1.015	—
Chapman-Enskog method:		
Cowling 1. approximation [66] (Ziman-Faber) [67]	0.299	0.299
Cowling 2. approximation [15]	0.299	0.587
Spitzer [15]	1.015	0.591
Grad's 13 moment approximation [6—8]	0.972	0.578
Kohler's variational method:		
Appel 1. approximation [35]	0.972	0.578
Appel 2. approximation [35]	1.014	0.580
quasi-equilibrium statistical operator, eq. (3.116)	0.972	0.578

electron-electron interaction the solution of the Boltzmann equation can be found only in an approximative manner. The results obtained by different methods are presented in table 3.1 in the nondegenerate limit.

For later discussions we included already the influence of the interaction between the electrons in table 3.1.

It is to be seen that the different approximation schemes converge relatively fast towards the exact result. However, the lowest approximation in which only a shifted Maxwell distribution is considered (FABER [67]) gives a result which differs from the exact one by a factor $(3\pi/32)$ (see also fig. 3.9).

3.5.2. Inclusion of Electron-Electron Interaction

While the Lorentz gas is a simple model for which the Boltzmann equation can be solved exactly, in a more realistic description of transport phenomena in a plasma one should also take into account electron-electron scattering. In the non-degenerate case different methods to solve the Boltzmann equation are known which give adequate results. Such are, for instance, the Grad method (sect. 3.2.2.), the Kohler variational principle and the Zubarev method (sect. 3.3.2.). In simplest approximation we get (see eqs. (3.69), (3.70), (3.96) and (3.124b)):

$$\sigma = f\sigma^* \frac{(k_B T)^{3/2}}{e^2 m_e^{1/2}}, \tag{3.202}$$

$$\sigma^* = \frac{3}{4\sqrt{2\pi}} \frac{\Phi_0}{1}, \qquad f = \frac{13 + 4\sqrt{2}}{4 + 4\sqrt{2}},$$

$$\Phi_0 = \frac{2}{Nn} \frac{(m_e k_B T)^{3/2}}{\hbar^3 (2\pi)^{3/2}} \int_0^\infty \mathrm{d}E(k) \left(-\frac{\mathrm{d}f}{\mathrm{d}E}\right) \int_0^{2k} \mathrm{d}q\, q^3 \left|\frac{V(q)^2}{4\pi e^2}\right|^2,$$

where the Coulomb logarithm is taken in first Born approximation for a statically screened potential. This expression follows from the formulae in sections 3.2.2. and 3.3.2. if the electron-electron and the electron-ion interactions are treated symmetrically, i.e., $t_e = t_{ee}$ in (3.69), so that $f = \text{const}$. With respect to the Lorentz gas model (3.201) $(\sigma = (13/4) \cdot \sigma^*$: Brooks-Herring formula$)$ the conductivity is reduced by a factor of about 0.6 (see also tab. 3.1).

In the strongly degenerate case (liquid metals, see also chap. 5.), in the first Born approximation the Ziman formula σ^* gives a correct result for the conductivity. In this limit it is not necessary to use further approximations for the charge carrier distribution function and we need not to include the effect of electron-electron collisions on the conductivity.

An expression for the conductivity valid at any degree of degeneration was given in section 3.3., equation (3.116). The evaluation in the first Born approximation for a statically screened potential gives an interpolation between the non-degenerate and the strongly degenerate limits, see figure 3.10.

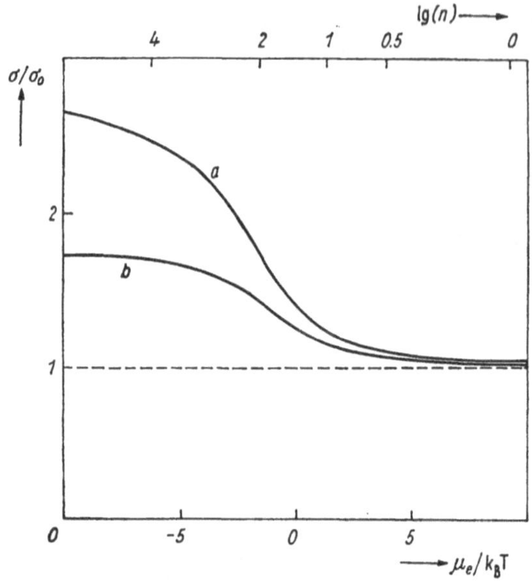

Fig. 3.10. Ratio of the conductivity without *ee* scattering σ_{ei} and including *ee* scattering, $\sigma_{ei,ee}$, to σ_0, eqs. (3.116), (3.118), (3.121), (3.122), in dependence on the degeneration parameter μ_e/k_BT, or, respectively, the electron density n_e (atomic units $\hbar = 2m = e^2/2 = 1$). $k_BT = 1$. a — $\sigma_{ei}(n_e, T)/\sigma_0(n_e, T)$; b — $\sigma_{ei,ee}(n_e, T)/\sigma_0(n_e, T)$

3.5.3. Equilibrium Correlation Effects

Structure factor effects are of increasing importance in dense plasmas. In the limiting case of liquid metals, the incorporation of the ion structure factor is a usual procedure (see chap. 5.). In the 1st Born approximation the influence of the structure factor is obtained in the Coulomb logarithm Φ_0, if the Fourier transform $V_{ei}^s(q)$ of the interaction of one electron with all ions located at r_{im} is decomposed into the single ion scattering part $v_{ei}^s(q)$ and the ion distribution factor so that (see sect. 3.3.3.)

$$|V_{ei}^s(q)|^2 = N \cdot S(q)\, |v_{ei}^s(q)|^2 ,$$

$$S(q) = \frac{1}{N} \langle \sum_{ij} e^{iq\cdot(R_i - R_j)} \rangle . \tag{3.203}$$

For randomly distributed scatterers we have $S(q) = 1$. In general, the ion structure factor is determined by the ion-ion interaction potential. For a hard core interaction we would obtain the Percus-Yevick expression for $S(q)$, whereas for the more realistic model potential of a superposition of a Coulombic and hard core contribution the structure factor may be given in mean spherical approximation (MSA), equation (3.127). The influence of the structure factor on the conductivity is shown in figure 3.11. For instance, in a Cs-plasma structure effects change the conductivity by more than 0% for parameter values $\mu^* > 2, 3$ at $T = 10^4$ K and for $\mu^* > 25$ at $T = 303$ K (plasma parameter $\mu^* = e^2\varkappa_e/k_BT = \mu/\sqrt{2}$).

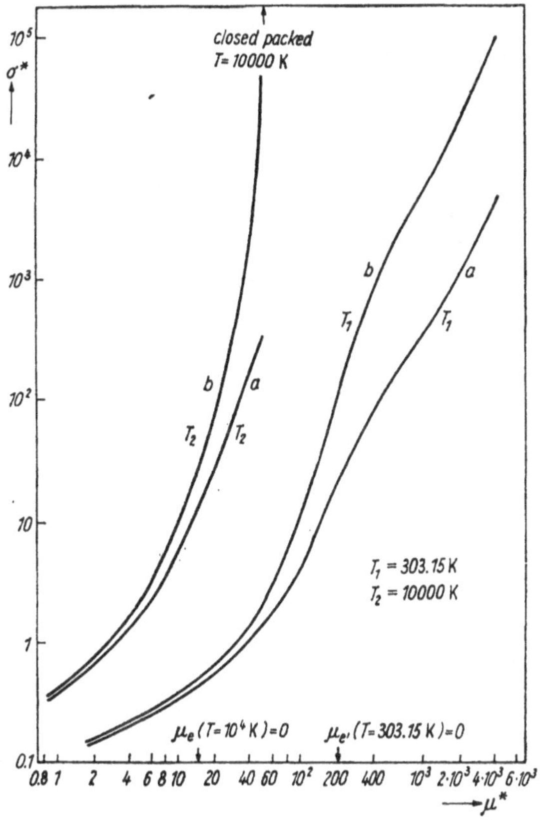

Fig. 3.11. Influence of the structure factor $S(q)$ in MSA, eq. (3.127), with $Ri = 1.69 \times \times 10^{-10}$ m (crystallografic radius of Cs) on the reduced conductivity $\sigma^*(n, T)$. Parameter values of the Cs electron ion pseudopotential (3.16): $A = 1.37$, $B = 1.037 \cdot 10^{10}\,\mathrm{m}^{-1}$ [68]. a — $S(q) = 1$; b — $S(q)$ in MSA. Degeneration of the electron gas at $\mu_e = 0$

3.5.4. Higher Order Scattering Effects

A central problem for Coulomb systems is the correct improvement of the Born approximation, which for the bare Coulomb potential is diverging. Especially, two effects are of importance if contributions of higher than the 1st Born approximation are considered: (i) The Coulomb interaction is replaced by a dynamically screened interaction (cf. sects. 3.4.; 3.3.5.) which can be approximated by a statically screened interaction (see above); (ii) The multiple scattering between two particles is taken into account if T-matrices are introduced (see 3.3.6.). In the well known single site approximation where the T-matrix is expressed by the single ion T-matrix $t^{\mathrm{s}}_{ei}(q)$ and the structure factor

$$|V^{\mathrm{s}}_{ei}(q)|^2 \rightarrow |T^{\mathrm{s}}_{ei}(q)|^2 = NS(q)\,|t^{\mathrm{s}}_{ei}(q)|^2\,, \tag{3.204}$$

the correspondingly extended Ziman formula has been applied to liquid transition metals, see [40]. In a non-degenerate plasma the improvement of the 1st Born approximation by introducing the single site T-matrix is of importance also for simple metals (Coulombic electron-ion interaction), whereas structure effects are less dominant.

We discuss two special cases: (i) The dynamically screened Coulomb potential is replaced by the statically screened one. The T-matrix $t^s_{ei}(q)$ is substituted by scattering phase shifts (see sect. 3.3.4.), which are obtained from the numerical integration of the Schrödinger equation. Using the conductivity formula (3.116) with the expressions (3.130), (3.133) for the correlation functions, the result is presented in figure 3.12 for two temperatures [36]. With decreasing density the Born approximation is corrected towards the Spitzer result which approximatively is arrived at $\mu^* \leqq 0.8$ for $T = 10^4$ K. We find that the statically screened T-matrix approximation interpolates between the Spitzer result at low densities and the 1st Born approximation at high densities, see also [59].

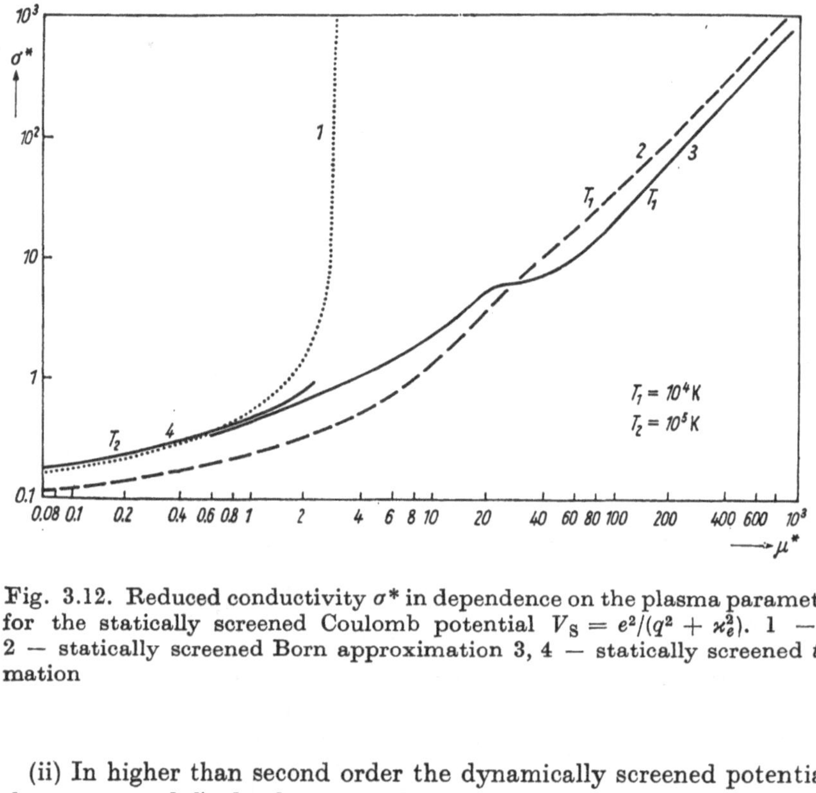

Fig. 3.12. Reduced conductivity σ^* in dependence on the plasma parameter $\mu^* = e^2\varkappa_e/k_BT$ for the statically screened Coulomb potential $V_S = e^2/(q^2 + \varkappa_e^2)$. 1 — Spitzer theory; 2 — statically screened Born approximation 3, 4 — statically screened t-matrix approximation

(ii) In higher than second order the dynamically screened potential is replaced by the unscreened Coulomb potential. Because for the Coulomb potential the 1st Born approximation always coincides with the correct T-matrix result, the higher order contributions cancel and the Born approximation remains.

We suppose that the results of a dynamically screened T-matrix will be found in an interval given by the Born approximation (which underestimates the effect of screening) and the statically screened T-matrix approximation (which overestimates screening effects). Experimental conductivity values for highly nonideal plasmas are not in contradiction with this proposition (see fig. 3.13). The correct treatment of dynamical screening is an essential topic to be investigated further in conductivity theory.

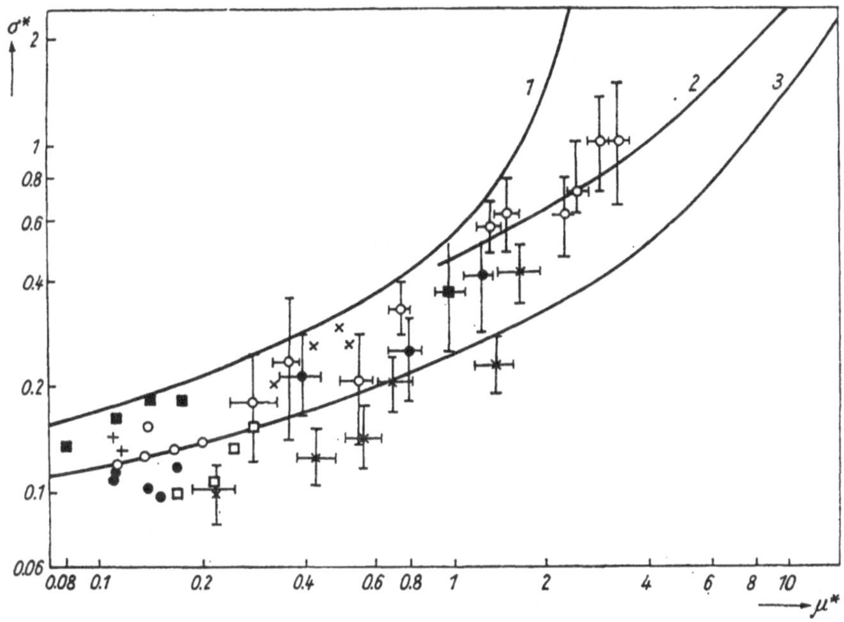

Fig. 3.13. Reduced conductivity σ^* in dependence on the plasma parameter μ^*. Comparison with experimental data. Theory: $T = 10^4$ K: 1 — Spitzer theory; 2 — statically screened t-matrix approximation; 3 — statically screened Born approximation; experiment: ■ Ar [69], ╋ Ar, ╋ Xe, ╋ Ne (with error bars) [70], + H [71], □ air [72], × C_2H_3Cl 7[3], ● Ar, ○ Xe [74], ╋ Cs (with error bars) [75]

3.5.5. Debye-Onsager Relaxation Effect

In nonideal plasmas the conductivity is further influenced by the coupling between the charged particle and its surrounding screening cloud. This effect tends to reduce the local electric field strength and was first discussed in the electrolyte theory [24] and us refered to as the Debye-Onsager relaxation effect. From the point of view of solid state physics this behaviour usually is taken into account introducing the quasiparticle concept.

As shown in sections 3.2.5., 3.2.6. the conductivity is reduced by a factor $(1 - Q)$ or in feedback theory by a factor $(1 + Q)^{-1}$ where in a simple approximation (static) screening [6, 20, 22]

$$Q = \mu/6 . \tag{3.205}$$

The same effect was treated also in the framework of the force-force-correlation function method [20]. The result of this approach coincides in the low density limit with (3.705). In solid state physics, interaction effects between the particles, usually are taken into account by introducing the quasi particle concept.

In [44, 60] was proposed to replace the electron masses by effective ones. For a nondegenerate plasma the effective mass reads [60]

$$m_e^{\text{eff}} = m_e \left(1 + \frac{1}{3} \frac{\hbar^2 \omega_p^2}{(k_B T)^2} + \frac{1}{3} \mu \left(1 - \frac{\pi}{8}\right)\right). \tag{3.206}$$

On insertion in the conductance expression (3.202) we get a correaction factor, which is of the same order in the density and has in the nondegenerate limit the following value

$$Q' = \frac{\mu}{6}\left(1 - \frac{\pi}{8}\right). \tag{3.207}$$

The corrections of the conductivity due to the Debye-Onsager relaxation effect (3.205) are of the ordér 15% for typical plasmas with a temperature of 10^4 K and plasma parameter $\mu = 1$.

3.5.6. Bound State Effects

The fully ionized plasma must be considered as a limiting case where the transport properties are determined only by the Coulombic scattering between charged particles. In a partially ionized plasma, the conductivity is reduced because of different effects.

(i) According to the Saha equation (mass action law) a part of the charged particles are bound to neutral complexes (atoms) and do not participate directly in the charge transport process. Roughly, in the conductivity formulae the total number of charged particles has to be replaced by the number of free particles.

(ii) Additional scattering processes between charged particles and neutrals (elastic, inelastic, ionization and recombination) occur which reduce the mobility of the free

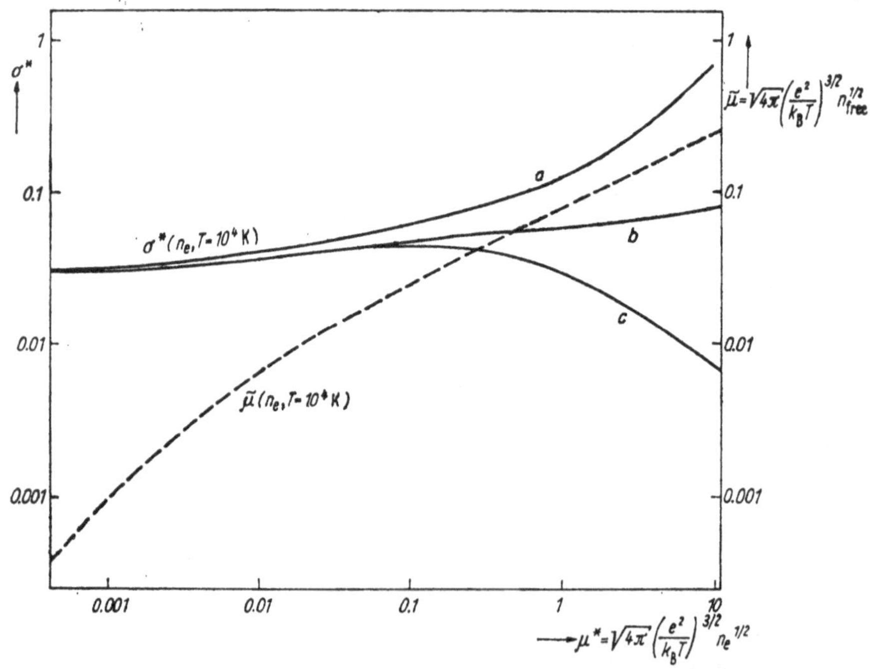

Fig. 3.14. Reduced conductivity σ^* in dependence on the plasma parameter $\mu^* = e^2\varkappa_e/k_BT$; $T = 10^4$ K [48]. a — no bound states present; b — hydrogen plasma, scattering processes with bound states neglected. Reduction of the free carrier dendity n_{free} due to the mass action law (3.160); c — hydrogen plasma, scattering processes with bound states included (eq. (3.152))

particles. Results of the electron-neutral scattering in 1st Born approximation (elastic, inelastic, ionization) and 2nd Born approximation (polarization potential, see sect. 3.3.7.), are presented in fig. 3.14, 3.15.

In figure 3.14, the reduced conductivity $\sigma_0^*(n, T)$ (because of the variational ansatz (3.150) the Ziman-Faber approximation (3.118) instead of (3.116) follows) is compared with the expression $\sigma^*(n_{\text{free}}, T)$ which takes into account the law of mass action and the conductivity $\sigma_0^*\ (n, T)$ which takes into account additionally the scattering by bound states. As shown in figure 3.15, the main contribution to the resistivity in addition to the electron-ion scattering stems from the polarization potential and elastic ground state scattering of atoms.

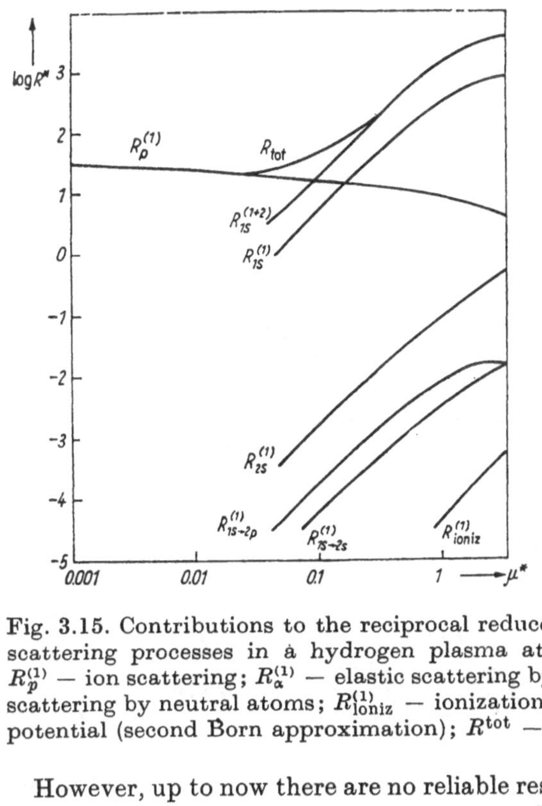

Fig. 3.15. Contributions to the reciprocal reduced conductivity $R^* = 1/\sigma^*$ from different scattering processes in a hydrogen plasma at $T = 10^4$ K. First Born approximation: $R_p^{(1)}$ – ion scattering; $R_\alpha^{(1)}$ – elastic scattering by neutral atoms, state α; $R_{\alpha\to\alpha'}^{(1)}$ – inelastic scattering by neutral atoms; $R_{\text{ioniz}}^{(1)}$ – ionization processes; $R_\alpha^{(1+2)}$ inclusion of polarization potential (second Born approximation); R^{tot} – total resistivity

However, up to now there are no reliable results in the region where the Mott transition occurs. Here a more refined theoretical approach must be developed. Furthermore, the Born approximations should be replaced by multi-particle scattering matrix analysis.

3.5.7. Deviations from Coulombic Interaction

Only in the simplest case, namely in the hydrogen plasma situation, the interaction between the constituents (e, p) is the Coulombic one. In other cases the electron-ion interaction must be described by a modified Coulomb potential (pseudo-potential)

which takes into account additional short range forces that are due to the core electrons. For instance, for alkali plasmas the following form of the pseudopotential may be applied [3, 21].

$$v_{ei}(r) = -e^2 \left[\frac{1}{r} - A \frac{e^{-Br}}{r} - C\,e^{-Br} \right]$$

where the parameters A, B, C are fitted to spectroscopic data. It is found that the conductivity slightly increases [21, 36]. The effects of deviations from the Coulombic behaviour are more essential at higher densities (see fig. 3.16).

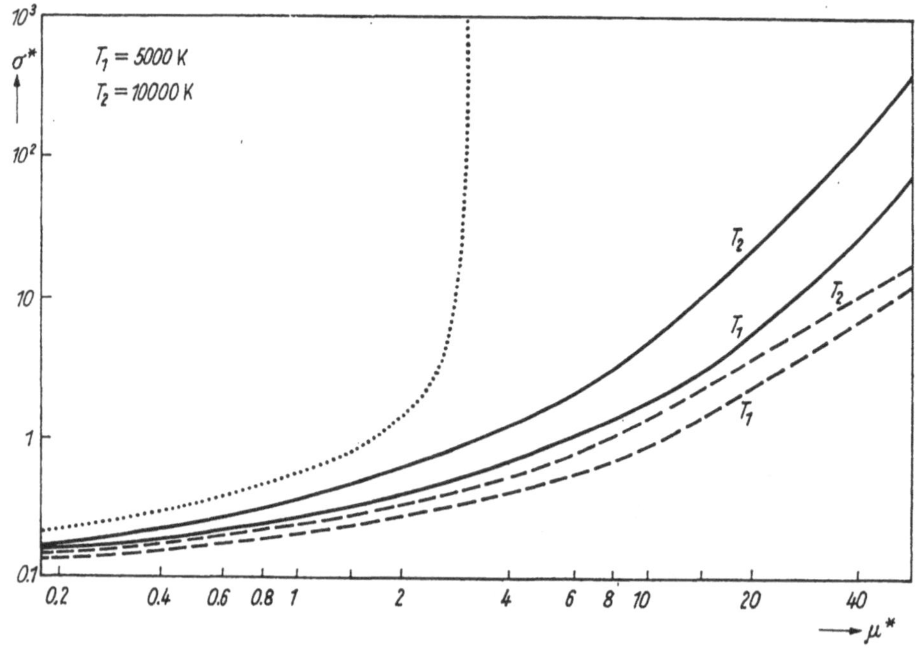

Fig. 3.16. Reduced conductivity σ^* in dependence on the plasma parameter $\mu^* = e^2\varkappa_e/k_BT$ for two different temperatures [36]. Full line: Hellmann potential, (3.1b) (parameter values: $A = 1.37$, $B = 1.037 \cdot 10^{10}\ \mathrm{m}^{-1}$ [68]), Born approximation; dashed line: Coulomb potential, Born approximation; dotted line: Spitzer theory, eq. (3.198)

In conclusion of this chapter we remark that it depends on density and temperature which of the different effects are the leading ones. In the non-degenerate limit (low densities) attention has to be paid essentially to the ionization equilibrium, bond state behaviour and scattering with neutrals, dynamical screening of the Coulomb potential, the appropriate solution of the Boltzmann equation and the inclusion of higher order collision processes. In the strongly degenerate limit (high densities) most important are the incorporation of quantum statistical effects, structure factors, pseudopotentials, dynamical screening and the Debye-Onsager relaxation effect.

We want to point out that it could not be our aim to achieve completeness in refering to the large number of theoretical and experimental papers on the subject in question. In addition to the references already quoted we mention the experimental papers [76] as well as the theoretical papers [77] and further references given in [1].

4. Bound States in the Quantum Kinetic Theory of Gases and Plasmas

Yu. L. Klimontovich, D. Kremp, M. Schlanges

4.1. Basic Equations

In this paper we will consider the kinetic properties of a quantum many particle system. The number of particles N and the volume are very large, that means we consider the thermodynamic limit

$$N \to \infty\,, \qquad V \to \infty\,, \qquad n = \frac{N}{V} = \text{finite}\,.$$

The Hamiltonian takes the form

$$H = \sum_i \frac{\boldsymbol{p}_i^2}{2m_i} + \sum_{i<j} V_{ij}(\boldsymbol{r}_i - \boldsymbol{r}_j)$$

V_{ij} represents the interaction between each pair of particles. The properties of the system are very interesting if this energy has an attractive part. In this case the formation of bound states is possible. If we interpret the bound states as new composite particles (atoms) the problem is connected with the important question of the statistical foundation of the reaction kinetics. Up to now exist only few papers concerning the question of bound states in the kinetic theory. Fundamental papers to this problem were given by Klimontovich [2, 4], Peletminski [5], Vashukov and Marusin [6], Ivanov [7], and Kolesnishenko [15].

The starting point of our considerations is the quantum BBGKY hierarchy [1a]. Usually a kinetic equation can be obtained from this by using the Bogoljubov condition on the total weakening of the initial correlation. As Peletminski has shown, this condition must be generalized for systems with bound states. We will give a generalization of the Bogoljubov condition for systems with two particles bound states.

Using this condition and cluster expansions for the nonequilibrium density operator we obtain kinetic equations for atoms and free particles taking into account three particle reactions. These equations can be considered as a generalization of the quantum Boltzmann equation for nonideal gases derived by Klimontovich and Ebeling [2a, 3]. Similar equations were given by Vashukov and Marusin [6].

Let us now consider these questions in detail. The quantum statistical state of such system is determined by the density operator ϱ_N with the normalization

$$\operatorname*{Sp}_{1\ldots N} \varrho_N = 1\,.$$

In allmost all cases of interest it is sufficient to know the reduced density operators given by

$$F_S = V^S \operatorname*{Sp}_{S+1\dots N} \varrho_N \tag{4.1}$$

with the normalization

$$\frac{1}{V^S} \operatorname*{Sp}_{S\dots 1} F_S = 1 . \tag{4.2}$$

Then we obtain the expectation value of the physical quantities A with the equation

$$\langle A \rangle = \frac{n^S}{s!} \operatorname*{Sp}_{1\dots S} F_S A_S . \tag{4.3}$$

Using a complet set of eigenvectors we get a matrix representation of these relations. Very useful are the eigenstates of the momentum operator $|\boldsymbol{p}\rangle$ with the properties

$$\langle \boldsymbol{p} \mid \overline{\boldsymbol{p}} \rangle = \delta(\boldsymbol{p} - \overline{\boldsymbol{p}}); \qquad \int \mathrm{d}\boldsymbol{p}\, |\boldsymbol{p}\rangle \langle \boldsymbol{p}| = 1 ,$$

because in homogeneous systems we have

$$[\boldsymbol{p}, \boldsymbol{F}(\boldsymbol{p})] = 0 .$$

Therefore in this case the one-particle density matrix in the momentum representation has the form

$$\langle \boldsymbol{p}_1 | \, F_1 \, | \boldsymbol{p}_1 \rangle = F(\boldsymbol{p}_1) \, \delta(\boldsymbol{p}_1 - \overline{\boldsymbol{p}}_1) . \tag{4.4}$$

$F(\boldsymbol{p}_1)$ is the quantum momentum distribution function. From (4.2) we get the normalization $\left(\delta(0) = V/(2\pi\hbar)^3\right)$

$$\int \frac{\mathrm{d}\boldsymbol{p}}{(2\pi\hbar)^3} F(\boldsymbol{p}) = 1 . \tag{4.5}$$

In order to consider the kinetic properties of the many particle system the equations of motion for the F_S are necessary. Such equations we get from the von Neumann equation for ϱ_N

$$\mathrm{i}\hbar \frac{\partial F_S}{\partial t} - [H_S, F_S] = n \operatorname*{Sp}_{S+1} \sum_{j=1}^{S} [V_{j,S+1}, F_{S+1}] . \tag{4.6}$$

That's the well known Bogoljubov hierarchy. H_S is the S-particle Hamiltonian.

The first of the equations takes the form

$$\mathrm{i}\hbar \frac{\partial F_1}{\partial t} - [H_1, F_S] = n \operatorname*{Sp}_{2} [V_{12}, F_{12}] . \tag{4.7}$$

In order to get a closed equation for F_1 (kinetic equation) it is necessary to determine F_{12} as a functional of F_1. To this purpose we write down the formal solutions of the Bogoljubov hierarchy:

$$F_S(t) = \mathrm{e}^{-\frac{\mathrm{i}}{\hbar} H_S \tau} F_S(t - \tau) \, \mathrm{e}^{\frac{\mathrm{i}}{\hbar} H_S \tau} + \frac{n}{\mathrm{i}\hbar} \int_0^\tau \mathrm{d}\tau' \, \mathrm{e}^{-\frac{\mathrm{i}}{\hbar} H_S \tau'} \operatorname*{Sp}_{S+1} \sum_{j \neq 1}^{S} \cdot$$

$$\cdot [V_{j,S+1}, F_S] \, \mathrm{e}^{\frac{\mathrm{i}}{\hbar} H_S \tau'} . \tag{4.8}$$

Especially we get for F_{12}

$$F_{12}(t) = e^{-\frac{i}{\hbar} H_{12}\tau} F_{12}(t-\tau)\, e^{\frac{i}{\hbar} H_{12}\tau} + \frac{n}{i\hbar} \int_0^\tau e^{-\frac{i}{\hbar} H_{12}\tau'} \cdot \operatorname*{Sp}_3 [V_{13} + V_{23}, F_{123}(\tau')]\, e^{\frac{i}{\hbar} H_{12}\tau'} . \tag{4.9}$$

The problem is the determination of $F_{12}(t-\tau)$.

Usually $F_{12}(t-\tau)$ can be obtained by a condition for $\tau \to \infty$ also by an asymptotic condition. A closed equation for F_1 we get if we use Bogoljubovs condition on the total weakening of the initial correlation

$$\lim_{\tau\to\infty} F_{12}(t-\tau) = F_1 \cdot F_2 . \tag{4.10}$$

Then we obtain the quantum Boltzmann equation under the following assumptions:

(i) In (4.9) contributions of the order n are dropped.
(ii) The retardation τ in (4.10) is totally neglected.
(iii) Bound states are not taken into account.

The result is

$$\frac{\partial f(\boldsymbol{p}_1)}{\partial t} = \frac{(2\pi)^3}{V} \hbar^2 \int d\boldsymbol{p}_2\, d\bar{\boldsymbol{p}}_1\, d\bar{\boldsymbol{p}}_2\, |\langle \boldsymbol{p}_1\boldsymbol{p}_2|\, T\, |\bar{\boldsymbol{p}}_2\bar{\boldsymbol{p}}_1\rangle|^2\, 2\pi\delta(E_{\boldsymbol{p}_1\boldsymbol{p}_2} - E_{\bar{\boldsymbol{p}}_1\bar{\boldsymbol{p}}_2}) \cdot \{f(\bar{\boldsymbol{p}}_1) f(\bar{\boldsymbol{p}}_2) - f(\boldsymbol{p}_1) f(\boldsymbol{p}_2)\} .$$

Here T is the two particle scattering operator and $f(\boldsymbol{p}) = nF(\boldsymbol{p})$.

Generalizations of the Boltzmann equation are possible in many directions.

First the Boltzmann kinetic equation is only valid for ideal systems. Indeed, in this equation the interaction between the particles is not taken into account completely. The interaction defines only dissipative processes. The thermodynamic functions are those of ideal systems. For example, the pressure and the internal energy are

$$P = nk_BT; \qquad U = \frac{2}{3} Nk_BT$$

and the conservation of the energy has the form

$$\frac{d}{dt} \left< \frac{p^2}{2m} \right> = 0 .$$

That means from the Boltzmann equation follows only the conservation of the kinetic energy. But in nonideal systems the conservation of the average potential and kinetic energy is necessary. Kinetic equations for nonideal systems follow if we take into account contributions of higher order with respect to the density. In this case the investigations carried out by Bogoljubov [1], Cohen and others [11] have shown that so called secular terms appear. It is possible to solve this problem e.g. following an idea of Klimontovich [2] namely that the secular terms can be eliminated by taking into account the retardation τ in a consistent manner. Using this idea generalizations of the quantum Boltzmann equations are given by Klimontovich and Ebeling [2a], Kadanoff and Baym [12], Baerwinkel [13] and Palzev [14].

Secondly, in many cases of physical and chemical interest it is necessary to take into account bound states and therefore chemical reactions. In systems with bound states the condition of the total weakening of the initial correlation is unsuitable. With this condition we obtain only the scattering solutions of the eigenvalue problem of H_{12} and so on, because the bound states have no free asymptotes. Therefore it is necessary to give a generalization of the condition (4.10).

Further for the formation of two particle bound states by a collision at least three particles are necessary. We must therefore take into account higher orders with respect to the density in (4.9) in order to get a kinetic theory for systems with two particle bound states.

4.2. Quantum Theory of Two- and Three-Particle Systems with Two-Particle Bound States. Generalization of the Asymptotic Condition

In the following we restrict ourselves to systems with only two particle bound states. The states of the two particle system are determined by the equation

$$H_{12}|\alpha_{12}\boldsymbol{P}_{12}\rangle = (E_{\alpha_{12}} + E_{P_{12}})\,|\alpha_{12}\boldsymbol{P}_{12}\rangle \tag{4.11}$$

with

$$\alpha_{12} = \begin{cases} \boldsymbol{p}_{12} = (m_1\boldsymbol{p}_1 - m_2\boldsymbol{p}_2)/M & \text{for scattering states,} \\ n_{12} & \text{for bound states} \end{cases}$$

and $\boldsymbol{P}_{12} = \boldsymbol{p}_1 + \boldsymbol{p}_2$ the center of mass momentum. The normalization of the eigenvectors is the following

$$\langle \boldsymbol{P}_{12}\alpha_{12} \mid \bar{\alpha}_{12}\bar{\boldsymbol{P}}_{12}\rangle = \begin{cases} \delta_{n\bar{n}}\delta(\boldsymbol{P}_{12} - \bar{\boldsymbol{P}}_{12})\,, & \alpha_{12} = n_{12}, \\ \delta(\boldsymbol{p}_{12} - \bar{\boldsymbol{p}}_{12})\,\delta(\boldsymbol{P}_{12} - \bar{\boldsymbol{P}}_{12})\,, & \alpha_{12} = \boldsymbol{p}_{12} \end{cases} \tag{4.12}$$

and the completeness is given by

$$1 = \sum_n \int \mathrm{d}\boldsymbol{P}_{12}\,|n\boldsymbol{P}_{12}\rangle\,\langle\boldsymbol{P}_{12}n| + \int \mathrm{d}\boldsymbol{P}_{12}\,\mathrm{d}\boldsymbol{p}_{12}\,|\boldsymbol{p}_{12}\boldsymbol{P}_{12}\rangle\,\langle\boldsymbol{P}_{12}\boldsymbol{p}_{12}|\,. \tag{4.13}$$

Further it is necessary to consider the quantum mechanics of three particle systems taking into account two particle bound states. The stationary states are given by

$$H_{123}\,|\varkappa\boldsymbol{p}\rangle^+ = E_{\varkappa\boldsymbol{p}}\,|\varkappa\boldsymbol{p}\rangle^+\,. \tag{4.14}$$

H_{123} is the three particle Hamiltonian

$$H_{123} = \frac{\boldsymbol{p}_1^2}{2m_1} + \frac{\boldsymbol{p}_2^2}{2m_2} + \frac{\boldsymbol{p}_3^2}{2m_3} + V_{12} + V_{13} + V_{23}\,. \tag{4.15}$$

By $\varkappa$ a classification of the solution is given with respect to the asymptotic initial states (channels) [8] and $\boldsymbol{p}$ is written for the dynamical variables. The channels are

$$\begin{aligned} &\varkappa = 0 \quad 1 + 2 + 3 \quad & |0\boldsymbol{p}\rangle^+ &= \Omega_{123}^0\,|\boldsymbol{p}_1\rangle\,|\boldsymbol{p}_2\rangle\,|\boldsymbol{p}_3\rangle\,, \\ &\varkappa = 1 \quad 1 + (2 + 3) \quad & |1\boldsymbol{p}\rangle^+ &= \Omega_{123}^1\,|\boldsymbol{p}_1\rangle\,|n_{23}\boldsymbol{P}_{23}\rangle\,, \\ &\varkappa = 2 \quad (1 + 3) + 2 \quad & |2\boldsymbol{p}\rangle^+ &= \Omega_{123}^2\,|\boldsymbol{p}_2\rangle\,|n_{13}\boldsymbol{P}_{13}\rangle\,, \\ &\varkappa = 3 \quad (1 + 2) + 3 \quad & |3\boldsymbol{p}\rangle^+ &= \Omega_{123}^3\,|\boldsymbol{p}_3\rangle\,|n_{12}\boldsymbol{P}_{12}\rangle\,. \end{aligned} \tag{4.16}$$

Here $\Omega^{\varkappa}_{123}$ are the Møller operators of the multi-channel scattering theory given by

$$\Omega^{\varkappa}_{123} = \lim_{t\to-\infty} e^{\frac{i}{\hbar} H_{123} t} e^{-\frac{i}{\hbar} H^{\varkappa}_{123} t} \tag{4.17}$$

$H^{\varkappa}_{123}$ is the Hamiltonian for the asymptotic initial state $\varkappa$. We can write

$$H^{\varkappa}_{123} = H_{123} - V^{\varkappa}_{123} . \tag{4.18}$$

$V^{\varkappa}_{123}$ is the interaction in the scattering channel $\varkappa$ which describes (for $\varkappa \neq 0$) the interaction between the free particle and a bound pair. Therefore we have

$$\begin{aligned} V^{1}_{123} &= V_{12} + V_{13} , \qquad V^{2}_{123} = V_{12} + V_{23} , \qquad V^{3}_{123} = V_{31} + V_{32} , \\ V^{0}_{123} &= V_{123} = V_{12} + V_{13} + V_{23} . \end{aligned} \tag{4.19}$$

Let us now consider the asymptotic condition in systems with two particle bound states. In generalization of Bogoljubovs condition (4.10) it is obvious to assume that the pair "1 and 2" moved independently of all other particles at $t \to -\infty$.

Then it is possible to write

$$\lim_{t\to-\infty} F_{12}(t) = \sum F_{\alpha}(\boldsymbol{P}) \, |\alpha \boldsymbol{P}\rangle \langle \boldsymbol{P} \alpha| = F^{\rm sc}_{12} + F^{12}_{\rm b} \tag{4.20}$$

$F^{\rm sc}_{12}$ is the contribution to $F_{12}(t)$ for pairs in scattering states and $F^{\rm b}_{12}$ is the part for pairs in bound states. Now in the sense of Bogoljubov for the scattering states we use the condition of the total weakening of the initial correlation

$$\lim_{t\to-\infty} F^{\rm sc}_{12} = F_1 \cdot F_2$$

but for $F^{\rm b}_{12}$ we do not have a weakening of the initial correlation. Because a total weakening of the initial correlation for $F^{\rm b}_{12}$ is absent, a system with two particle bound states can not be described only by F_1 but by F_1 and $F^{\rm b}_{12}$.

By this consideration a chemical picture is introduced. We have for $t \to -\infty$ a system with free elementary particles and "atoms" (bound pairs).

For the density operator F_{123} the asymptotic condition takes then the form

$$\lim_{t\to-\infty} F_{123}(t) = F^{\rm b}_{12} F_3 + F^{\rm b}_{13} F_2 + F^{\rm b}_{23} F_1 + F_1 F_2 F_3 = \sum_{\varkappa=0}^{3} F^{\varkappa}_{123} .$$

Summarizing, the generalization of the asymptotic condition for systems with two particle bound states is

$$\lim_{t\to-\infty} F_{12} = F^{\rm b}_{12} + F^{\rm sc}_{12} , \qquad \lim_{t\to-\infty} F^{\rm sc}_{12} = F_1 \cdot F_2 , \qquad \lim_{t\to-\infty} F_{123} = \sum_{\varkappa=0}^{3} F^{\varkappa}_{123} . \tag{4.21}$$

If bound states do not appear in the system we get again the condition (4.10).

4.3. Binary Density Operator. Definition of Distribution Functions for Free Particles and Atoms

In order to obtain a kinetic equation which describes the decay and the formation of two particle bound states it is necessary to determine F_{12} taking into account three particle collisions. This is possible if we determine F_{12} with the help of the equations

(4.8) and (4.9) up to the order n. In this case we can use for F_{123} and F_{12} the approximation

$$F_{123}(t) = e^{-\frac{i}{\hbar}H_{123}\tau} F_{123}(t-\tau)\, e^{\frac{i}{\hbar}H_{123}\tau},$$

$$F_{12}(t) = e^{-\frac{i}{\hbar}H_{12}\tau} F_{12}(t-\tau)\, e^{\frac{i}{\hbar}H_{12}\tau} +$$

$$+ \frac{n}{i\hbar}\int_0^\tau d\tau'\, e^{-\frac{i}{\hbar}H_{12}\tau'} \operatorname*{Sp}_3 [V_{13} + V_{23}, F_{123}(t)]\, e^{\frac{i}{\hbar}H_{12}\tau'}\, d\tau'\,. \tag{4.22}$$

Using now the identity $V_{12} + V_{23} = H_{123} - H_{12} - H_3$ the integration can be carried out. The result is

$$F_{12}(t) = e^{-\frac{i}{\hbar}H_{12}\tau} F_{12}(t-\tau)\, e^{\frac{i}{\hbar}H_{12}\tau} + n \operatorname*{Sp}_3 \{F_{123}(t) - e^{-\frac{i}{\hbar}H_{12}\tau} \cdot$$

$$\cdot F_{123}(t-\tau)\, e^{\frac{i}{\hbar}H_{12}\tau}\}\,. \tag{4.23}$$

From the asymptotic condition (4.21) for F_{123} and the relation $H_{123}\Omega^{\varkappa}_{123} = \Omega^{\varkappa}_{123}H^{\varkappa}_{123}$ and with the asymptotic condition for F_{12} finally we have for F_{12}

$$F_{12}(t) = e^{-\frac{i}{\hbar}H_{12}\tau} F_1(t-\tau)\, F_2(t-\tau)\, e^{\frac{i}{\hbar}H_{12}\tau} + F^{\mathrm{b}}_{12} +$$

$$+ n \sum_{\varkappa} \operatorname*{Sp}_3 \{\Omega^{\varkappa}_{123}F^{\varkappa}_{123}(t)\, \Omega^{\varkappa +}_{123} - \Omega_{12}F^{\varkappa}_{123}(t)\, \Omega^{+}_{12}\}\,. \tag{4.24}$$

In connection with this expression occurs the problem of secular divergencies. It is well known from the papers [1, 11, 2] that this equation contains terms which grow with $\tau/\tau_{\mathrm{coll}}$ by $\tau \to \infty$. The physical origin of the terms are succesive binary collisions. In order to analyse this problem we consider the formal solution for F_1 (4.8). Using the binary collision approximation for F_{12} similarly as before the τ integration can be carried out. The result for homogeneous systems is

$$F_1(t) = F_1(t-\tau) + n \operatorname*{Sp}_2 \{\Omega_{12}F_1(t)\, F_2(t)\, \Omega^{+}_{12} - F_1(t)\, F_2(t)\} + n \operatorname*{Sp}_2 F^{\mathrm{b}}_{12}(t)\,. \tag{4.25}$$

In this equation the second term grows with $\tau/\tau_{\mathrm{coll}}$. As can be seen the difference between $F_1(t)$ and $F_1(t-\tau)$ is of the order n. If we take into account three particle collisions it is therefore not possible to neglect the retardation. Following an idea of Klimontovich [2] we expand equation (4.25) and (4.24) with respect to τ. Thus we get in first order

$$F_1(t) = F_1(t) - \tau \frac{d}{dt} F(t) + n \operatorname*{Sp}_2 \{\Omega_{12}F_1F_2\Omega^{+}_{12} - F_1F_2\} + n \operatorname*{Sp}_2 F^{\mathrm{b}}_{12}\,,$$

$$F_{12}(t) = \Omega_{12}F_1F_2\Omega^{+}_{12} - \tau \frac{d}{dt}\Omega_{12}F_1F_2\Omega^{+}_{12} + F^{\mathrm{b}}_{12} +$$

$$+ n \sum_{\varkappa} \operatorname*{Sp}_3 \{\Omega^{k}_{123}F^{\varkappa}_{123}\Omega^{\varkappa +}_{123} - \Omega_{12}F^{\varkappa}_{123}(t)\, \Omega^{+}_{12}\}\,. \tag{4.26}$$

From this equation the retardation terms $\tau \frac{\mathrm{d}}{\mathrm{d}t}$ ... in the equation for $F_{12}(t)$ can be eliminated. Thus we obtain the expression

$$F_{12}(t) = \Omega_{12}F_1F_2\Omega_{12}^{+} + F_{12}^{\mathrm{b}} + n\,\mathop{\mathrm{Sp}}_{3}\,\{\sum_{\varkappa=0}^{3}(\Omega_{123}^{\varkappa}F_{123}^{\varkappa}\Omega_{123}^{\varkappa+} - \Omega_{12}F_{123}^{\varkappa}\Omega_{12}^{+}) - \\ - \Omega_{12}(\Omega_{13}F_1F_2F_3\Omega_{13}^{+} - F_1F_2)\,\Omega_{12}^{+} - \Omega_{12}(\Omega_{23}F_1F_2F_3\Omega_{23}^{+} - F_1F_2)\Omega_{12}^{+}\}\,. \tag{4.27}$$

It is easy to be seen, that the last terms vanish for two succesive binary collisions. That means F_{12} exists in this approximation.

Due to the fact that we have taken into account bound states in the binary density operator it is possible to consider chemical reactions in the kinetic theory. It is well known, that there is no essential difference between a bound state and an elementary particle. For many physical properties e.g. the pressure, the difference between a bound state and elementary particles does not play any role [10]. For this reason, in statistical mechanics it is convenient too, to use the following interpretation:
Bound states $\doteq$ new composite particles (atoms).
Therefore we have now a two component system consisting of free particles and atoms. For the description of this system it is necessary to introduce one particle distribution functions for free particles and atoms. It seems to be obvious, that the distribution function of the atoms is connected with the bound state contribution of F_{12}. We consider the normalization condition for F_{12}. From (4.27) and (4.1) follows in the thermodynamic limit

$$\frac{1}{V^2}\,\mathrm{Sp}\,F_{12} = \frac{1}{V^2}\left\{\int \mathrm{d}\boldsymbol{p}\,\mathrm{d}\boldsymbol{P}\,\langle \boldsymbol{pP}|\,F_{12}\,|\boldsymbol{Pp}\rangle + \sum_n \int \mathrm{d}\boldsymbol{P}\,\langle n\boldsymbol{P}|\,F_{12}\,|\boldsymbol{P}n\rangle\right\} \\ = \frac{1}{V}\int \mathrm{d}\boldsymbol{p}_1\,\langle \boldsymbol{p}_1|\,F_1\,|\boldsymbol{p}_1\rangle + \frac{n}{V}\sum_n \int \mathrm{d}\boldsymbol{P}\,\langle n\boldsymbol{P}|\,F_{12}\,|\boldsymbol{P}n\rangle = 1\,. \tag{4.28}$$

Now we interpret the first term as contribution of the free particles

$$\frac{1}{V}\int \mathrm{d}\boldsymbol{p}_1\,\langle \boldsymbol{p}_1|\,F_1\,|\boldsymbol{p}_1\rangle = \frac{N_1}{N}\,. \tag{4.29}$$

N_1 = number of the free particles, and the second term as the contribution of the atoms

$$\frac{n}{V}\sum_{n_{12}} \int \mathrm{d}\boldsymbol{P}_{12}\,\langle n_{12}\boldsymbol{P}_{12}|\,F_{12}\,|\boldsymbol{P}_{12}n_{12}\rangle = \frac{2N_{12}}{N}\,. \tag{4.30}$$

N_{12} = number of the atoms.
Then from (4.28) follows the condition

$$N_1 + 2N_{12} = N\,. \tag{4.31}$$

Therefore we can give in consistent manner the following definitions for the distribution functions

$$n\langle \boldsymbol{p}_1|\,F_1\,|\boldsymbol{p}_1\rangle = \frac{V}{(2\pi\hbar)^3}\,f(\boldsymbol{p}_1) \tag{4.32}$$

for the free particles and

$$\frac{n^2}{2}\,\langle n_{12}\boldsymbol{P}_{12}|\,F_{12}\,|\boldsymbol{P}_{12}n_{12}\rangle = \frac{V}{(2\pi\hbar)^3}\,f_{n_{12}}(\boldsymbol{P}) \tag{4.33}$$

for the atoms.

The normalization of these functions is given by

$$\int f(\boldsymbol{p}_1) \frac{d\boldsymbol{p}_1}{(2\pi\hbar)^3} = n_1 , \tag{4.33a}$$

$$\sum_{n_{12}} \int \frac{d\boldsymbol{P}}{(2\pi\hbar)^3} f_{n_{12}}(\boldsymbol{P}) = n_{12} ,$$
$$2n_{12} + n_1 = n . \tag{4.33}$$

4.4. Kinetic Equations for Free Particles

With the expression (4.27) for F_{12} it is possible to write down kinetic equations which describe the formation and the decay of two particle bound states in three particle collisions. Introducing (4.27) into the first of the eqs. (4.7) and using the momentum representation we can write the kinetic equation in the form

$$\frac{\partial}{\partial t} f(\boldsymbol{p}_1) = I_2^B + I_3^R + I_3 . \tag{4.34}$$

Here is

$$I_2^B = \frac{1}{i\hbar} \frac{(2\pi\hbar)^3}{V} n^2 \int d\boldsymbol{p}_2 \langle \boldsymbol{p}_1\boldsymbol{p}_2| [V_{12}, \Omega_{12}F_1F_2\Omega_{12}^+] |\boldsymbol{p}_1\boldsymbol{p}_2\rangle \tag{4.34a}$$

the Boltzmann collision term for two particle collisions and I^R the retardation term

$$I^R = i \frac{(2\pi\hbar)^3}{V\hbar} n^3 \int d\boldsymbol{p}_2 \, d\boldsymbol{p}_3 \langle \boldsymbol{p}_1\boldsymbol{p}_2\boldsymbol{p}_3| [V_{12}, \Omega_{12}(\Omega_{13}F_1F_2F_3\Omega_{13}^+ - F_1F_2F_3) \Omega_{12}^+$$
$$+ \Omega_{12}(\Omega_{23}F_1F_2F_3\Omega_{23}^+ - F_1F_2F_3) \Omega_{12}^+ + \Omega_{12}F_1F_2F_3\Omega_{12}^+] |\boldsymbol{p}_1\boldsymbol{p}_2\boldsymbol{p}_3\rangle \cdot \tag{4.34b}$$

Further it is useful to write down I_3 in the form

$$iI_3 = \frac{(2\pi\hbar)^3}{V\hbar} \frac{n^3}{2} \int d\boldsymbol{p}_2 \, d\boldsymbol{p}_3 \langle \boldsymbol{p}_1\boldsymbol{p}_2\boldsymbol{p}_3| [V_{12} + V_{23} + V_{13}, \sum_{\varkappa=0}^{3} \Omega_{123}^{\varkappa} F_{123}^{\varkappa} \Omega_{123}^{\varkappa+}]$$
$$\cdot |\boldsymbol{p}_1\boldsymbol{p}_2\boldsymbol{p}_3\rangle + \sum_{n_{23}} \frac{(2\pi\hbar)^3}{V\hbar} \frac{n^3}{2} \int d\boldsymbol{P}_{23} \langle \boldsymbol{p}_1\boldsymbol{P}_{23}n_{23}| [V_{12} + V_{23}, \sum_{\varkappa=0}^{3} \Omega_{123}^{\varkappa} F_{123}^{\varkappa} \Omega_{123}^{\varkappa+}]$$
$$\cdot |n_{23}\boldsymbol{P}_{23}\boldsymbol{p}_1\rangle . \tag{4.34c}$$

This contribution describes the formation and the decay of atoms by three particle reactions. Because I_3 contains the distribution function of the atoms, we do not get a closed equation for $f_1(\boldsymbol{p}_1)$.

First we consider I_2. With the help of the scattering theory it is possible to express I_2 by the two particle T-matrix. For this purpose we use the following relations from the scattering theory [8]:

1. the definition of the T-matrix: $T = V\Omega$,
2. the Lippmann-Schwinger equation:

$$\Omega = 1 + G^0T; \qquad G_0 = (z - H_0)^{-1} , \tag{4.35}$$

3. the optical theorem

$$T - T^+ = -2\pi \mathrm{i} T \delta(E - H_0)\, T^+ \,. \tag{4.36}$$

With these relations it is easy to get the usual form of the Boltzmann collision integral

$$\begin{aligned} I_2^B = \frac{(2\pi\hbar)^3}{V\hbar} \int & \mathrm{d}\boldsymbol{p}_2\, \mathrm{d}\overline{\boldsymbol{p}}_1\, \mathrm{d}\overline{\boldsymbol{p}}_2\, |\langle \boldsymbol{p}_1\boldsymbol{p}_2|\, T\, |\overline{\boldsymbol{p}}_1\overline{\boldsymbol{p}}_2\rangle|^2\, \delta(E_{\boldsymbol{p}_1\boldsymbol{p}_2} - E_{\overline{\boldsymbol{p}}_1\overline{\boldsymbol{p}}_2}) \\ & \cdot 2\pi\, \{f(\overline{\boldsymbol{p}}_1)\, f(\overline{\boldsymbol{p}}_2) - f(\boldsymbol{p}_1)\, f(\boldsymbol{p}_2)\} \,. \end{aligned} \tag{4.37}$$

The second term I^{R} guarantees the conservation of the potential and the kinetic energy in the binary collision approximation. In the same way it is possible to introduce the T-matrix in the contribution $I_3(\boldsymbol{p}_1)$ using following relations from the many-channel-scattering theory [8]:

1. the Lippmann-Schwinger equation for

$$\begin{aligned} &\Omega^\varkappa = 1 + G V^\varkappa = 1 + G^\varkappa T^\varkappa \,, \\ &G^\varkappa = (z - H^\varkappa)^{-1} \,, \qquad G = (z - H)^{-1} \,, \end{aligned} \tag{4.38}$$

2. the definition of the T-operator by

$$\begin{aligned} &T^{\varkappa\varkappa'} = V^\varkappa - V^{\varkappa'} G V^\varkappa = V^\varkappa + V^{\varkappa'} + V^{\varkappa'}\Omega^\varkappa = V^\varkappa - V^{\varkappa'} + \tilde{T}^{\varkappa\varkappa'} \,, \\ &\tilde{T}^{\varkappa'} = V^{\varkappa'}\Omega^\varkappa \,, \end{aligned} \tag{4.39}$$

3. the optical theorem of the multi channel scattering theory

$$2\,\mathrm{Im}\, \langle \varkappa\boldsymbol{p}|\, T^{\varkappa\varkappa}\, |\boldsymbol{p}\varkappa\rangle = -2\pi \sum_{\bar{\varkappa}} \int \mathrm{d}\overline{\boldsymbol{p}}\, \langle \varkappa\boldsymbol{p}|\, T^{\varkappa\bar{\varkappa}'}\, |\overline{\boldsymbol{p}\varkappa}\rangle\, \delta(E_{\bar{p}\bar{\varkappa}} - E_{p\varkappa})\, \langle \overline{\varkappa\boldsymbol{p}}\,|\, \overset{+}{T^{\bar{\varkappa}\varkappa}}\, |\boldsymbol{p}\varkappa\rangle \tag{4.40}$$

We notice that $T^{\varkappa'\varkappa}$ and $\tilde{T}^{\varkappa'\varkappa}$ are equivalent on the energy shell. Starting from (4.23) and using the relations (4.38) up to (4.40) and the completeness of the states $|\varkappa\boldsymbol{p}\rangle$ it is easy to get an explicit expression for I_3:

$$\begin{aligned} I_3 = \frac{(2\pi\hbar)^3}{V\hbar}\, \frac{1}{2} \sum_{\varkappa=0}^{3} & \{\mathrm{d}\boldsymbol{p}_2\, \mathrm{d}\boldsymbol{p}_3\, \mathrm{d}\overline{\boldsymbol{p}}\, |\langle \boldsymbol{p}_1\boldsymbol{p}_2\boldsymbol{p}_3|\, \tilde{T}^{0\varkappa}\, |\overline{\varkappa\boldsymbol{p}}\rangle|^2\, 2\pi\delta(E_{123} - \tilde{E}_{123}) \\ & \cdot \{f(\overline{\varkappa\boldsymbol{p}}) - f(\boldsymbol{p}_1)\, f(\boldsymbol{p}_2)\, f(\boldsymbol{p}_3)\} + \sum_{n_{23}} \int \mathrm{d}\boldsymbol{P}_{23}\, \mathrm{d}\overline{\boldsymbol{p}}\, |\langle \boldsymbol{p}_1\boldsymbol{P}_{23}n_{23}|\, \tilde{T}^{1k}\, |\overline{\boldsymbol{p}\varkappa}\rangle|^2 \\ & \cdot 2\pi\delta(E_{1(23)} - E_{\bar{p}\bar{\varkappa}})\, \{f(\overline{\varkappa p}) - f(\boldsymbol{p}_1)\, f_{n_{23}}(\boldsymbol{P}_{23})\} \,. \end{aligned} \tag{4.41}$$

Here is

$$f(\varkappa\boldsymbol{p}) = \begin{cases} f(\boldsymbol{p}_1)\, f(\boldsymbol{p}_2)\, f(\boldsymbol{p}_3) & \varkappa = 0 \,, \\ f(\boldsymbol{p}_1)\, f_{n_{23}}(\boldsymbol{P}_{23}) & \text{for} \quad \varkappa = 1 \,, \\ f(\boldsymbol{p}_2)\, f_{n_{13}}(\boldsymbol{P}_{13}) & \varkappa = 2 \,, \\ f(\boldsymbol{p}_3)\, f_{n_{12}}(\boldsymbol{P}_{12}) & \varkappa = 3 \,. \end{cases}$$

In order to discuss this expression in a transparent way we use the following form for I_3 [2]:

$$I_3 = [I]_1 + [I]_2 \,. \tag{4.42}$$

$[I]_1$ is given by the terms with $\varkappa = 0, 1$, that means we have bound states only between the particles 2 and 3. $[I]_2$ is given by the contribution with $\varkappa = 2, 3$; in this term are also bound states with the particle 1. Before we consider the explicit form of this term we will give a little generalization. So far we have considered a one-component system, but now we will deal with a many component system. In this case it is necessary to indicate the particles by the number and the index of the species e.g. $a, b, \ldots$ For simplification we assume that bound states are possible only between particles of different species.

First we will write down $[I]_1$ for many component systems. $[I]_1$ consists of 4 parts which describe 4 processes:

$$[I]_1 = [I^1]_1 + [I^2]_1 + [I^3]_1 + [I^0]_1 .$$

For $\varkappa = 0$ we have the two contributions $[I^0]_1$ and $[I^1]_1$. We get the three particle Boltzmann integral

$$[I^0]_1 = \frac{(2\pi\hbar)^3}{V\hbar} \sum_{c,b} \int \mathrm{d}\boldsymbol{p}_a \, \mathrm{d}\boldsymbol{p}_b \, \mathrm{d}\overline{\boldsymbol{p}}_a \, \mathrm{d}\overline{\boldsymbol{p}}_b \, \mathrm{d}\overline{\boldsymbol{p}}_c \, |\langle \boldsymbol{p}_a \boldsymbol{p}_b \boldsymbol{p}_c | \, T^{00} \, | \overline{\boldsymbol{p}}_a \overline{\boldsymbol{p}}_b \overline{\boldsymbol{p}}_c \rangle|^2$$
$$\cdot 2\pi\delta(E_{abc} - \overline{E}_{abc}) \frac{1}{2} \{f(\overline{\boldsymbol{p}}_a) f(\overline{\boldsymbol{p}}_b) f(\overline{\boldsymbol{p}}_c) - f(\boldsymbol{p}_a) f(\boldsymbol{p}_b) f(\boldsymbol{p}_c)\} . \tag{4.43}$$

This integral describes the scattering between three particles

$$a + b + c \rightarrow \bar{a} + \bar{b} + \bar{c} .$$

Without the following reaction terms we get with $[I_0]_1$ the quantum version of the kinetic equation of UHLENBECK, COHEN [11], KLIMONTOIVCH and others. With other methods this equation was also given by PALZEV [14].

The second contribution for $\varkappa = 0$ is given by

$$[I^1]_1 = \frac{(2\pi\hbar)^3}{V\hbar} \sum_{\substack{n_{bc} \\ c,b}} \int \mathrm{d}\boldsymbol{P}_{bc} \, \mathrm{d}\overline{\boldsymbol{p}}_a \, \mathrm{d}\overline{\boldsymbol{p}}_b \, \mathrm{d}\overline{\boldsymbol{p}}_c \, |\langle \boldsymbol{p}_a \boldsymbol{P}_{bc} n_{bc} | \, T^{10} | \overline{\boldsymbol{p}}_a \overline{\boldsymbol{p}}_b \overline{\boldsymbol{p}}_c \rangle|^2$$
$$\cdot 2\pi\delta(E_{a(bc)} - \overline{E}_{abc}) \frac{1}{2} \{f(\overline{\boldsymbol{p}}_a) f(\overline{\boldsymbol{p}}_b) f(\overline{\boldsymbol{p}}_c) - f(\boldsymbol{p}_a) f_{n_{bc}}(\boldsymbol{P}_{bc})\} . \tag{4.44}$$

This collision integral describes the reaction

1. $a + (bc) \rightarrow \bar{a} + \bar{b} + \bar{c}$.

For $\varkappa = 1$ we have the following contribution

$$[I^2]_1 = \frac{(2\pi\hbar)^3}{V\hbar} \sum_{\substack{n_{bc} \\ c,b}} \int \mathrm{d}\overline{\boldsymbol{P}}_{bc} \, \mathrm{d}\overline{\boldsymbol{p}}_a \mathrm{d}\boldsymbol{p}_b \mathrm{d}\boldsymbol{p}_c \, |\langle \boldsymbol{p}_a \boldsymbol{p}_b \boldsymbol{p}_c | \, T^{01} \, | \overline{n}_{bc} \overline{\boldsymbol{P}}_{bc} \overline{\boldsymbol{p}}_a \rangle|^2$$
$$\cdot 2\pi\delta(E_{abc} - \overline{E}_{a(bc)}) \frac{1}{2} \{f(\overline{\boldsymbol{p}}_a) f_{\overline{n}_{bc}}(\overline{\boldsymbol{P}}_{bc}) - f(\boldsymbol{p}_a) f(\boldsymbol{p}_b) f(\boldsymbol{p}_c)\} \tag{4.45}$$

which corresponds to the reaction

2. $a + b + c \rightarrow \bar{a} + (\overline{bc})$

and for

$$[I^3]_1 = \frac{(2\pi\hbar)^3}{V\hbar} \sum_{\substack{n_{bc} \\ c,b}} \sum_{\overline{n}_{bc}} \int \mathrm{d}\boldsymbol{P}_{bc} \, \mathrm{d}\overline{\boldsymbol{P}}_{bc} \, \mathrm{d}\overline{\boldsymbol{p}}_a \, |\langle \boldsymbol{p}_a \boldsymbol{P}_{bc} n_{bc} | \, T^{11} \, | \overline{n}_{bc} \overline{\boldsymbol{P}}_{bc} \overline{\boldsymbol{p}}_a \rangle|^2 \cdot$$
$$\cdot 2\pi\delta(E_{a(bc)} - \overline{E}_{a(bc)}) \frac{1}{2} \{f(\overline{\boldsymbol{p}}_a) f_{\overline{n}_{bc}}(\overline{\boldsymbol{P}}_{bc}) - f(\boldsymbol{p}_a) f_{n_{bc}}(\boldsymbol{P}_{bc})\}. \tag{4.46}$$

This term describes the scattering between atoms and free particles.

3. $a + (bc) \to \bar{a} + (\bar{b}\bar{c})$.

In all these processes the number of particles with momentum $\boldsymbol{p}_a$ remains unchanged. Therefore the collision integral has the important property

$$\int [I(\boldsymbol{p}_a)]_1 \, d\boldsymbol{p}_a = 0 \quad \text{(conservation of } n_a\text{)} . \tag{4.47}$$

Now we consider $[I]_2$. This contribution splits up into four parts. Two parts we get for $\varkappa = 2$, the first of which is

$$[I^4]_2 = \frac{(2\pi\hbar)^3}{V\hbar} \sum_{\substack{\bar{n}_{ac} \\ cb}} \int d\boldsymbol{p}_b \, d\boldsymbol{p}_c \, d\bar{\boldsymbol{p}}_b \, d\bar{\boldsymbol{P}}_{ac} \, |\langle \boldsymbol{p}_a \boldsymbol{p}_b \boldsymbol{p}_c| \, T^{02} \, |\bar{n}_{ac} \bar{\boldsymbol{P}}_{ac} \bar{\boldsymbol{p}}_b\rangle|^2 \cdot$$
$$\cdot \, 2\pi\delta(E_{abc} - E_{b(ac)}) \, \{f(\bar{\boldsymbol{p}}_b) \, f(\bar{\boldsymbol{P}}_{ac}) - f(\boldsymbol{p}_a) \, f(\boldsymbol{p}_b) \, f(\boldsymbol{p}_c)\} \tag{4.48}$$

with the reaction

4. $a + b + c \to \bar{b} + (\bar{a}\bar{c})$.

The second contribution reads

$$[I^5]_2 = \frac{(2\pi\hbar)^3}{V\hbar} \sum_{n_{bc}} \sum_{\substack{\bar{n}_{ac} \\ cb}} \int d\boldsymbol{P}_{cb} \, d\bar{\boldsymbol{P}}_{ac} d\bar{\boldsymbol{p}}_b \, |\langle \boldsymbol{p}_a \boldsymbol{P}_{cb} n_{cb}| \, T^{12} \, |\bar{n}_{ac} \bar{\boldsymbol{P}}_{ca} \bar{\boldsymbol{p}}_b\rangle| \cdot$$
$$\cdot \, 2\pi\delta(E_{a(bc)} - \bar{E}_{b(ac)}) \, \{f(\bar{\boldsymbol{p}}_b) \, f_{\bar{n}_{ac}}(\bar{\boldsymbol{P}}_{ac}) - f(\boldsymbol{p}_a) \, f_{n_{bc}}(\boldsymbol{P}_{bc})\} .$$

with the exchange reaction

5. $a + (bc) \to \bar{b} + (\bar{a}\bar{c})$.

Furthermore we get two parts for $\varkappa = 3$ namely

$$[I^6]_2 = [I^4]_2 \, (\bar{b} \to \bar{c}) \tag{4.49}$$

with

6. $a + b + c \to \bar{c} + (\bar{b}\bar{a})$

and

$$[I^7]_2 = [I^5]_2 \, (\bar{c} \to \bar{b}) \tag{4.50}$$

with the exchange reaction

7. $a + (bc) \to \bar{c} + (\bar{b}\bar{a})$.

As can be seen the processes 4. up to 7. changes the number of free particles. Therefore we have

$$\int d\boldsymbol{p}_a \, [I]_2 \neq 0 . \tag{4.51}$$

4.5. Kinetic Equation for Atoms

The collision integral I^{reac} contains the distribution function for the atoms $f_n(\boldsymbol{P})$. Therefore the kinetic equation (4.34) is not closed. It is also necessary to get a kinetic equation for the atoms. Such equation can be derived from the second equation of the hierarchy (4.6). Using (4.21) we get

$$\begin{aligned} &i\hbar \frac{\partial}{\partial t} \langle n_{ab}\boldsymbol{P}_{ab}| F_{ab} |\boldsymbol{P}_{ab} n_{ab}\rangle + \langle n_{ab}\boldsymbol{P}_{ab}| [H_{ab}, F_{ab}] |n_{ab}\boldsymbol{P}_{ab}\rangle = \\ &= \sum_{c\varkappa} n_c \int d\boldsymbol{p}_c \langle n_{ab}\boldsymbol{P}_{ab}\boldsymbol{p}_c| [V_{13} + V_{23}, \Omega^{\varkappa}_{123} F^{\varkappa}_{123} \Omega^{\varkappa+}_{123}] |n_{ab}\boldsymbol{P}_{ab}\boldsymbol{p}_c\rangle \,. \end{aligned} \tag{4.52}$$

Similar as before it is possible to write this equation in the form

$$\frac{\partial}{\partial t} f_{n_{ab}}(\boldsymbol{P}_{ab}) = I^{\mathrm{sc}}_{n_{ab}} + I^{\mathrm{reac}}_{n_{ab}} \tag{4.53}$$

$I^{\mathrm{sc}}_{n_{ab}}$ is given by the contribution with $\varkappa = 3$ in (4.52) and describes the collisions between free particles and atoms

$$\begin{aligned} I^{\mathrm{sc}}_{n_{ab}}(\boldsymbol{P}_{ab}) = &\frac{(2\pi\hbar)^3}{\hbar V} \sum_c \int d\boldsymbol{p}_c \sum_{n_{ab}} \int d\overline{\boldsymbol{P}}_{ab}\, d\overline{\boldsymbol{p}}_c\, |\langle \boldsymbol{P}_{ab} n_{ab} \boldsymbol{p}_c| T^{33} |\overline{\boldsymbol{p}}_c \overline{n}_{ab} \overline{\boldsymbol{P}}_{ab}\rangle|^2 \cdot \\ &\cdot 2\pi\delta(E_{(ab)c} - \overline{E}_{(ab)c}) \{f(\overline{\boldsymbol{p}}_c) f_{\overline{n}_{ab}}(\overline{\boldsymbol{P}}_{ab}) - f(\boldsymbol{p}_c) f_{n_{ab}}(\boldsymbol{P}_{ab})\} \end{aligned} \tag{4.54}$$

with

$$(ab) + c \to (\overline{a}\overline{b}) + \overline{c}\,.$$

I^{reac} is determined by the contribution with $\varkappa = 0, 1, 2$. For instance follows for $\varkappa = 0$

$$\begin{aligned} [I_{n_{ab}}(\boldsymbol{P}_{ab})]^{\mathrm{reac}}_0 = &\frac{(2\pi\hbar)^3}{\hbar V} \sum_c \int d\boldsymbol{p}_c\, d\overline{\boldsymbol{p}}_c\, d\overline{\boldsymbol{p}}_b\, d\overline{\boldsymbol{p}}_a\, |\langle n_{ab}\boldsymbol{P}_{ab}\boldsymbol{p}_c| T^{30} |\overline{\boldsymbol{p}}_a \overline{\boldsymbol{p}}_b \overline{\boldsymbol{p}}_c\rangle|^2 \cdot \\ &\cdot 2\pi\delta(E_{(ab)c} - \overline{E}_{abc}) \{f(\overline{\boldsymbol{p}}_a) f(\overline{\boldsymbol{p}}_b) f(\overline{\boldsymbol{p}}_c) - f(\boldsymbol{p}_c) f_{n_{ab}}(\boldsymbol{P}_{ab})\} \end{aligned} \tag{4.55}$$

corresponding to the process

$$(ab) + c \to \overline{a} + \overline{b} + \overline{c}\,.$$

In a similar manner we can write down the contribution for $\varkappa = 1, 2$.

The basic equations for the kinetic theory of systems with two particle bound states are therefore given by

$$\begin{aligned} &\frac{\partial}{\partial t} f(\boldsymbol{p}_a) = I^{\mathrm{B}}_2 + I^{\mathrm{R}}_2 + I_3\,, \\ &\frac{\partial}{\partial t} f_{n_{ab}}(\boldsymbol{P}_{ab}) = I^{\mathrm{sc}}_{n_{ab}}(\boldsymbol{P}_{ab}) + I^{\mathrm{reac}}_{n_{ab}}(\boldsymbol{P}_{ab})\,. \end{aligned} \tag{4.56}$$

4.6. Equilibrium Solutions and Conservation Laws

In the equilibrium case the equations (4.56) has the form

$$I^{\mathrm{sc}}(\boldsymbol{p}_a) + I^{\mathrm{reac}}(\boldsymbol{p}_a) = 0; \qquad I^{\mathrm{sc}}_{n_{ab}}(\boldsymbol{P}_{ab}) + I^{\mathrm{reac}}_{n_{ab}}(\boldsymbol{P}_{ab}) = 0\,.$$

Neglecting the terms of retardation we obtain the solutions

$$f^0(\boldsymbol{p}_a) = \exp\left[-\beta\left(\frac{p_a^2}{2m_a} - \mu_{ab}\right)\right],$$
$$f_{n_{ab}}(\boldsymbol{P}_{ab}) = \exp\left[-\beta\left(E_{n_{ab}} + \frac{P_{ab}^2}{2M} - \mu_{ab}\right)\right] \tag{4.57}$$

as can be seen easily from the explicit form of the collision integrals. We can also see that (4.57) is only a solution if the condition.

$$\mu_{ab} = \mu_a + \mu_b \tag{4.58}$$

is fulfilled. Therefore the μ have the meaning of the chemical potentials and (4.58) is the condition of chemical equilibrium.

If we take into account the retardation in I_2, i.e. that we use the complete expression (4.34), we get additional terms in (4.57). It is possible to determine this term from F_{12} without the bound state contributions. The result is the cluster expansion of F_1 [2]

$$F_1 = F_1^0 + n \operatorname*{Sp}_2 (\Omega_{12} F_1^0 F_2^0 \Omega_{12}^+ - F_1^0 F_2^0) . \tag{4.59}$$

Using (4.57) and $H_{12}^{sc} = \Omega_{12} H_{12}^0 \Omega_{12}^+$ we obtain in the case of equilibrium [2]

$$f^{eq} = f^0(\boldsymbol{p}_a) + (2\pi)^3 \int d\boldsymbol{p}_b \langle \boldsymbol{p}_a \boldsymbol{p}_b | \, e^{-\beta(H_{ab} - \mu_a - \mu_b)} - e^{-\beta(H_{ab}^0 - \mu_a - \mu_b)} | \boldsymbol{p}_b \boldsymbol{p}_a \rangle . \tag{4.60}$$

We can determine μ from the normalization conditions (4.33) with the result

$$\mu_a = \ln(n_a \lambda_a^3) - 2 \sum_c n_a n_c b_{ac}^{sc} ,$$
$$\mu_{ab} = \ln(n_{ab} \lambda_{ac}^3) - \ln \sum_n e^{-\beta E_n} . \tag{4.61}$$

Here is

$$b_{ab}^{sc} = \operatorname{Sp}(\Omega_{ab} F_a^0 F_b^0 \Omega_{ab}^+ - F_a^0 F_b^0) = \operatorname{Sp}(e^{-\beta H_{ab}} - e^{-\beta H_{ab}^0}) \tag{4.62}$$

the scattering part of the second virial coefficient [9]. $\lambda_a = \dfrac{\hbar}{\sqrt{2\pi m_a k_B T}}$ thermal wave length. Now we return to the chemical equilibrium. Introducing (4.61) in the equilibrium condition we get the mass action law (MAL) for nonideal systems

$$\frac{n_{ab}}{n_a n_{ab}} = \frac{\lambda_{ab}^3}{\lambda_a^3 \lambda_b^3} \sum_n e^{-\beta E_n} \, e^{4 n_a b_{ab}^{sc}} . \tag{4.63}$$

This result is in agreement with the statistical theory of the chemical equilibrium which was given in [10].

In the non-equilibrium situation the densities n_a, n_b, n_{ab} do not satisfy the MAL but are functions to the time. In order to get equations for the density we multiply equation (4.56) with $(2\pi\hbar)^{-3}$ and integrate with respect to p_a. Using the relation (4.47) we get

$$\frac{dn_a}{dt} = \int \frac{d\boldsymbol{p}_a}{(2\pi\hbar)^3} [I(\boldsymbol{p}_a)]_1 + \int \frac{d\boldsymbol{p}_a}{(2\pi\hbar)^3} [I(\boldsymbol{p}_a)]_2 = \int \frac{d\boldsymbol{p}_a}{(2\pi\hbar)^3} [I(\boldsymbol{p}_a)]_2 . \tag{4.64}$$

For simplification we neglect the exchange reactions and obtain the rate equation

$$\frac{dn_a}{dt} = \alpha n_a n_{ab} - \beta n_a^3 \tag{4.65}$$

Here α is the ionization coefficient given by

$$\alpha = \frac{(2\pi)}{V\hbar} \sum_{\substack{\bar{n}_{ac} \\ ca}} \int \mathrm{d}\boldsymbol{p}_b \, \mathrm{d}\boldsymbol{p}_c \, \mathrm{d}\bar{\boldsymbol{p}}_b \, \mathrm{d}\bar{\boldsymbol{P}}_{ca} \, |\langle \boldsymbol{p}_a \boldsymbol{p}_b \boldsymbol{p}_c | \, T^{02} \, |\bar{n}_{ca} \bar{\boldsymbol{P}}_{ca} \boldsymbol{p}_b \rangle|^2 \cdot \delta(E_{abc} - E_{(ac)b}) \frac{f_{\bar{n}_{ac}}(\bar{\boldsymbol{P}}_{ca})}{n_{ac}} \frac{f(\bar{\boldsymbol{p}}_b)}{n_b} \, \mathrm{d}\boldsymbol{p}_a \, . \tag{4.66}$$

β — the tripel-recombination coefficient

$$\beta = \frac{(2\pi)}{V\hbar} \sum_{\bar{n}_{ca}} \int \mathrm{d}\boldsymbol{p}_b \, \mathrm{d}\boldsymbol{p}_c \, \mathrm{d}\bar{\boldsymbol{p}}_b \, \mathrm{d}\bar{\boldsymbol{P}}_{ca} \, |\langle \boldsymbol{p}_a \boldsymbol{p}_b \boldsymbol{p}_c | \, T^{02} \, |\bar{n}_{ca} \bar{\boldsymbol{P}}_{ca} \bar{\boldsymbol{p}}_b \rangle|^2 \cdot \delta(E_{abc} - E_{(ac)b}) \frac{f(\boldsymbol{p}_a)}{n_a} \frac{f(\boldsymbol{p}_b)}{n_b} \frac{f(\boldsymbol{p}_c)}{n_c} \, \mathrm{d}\boldsymbol{p}_c \, . \tag{4.67}$$

In the equilibrium case we get from (4.65)

$$n_{ab}/n_a n_b = \beta/\alpha \, .$$

Using (4.66) and (4.67) we obtain again the MAL. Further the properties of the collision integral are such that we have the conservation of the total moment, total number of particles $n_a + n_b = n$, and the energie in the binary collision approximation of the free particles. That means the collision integrals obey the relations

$$\sum_a \left\{ \int \frac{\mathrm{d}\boldsymbol{p}_a}{(2\pi\hbar)^3} \varphi(\boldsymbol{p}_a) \left[I_2^B(\boldsymbol{p}_a) + [I_3]_1 + [I_3]_2 \right] + \sum_{\substack{n_{ab} \\ b}} \int \frac{\mathrm{d}\boldsymbol{P}_{ab}}{(2\pi\hbar)^3} I_{n_{ab}}(\boldsymbol{P}_{ab}) \, \varphi_{n_{ab}}(\boldsymbol{P}_{ab}) = 0 \right. \tag{4.68}$$

for

$$\varphi_a(\boldsymbol{p}_a) = 1, \boldsymbol{p}_a, \frac{p_a^2}{2m} \quad \text{and} \quad \varphi_n(P_{ab}) = E_{n_{ab}} + \frac{P_{ab}^2}{2M} \, . \tag{4.69}$$

Using this relation from equations (4.56) follow the conservation law

$$\frac{\mathrm{d}}{\mathrm{d}t} \sum_{ab} \{\langle \varphi_a \rangle + \langle \varphi_{n_{ab}} \rangle\} = \sum_a \int \frac{\mathrm{d}\boldsymbol{p}_a}{(2\pi\hbar)^3} I_a^{\mathrm{R}}(\boldsymbol{p}_a) \, \varphi_a(\boldsymbol{p}_a) \tag{4.70}$$

for $\varphi(\boldsymbol{p}_a) = 1, \boldsymbol{p}_a$ the integral at the right hand side is zero and we get the conservation of the total density and the total momentum. For $\varphi_a(\boldsymbol{p}_a) = \frac{p_a^2}{2m_a}$ it is possible to show that

$$\sum_a \int \frac{\mathrm{d}\boldsymbol{p}_a}{(2\pi\hbar)^3} I_{\mathrm{R}}^a(\boldsymbol{p}_a) \frac{p_a^2}{2m_a} = -\frac{\mathrm{d}}{\mathrm{d}t} \mathrm{Sp}\,(\Omega_{12} F_1 F_2 \Omega_{12}^+ V_{12}) = -\frac{\mathrm{d}}{\mathrm{d}t} \mathrm{Sp}\,(F_{12} V_{12}) = -\frac{\mathrm{d}}{\mathrm{d}t} \langle V \rangle \tag{4.71}$$

is the average value of the potential energy in the binary collision approximation. Therefore the conservation of the full energy

$$\sum_a \frac{\mathrm{d}}{\mathrm{d}t} (\langle E_{p_a} \rangle + \langle E_{Pn} \rangle + \langle V_{ab} \rangle) = 0 \, , \tag{4.72}$$

is ensured by the contribution of the retardation $I^{R}(\boldsymbol{p}_a)$. The discussion of the equilibrium properties and the conservation laws shows that we have obtained a kinetic theory for nonideal gases with chemical reactions.

4.7. Kinetic Equations for Partially Ionized Plasmas

Now we consider the kinetic theory of partially-ionized plasmas. In the simplest case partly-ionized plasmas consist of three components e.g. a H-plasma consists of electrons, protons and H-atoms. We can use the basis equation (4.34). Then the kinetic equation for the electrons is given by

$$\frac{\partial}{\partial t} f_e(\boldsymbol{p}_e) = I_2(\boldsymbol{p}_e) + I_3(\boldsymbol{p}_e) + I^{R}(\boldsymbol{p}_e) \,. \tag{4.73}$$

For simplification we consider the collision terms in the first Born approximation that means

$$\begin{aligned} \langle \boldsymbol{p}_e\boldsymbol{p}_b| \, T_{eb} \, |\overline{\boldsymbol{p}}_e\overline{\boldsymbol{p}}_b\rangle &= \langle \boldsymbol{p}_e\boldsymbol{p}_b| \, V_{eb} \, |\overline{\boldsymbol{p}}_e\overline{\boldsymbol{p}}_b\rangle \,, \\ \langle \boldsymbol{p}\varkappa| \, T_{ebc} \, |\overline{\varkappa\boldsymbol{p}}\rangle &= \langle \boldsymbol{p}\varkappa| \, V^{\varkappa}_{ebc} \, |\overline{\varkappa\boldsymbol{p}}\rangle \,. \end{aligned} \tag{4.74}$$

Moreover we do not take into account effects which are connected with the identity of the particles, that means we neglect exchange terms.

Using (4.74) in $I_2(\boldsymbol{p}_e)$ we get the quantum version of the Landau collision integral, which is well known (see e.g. KADANOFF, BAYM [12] and KLIMONTOVICH [2]).

Let us consider $I_3(\boldsymbol{p}_e)$ more in detail. It is easy to show that $[I]_1$ can be written in the form

$$\begin{aligned} [I]_1 = \sum_{\substack{c,b \\ =e,p}} \sum_{\substack{\bar\alpha_{eb} \\ \alpha_{eb}}} \int \frac{\mathrm{d}\boldsymbol{P}_{bc}\,\mathrm{d}\overline{\boldsymbol{P}}_{bc}}{(2\pi)^6} \,\mathrm{d}\overline{\boldsymbol{p}}_e \left| \frac{4\pi}{k^2} \, e P_{\alpha_{ab}\bar\alpha_{ab}}(\boldsymbol{k}) \right|^2 \delta(\overline{\boldsymbol{p}}_e - \boldsymbol{p}_e + \boldsymbol{k}) \\ \cdot 2\pi\delta(E_{e(bc)} - \overline{E}_{ebc}) \, \frac{1}{2} \, \{f(\overline{\boldsymbol{p}}_e) \, f_{\bar\alpha_{cb}}(\overline{\boldsymbol{P}}_{bc}) - f(\boldsymbol{p}_e) \, f_{\alpha_{cb}}(\boldsymbol{P}_{cb})\} \end{aligned} \tag{4.75}$$

with $\hbar\boldsymbol{k} = \boldsymbol{P}_{bc} - \overline{\boldsymbol{P}}_{bc}$ and

$$f_{\bar\alpha_{bc}} = \begin{cases} f(\boldsymbol{p}_e) \, f(\boldsymbol{p}_b) & \text{for} \quad \alpha = \boldsymbol{p} \,, \\ f_{n_{bc}}(\boldsymbol{P}_{cb}) & \qquad\;\; \alpha = n \,. \end{cases}$$

The matrix elements $P_{\alpha\bar\alpha}$ are determined by the expression

$$P_{\alpha_{bc}\bar\alpha_{bc}}(\boldsymbol{k}) = \int \mathrm{d}\boldsymbol{r} \, \psi_{\bar\alpha_{bc}}(\boldsymbol{r}) \, \psi^{*}_{\alpha_{cb}}(\boldsymbol{r}) \left\{ e_b \, \mathrm{e}^{\mathrm{i}\frac{m_c}{M}\boldsymbol{kr}} + e_c \, \mathrm{e}^{-\mathrm{i}\frac{m_b}{M}\boldsymbol{kr}} \right\}. \tag{4.76}$$

In the case $\alpha = \boldsymbol{p}$, $\bar\alpha = \overline{\boldsymbol{p}}$ and with the free particle wave function ψ_0 we get approximately

$$P_{\boldsymbol{p}_{bc}\overline{\boldsymbol{p}}_{bc}}(\boldsymbol{k}) = e_b\delta\left(\overline{\boldsymbol{p}}_{bc} - \boldsymbol{p}_{bc} - \mathrm{i}\,\frac{m_c}{M}\,\boldsymbol{k}\right) + e_c\delta(\boldsymbol{p}_{bc} - \overline{\boldsymbol{p}}_{bc} - \mathrm{i}\,\frac{m_b}{M}\,\boldsymbol{k}). \tag{4.77}$$

Further we have taken into account that bound states are possible only between e and p, that means for the bound-free and bound-bound parts in (4.75) the sum over the indices c, b is restricted by $c \neq b$. Now we consider the contribution $[I]_2$.

If we do not take into account exchange terms we get only

$$[I]_2 = \sum_{\bar{n}_{ep}} \int \mathrm{d}\bar{\boldsymbol{P}}_{ep}\, \mathrm{d}\bar{\boldsymbol{p}}_p\, \mathrm{d}\boldsymbol{p}_p\, \mathrm{d}\boldsymbol{p}_p \left| \frac{4\pi}{k^2} e P_{p_{ep}\bar{n}_{ep}}(\boldsymbol{k}) \right|^2 \delta(\bar{\boldsymbol{p}}_p - \boldsymbol{p}_p + \boldsymbol{k}) \cdot \\ \cdot \pi\delta(E_{epp} - \bar{E}_{(ep)e}) \{ f(\bar{\boldsymbol{p}}_p) f_{n_{ep}}(\bar{\boldsymbol{P}}_{ep}) - f(\boldsymbol{p}_e) f(\boldsymbol{p}_p) f(\boldsymbol{p}_p) \} . \tag{4.78}$$

This contribution determines the ionization and recombination coefficient.

Up to now we have not taken into account the long range behaviour of the Coulomb interaction. Because this long range behaviour the Landau collision integral is divergent. The correct equation for the description of the collective behaviour of the plasma is the Lennard Balescu collision integral. Formal we get this collision integral with the substitution

$$V_{ab}(\boldsymbol{k}) = \frac{4\pi e^2}{k^2} \Rightarrow V_{ab}(\boldsymbol{k})/\varepsilon(\omega \boldsymbol{k}) . \tag{4.79}$$

$\varepsilon(\omega\boldsymbol{k})$ is the dielectric function that means the polarization of the plasma is taken into account. In the RPA it is given by

$$\varepsilon(\omega\boldsymbol{k}) = \int \frac{\mathrm{d}\boldsymbol{q}}{(2\pi\hbar)^3} \frac{f(\boldsymbol{k}-\boldsymbol{q}) - f(\boldsymbol{k}+\boldsymbol{q})}{\omega - E_{\boldsymbol{k}-\boldsymbol{q}} - E_{\boldsymbol{k}+\boldsymbol{q}}} .$$

In the static approximation $\varepsilon(0\boldsymbol{k})$ describes the Debye screening

$$\varepsilon(0\boldsymbol{k}) = \frac{1}{1 + \frac{1}{k^2}\varkappa^2} .$$

If we use this substitution in the equations (4.75) and (4.78) it is possible to compare them with the kinetic equation which is given by Klimontovich in [2].

It is easy to see, that these equations are in agreement if collisions between four particles are neglected.

5. Theory of Equilibrium and Kinetic Properties of Liquid Metals

N. P. Kowalenko, Yu. P. Krasny

5.1. Introduction and Concept of Pseudopotential

If dense plasmas in the region of about 10^{21} particles per cm^3 are cooled below the critical temperature $T_c \approx 10^4$ K the uniform state becomes unstable and the system divides into two phases, a liquid and a gaseous one. In metallic systems the liquid phase is highly conducting and shows many similarities with the plasma state. Because the theoretical description of dense gaseous plasmas should be consistent with the theory of liquid metals, in this chapter some aspects of the liquid metal theory are presented.

In recent years liquid metals and alloys are more and more used, especially in atomic energetics and in other fields of practical importance. The increase in the practical use of liquid metals is accompanied by an intensive growth of the theory of their equilibrium and kinetic properties. The progress reached in the theoretical description of liquid metals is mainly due to the method of pseudopotential (PP). With its help it is possible to calculate thermodynamic properties as well as transport effects (electrical conductivity, thermo-diffusion, etc.). The present article is based on the PP-method and the Green's function method. Many aspects of these methods have been explained already in the former chapters (see e.g. 3.1., 3.3. and 3.4.). The first part of this chapter (5.1. up to 5.3.) is devoted to the evaluation of the thermodynamic potential with consequent account for the conduction electron influence on the formation of the interionic interaction (many-particle interaction included). The second part (5.4. up to 5.6.) deals with the theory of transport processes. Special stress is laid on the finding of the electric conductivity and the elucidation of the applicability of the Born approximation.

It is common knowledge that both liquid and solid metals are good conductors of the electric current. The fact was explained way back in 1900 by Drude, who supposed that in metals valence electrons behave as free ones and may move all over the metal, and the rest of the electrons gather around ions and make a stable electron configuration. At first glance this model seems quite unrealistic, as in reality valence electrons interact strongly with each other and with the ions of the metal — the potential energy of these interactions is of the order of the kinetic energy of the electrons. At present, however, there is a strict foundation of the Drude model. Firstly, it has been shown that there is an analogy in the behaviour of strongly interacting electrons (electron liquid) and that of the system of noninteracting particles [1]. Secondly, valence electrons may be in some sense treated on the basis of weak bond approximation. The reason why the field acting on the electron is so small was established in 1940 by HERRING [2]. It consists in the fact that conductivity electrons suffer an effective ion repulsion which compensates their large Coulomb potential since wave functions of the conductivity electrons are orthogonal to the wave functions of inner shell electrons. In this connection the concept of PP was introduced as an effective potential acting on the conductivity electrons; this potential remains fairly small even when the momentum is large. This fact is very important as due to it a perturbation theory may be used.

The PP-method is now a well developed theory, on the basis of which various properties of metals may be calculated. It is also important that the idea of a weak PP electron-ion interaction is equally applicable both to ordered systems (solid metals) and disordered systems (liquid metals). But while the calculation of physical properties of metals in crystal state is based mainly on the symmetry properties of the system (long range order, the existence of the reciprocal lattice etc.), to determine the physical properties of liquid metals one must take into consideration the nonregular character of the system and use methods of the liquid state theory.

Let us give the main forms of the PP, which are used to calculate various characteristics of metals. One of the possible ways of constructing the PP of the electron-ion interaction is the calculation on the basis of "first principles", i.e. on a microscopic theory, which does not use any phenomenological parameter. These calculations are very complicated. To carry them out it is necessary to introduce a lot of simplifying approximations, which in the end cuts the accuracy of the final results [3]. It is therefore a widely accepted practice to take some model PP and determine its parameters from the agreement between calculated and measured values. The model PP-method is mainly due to HEINE, ABARENKOV, ANIMALU [4—6]. HEINE and ABARENKOV offered a PP of the following type [4]:

$$w_0(r) = \begin{cases} -\sum\limits_l A_l(E)\, \hat{P}_l\,, & r < r_0\,, \\ -ze^2/r\,, & r > r_0\,. \end{cases} \tag{5.1}$$

The symbols denote here: z — valency of ion, e — electron charge, r — distance between electron and ion, $\hat{P}_l$ — projection operator, which singles out (from the wave function on which it acts) components with an azimuth quantum number l, and E — energy eigenvalue corresponding to the quantum number, $A_l(E)$ and r_0 — adjustable parameters. Information on coefficients A_l and parameter r_0 is taken from spectroscopic data of energy levels of a free ion. ANIMALU and HEINE [5], considering $A_l = A_2$ for $l \geqq 2$, suggested a simplified model PP

$$w_0(r) = \begin{cases} -A_2 - (A_0 - A_2)\, \hat{P}_2 - (A_1 - A_2)\, \hat{P}_1\,, & r < r_0\,, \\ -ze^2/r\,, & r > r_0 \end{cases} \tag{5.2}$$

and applied it to many metals. The formfactor of this PP is tabulated for many non-transition metals in [3].

There are other much simpler, but often used model PP, among them the point ion PP, suggested by HARRISON [3], and the ASHCROFT PP [7]. The point ion PP

$$w_0(r) = -ze^2/r + \beta\delta(r) \tag{5.3}$$

consists of the Coulomb attraction potential and a δ-like repulsive part describing the repulsion of an electron from the inner shell of the ion.

The only parameter of this PP is found from the condition of finding correct values of the PP form-factor when $q = 2k_F$ (k_F is Fermi wave number). The Ashcroft PP is one of the most simplified variants of the Heine-Abarenkov PP, it looks as follows:

$$w_0(r) = \begin{cases} 0\,, & r < r_0\,, \\ -ze^2/r\,, & r > r_0\,. \end{cases} \tag{5.4}$$

The adjustable parameter r_0 is determined from experimental data on liquid metal resistivity or the Fermi surface.

Let us take another frequently used PP — the Cohen PP:

$$w_0(r) = \begin{cases} ze^2U/r_0\,, & r < r_0\,, \\ -ze^2/r\,, & r > r_0\,. \end{cases} \tag{5.5}$$

This PP formfactor is as follows:

$$\begin{aligned} w_0(q) &\equiv \frac{1}{v_0}\int w_0(r)\exp\left(-\mathrm{i}\boldsymbol{q}\boldsymbol{r}\right)\mathrm{d}^3r = \\ &= \frac{4\pi ze^2}{v_0q^2}\left[(1+U)\cos\left(qr_0\right) - U\frac{\sin\left(qr_0\right)}{qr_0}\right]. \end{aligned} \tag{5.6}$$

Where $v_0 = V/N$ is the volume per ion.

The adjustable parameters U and r_0, as a rule, are determined from the comparison of theoretical and experimental data on static properties of metal. For example, in [9] they use the condition of zero pressure at zero temperature in an equilibrium crystal lattice and coincidence of expected and experimental values of shear modulus C_{44}. Such adjustment of parameters in (5.6) turned out to be very effective: one set of U and r_0 is enough to describe fairly well the binding energy, the equation of state at $T = 0$, the phonon spectrum, results of measurements of Fermi surface parameters in the case of Na and K.

All the above mentioned PP are described by a broken function in R-space. As a result, their formfactors oscillate in case of great q and do not give a fairly quick convergence of series or integrals when estimating atomic properties. In this connection Gursky and Krasko [10—12] made a successful effort of building a model PP unruptured in r-space. This PP may be used for description of not only alkaline metals but various other simple metals (Mg, Zn, Pb, Sn). The Gursky-Krasko PP is as follows:

$$w_0(r) = ze^2\left\{\frac{e^{-r/r_0} - 1}{r} + \frac{a}{r_0}e^{-r/r_0}\right\} \tag{5.7}$$

To determine the parameters the following data are taken: the experimental values of the first ionization potential of the ion of valence z and the value of the formfactor of the screened PP in the first points of reciprocal lattice.

On the basis of the concept of PP the present-day knowledge of the electron system of nontransition metals may be formulated in the following way:

1. All valence electrons in metals collectivize and make a quasifree electron subsystem. (This may be easily verified experimentally from the volume occupied by the electrons in the momentum space, i.e. the value of Fermi momentum.)

2. The remaining ions are of comparatively small size (about 10% of the whole volume), they are not blocked and retain practically all properties of isolated ions. Their interaction is of Coulomb character.

3. In the main part of the volume a separate electron interacts with an ion in a purely Coulombic way. Within the ion the true wave function of the electron suffers strong oscillations as it must be orthogonal to the eigenfunctions of the ion core electrons. This leads to a sharp decrease of the electron-ion interaction and consequently of the amplitude of the electron scattering by the ion. The decrease of the interaction is equivalent to the appearance of some effective repulsive potential within the ion which compensates the selfconsistent potential of the ion core up to a considerable degree. As a result it becomes possible to substitute the true potential by some effective PP (nonlocal in a general case), giving a value close to the true value of scattering amplitude and close to a plane wave behaviour of the electron wave function in the entire volume.

4. The effective weakening of the interaction within the ion causes the appearance of a small parameter in the theory of metals. This parameter is the ratio of the Fourier component of the PP $w_0(q)$ to the Fermi energy E_F (for simplicity the PP will be from now considered local). The very presence of the small parameter paved the way for the model of quasifree electrons in describing the electron properties of metals.

5. The movement of ions is supposed to be so slow and their mass so large in comparison with the mass of electrons that the adiabatic approximation is justified.

5.2. Thermodynamic Potential of Liquid Metals

On the basis of the above formulated ideas of the structure of simple metals we may choose the following Hamiltonian of the metal system [13, 14]:

$$H = \sum_{\nu=1}^{N} \frac{\boldsymbol{P}_\nu^2}{2M} + \frac{1}{2} \sum_{\nu\nu'}^{N} \frac{z^2 e^2}{|\boldsymbol{R}_\nu - \boldsymbol{R}_{\nu'}|} + \sum_{i=1}^{zN} \frac{\boldsymbol{p}_i^2}{2m} + \frac{1}{2} \sum_{ii'}^{zN} \frac{e^2}{|\boldsymbol{r}_i - \boldsymbol{r}_{i'}|} + \sum_{i\nu} w_0(\boldsymbol{r}_i - \boldsymbol{R}_\nu) \,. \tag{5.8}$$

Here $w_0(r)$ is the local PP of electron-ion interaction, z is the valence of the ion, N is the number of ions, V is the volume of the metal, $\boldsymbol{r}$ and $\boldsymbol{R}$, $\boldsymbol{p}$ and $\boldsymbol{P}$, m and M are radius-vectors, momentum and mass of electrons and ions correspondingly.

Let us proceed in (5.8) to the momentum representation accordingly to the formula

$$f(\boldsymbol{r}) = \frac{1}{V} \sum_{\boldsymbol{q}} f(\boldsymbol{q})\, e^{i\boldsymbol{q}\boldsymbol{r}} \,. \tag{5.9}$$

Then the Hamiltonian (5.8) will take the form:

$$H = \sum_\nu \frac{\boldsymbol{P}_\nu^2}{2M} + \frac{N}{2} \sum_{\boldsymbol{q}}{}' \left(|\varrho(\boldsymbol{q})|^2 - \frac{1}{N} \right) \frac{4\pi z^2 e^2}{v_0 q^2} + \sum_i \frac{\boldsymbol{p}_i^2}{2m} + \frac{1}{2} \sum_{ii'}{}' \frac{1}{V} \sum_{\boldsymbol{q}}{}' \frac{4\pi e^2}{q^2} e^{i\boldsymbol{q}(\boldsymbol{r}_i - \boldsymbol{r}_{i'})} + \sum_{i\nu} \frac{1}{N} \sum_{\boldsymbol{q}}{}' w_0(q)\, e^{i\boldsymbol{q}(\boldsymbol{r}_i - \boldsymbol{R}_\nu)} + Nz\beta/v_0 \,, \tag{5.10}$$

where

$$\varrho(\boldsymbol{q}) = \frac{1}{N} \sum_{\nu} \mathrm{e}^{-i\boldsymbol{q}\boldsymbol{R}_\nu}, \qquad v_0 = \mathrm{N}/N\,.$$

The prime at the sum symbol means that the terms with coinciding indices or with $q = 0$ are omitted; the latter were separated in the last term of (10). Here β is the non-Coulombic part of the PP of the electron-ion interaction:

$$\beta = \lim_{q \to 0} \left[v_0 w_0(q) + 4\pi z e^2/q^2\right].$$

Let us consider furtheron temperatures which satisfy the inequality:

$$\frac{\hbar^2}{2k_\mathrm{B}M}\left(\frac{3\pi^2 N}{V}\right)^{2/3} \ll T \ll \frac{\hbar^2}{2k_\mathrm{B}m}\left(\frac{3\pi^2 zN}{V}\right)^{2/3} \tag{5.11}$$

(k_B is the Boltzmann constant). The interval of temperatures, determined by this inequality is rather wide and includes practically the entire region of liquid metal state. At such temperatures the movement of ions may be considered classically, the electrons present a highly degenerate electron gas, due to the laws of quantum mechanics. It is therefore convenient to proceed in (5.10) to the second quantization representation for the electron subsystem of the metal. Taking plane waves as basis functions the Hamiltonian H may be presented as follows:

$$H = H_i + H_{ie} \tag{5.12}$$

where

$$H_i = \sum_{\nu}\left(\frac{P_\nu^2}{2M} + \frac{z\beta}{v_0}\right) + \frac{N}{2}\sum_{\boldsymbol{q}}{}'\left(|\varrho(\boldsymbol{q})|^2 - \frac{1}{N}\right)\frac{4\pi z^2 e^2}{v_0 q^2} \tag{5.13}$$

— the Hamiltonian of the ion subsystem,

$$H_{ie} = \sum_{\boldsymbol{k}\sigma}\left(\frac{\hbar^2 k^2}{2m} a_{\boldsymbol{k}\sigma}^+ a_{\boldsymbol{k}\sigma} + \sum_{\substack{\boldsymbol{k}\boldsymbol{q}\sigma\\(\boldsymbol{q}\neq 0)}} w_0(q)\,\varrho(\boldsymbol{q})\,a_{\boldsymbol{k}-\boldsymbol{q}\sigma}^+ a_{\boldsymbol{k}\sigma}\right)$$
$$+ \frac{1}{2V}\sum_{\substack{\boldsymbol{k}\boldsymbol{k}'q\sigma\sigma'\\(q\neq 0)}} \frac{4\pi e^2}{q^2}\, a_{\boldsymbol{k}-\boldsymbol{q}\sigma}^+ a_{\boldsymbol{k}'+\boldsymbol{q}\sigma'}^+ a_{\boldsymbol{k}'\sigma'} a_{\boldsymbol{k}\sigma} \tag{5.14}$$

— the Hamiltonian of the electron subsystem, $a_{\boldsymbol{k}\sigma}$ and $a_{\boldsymbol{k}\sigma}^+$ are annihilation and creation operators in the state with momentum $\hbar\boldsymbol{k}$ and spin σ.

To find thermodynamic characteristics of liquid metals it is sufficient to calculate the thermodynamic potential Ω:

$$\Omega = -k_\mathrm{B}T \ln \mathrm{Sp}\left\{\exp\left[\frac{-H + \mu_e N_e + \mu_i N_i}{k_\mathrm{B}T}\right]\right\} \tag{5.15}$$

where μ_i and μ_e are chemical potentials of ions and electrons. They must be chosen so as to satisfy the condition of electric neutrality:

$$\frac{\partial\Omega}{\partial\mu_e} = z\frac{\partial\Omega}{\partial\mu_i}. \tag{5.16}$$

It is convient to perform the calculation of Ω in two steps: first to find the average over the states of the electron subsystem with fixed positions of the ions and then over the states of the ion subsystem:

$$\Omega = -k_B T \ln \mathrm{Sp}^{(i)} \left\{ \exp \frac{-H_i + \mu_i N_i}{k_B T} \mathrm{Sp}^{(e)} \left[\exp \frac{-H_{ie} + \mu_e N_e}{k_B T} \right] \right\}. \tag{5.17}$$

Further transformations of (5.17) will be done in accordance with [15]. While finding the electron average, the H_{ie} part, corresponding to the electron interaction, will be singled out:

$$H_{ie} = H_0 + H_i \,, \qquad H_i = H_i' + H_i'' \,, \tag{5.18}$$

where the operators H_0, $H_{i'}$ correspond to the kinetic and the potential energy of the electrons, $H_{i''}$ is the interaction operator of electrons and ions. For the interaction representation in (5.17) let us choose as a perturbation the operator H_i. Then by the methods of quantum field theory for non-zero temperatures [16], we shall get:

$$\exp \left[\frac{-H_{ie} + \mu_e N_e}{k_B T} \right] = \exp \left[\frac{-H_0 + \mu_e N_e}{k_B T} \right] S \,, \tag{5.19}$$

$$S = T_\tau \exp \left[- \int_0^{1/k_B T} H_i(\tau) \, \mathrm{d}\tau \right], \tag{5.20}$$

$$H_i(\tau) = \exp \left[\tau (H_0 - \mu_e N_e) \right] H_i \exp \left[-\tau (H_0 - \mu_e N_e) \right]. \tag{5.21}$$

Hence

$$\mathrm{Sp}^{(e)} \left[\exp \frac{-H_{ie} + \mu_e N_e}{k_B T} \right] = \exp \left(\frac{-\Omega_0}{k_B T} \right) \langle S \rangle \,, \tag{5.22}$$

where Ω_0 is the thermodynamic potential of the ideal electron gas and

$$\langle S \rangle = \mathrm{Sp}^{(e)} \left[\exp \frac{\Omega_0 - H_0 + \mu_e N_e}{k_B T} S \right]. \tag{5.23}$$

Expanding the operator S in the last equation into powers of H_i, we obtain a perturbation series for $\langle S \rangle$. Let us transform it by Wick's theorem. If every part of the series is represented by Feynman diagrams, connected and disconnected diagrams may be singled out. Let us designate by $\langle S \rangle_c$ the sum of connected diagrams of all orders that enter $\langle S \rangle$. Then it is possible to show [16] that

$$\langle S \rangle = \exp \left[\langle S \rangle_c - 1 \right]. \tag{5.24}$$

Using this result in (5.22) we obtain:

$$\Delta\Omega \equiv -k_B T \ln \mathrm{Sp}^{(e)} \left[\exp \frac{-H_{ie} + \mu_e N_e}{k_B T} \right] = \Omega_0 - k_B T [\langle S \rangle_c - 1]. \tag{5.25}$$

The perturbation series for $\langle S \rangle_c$ contains members of all orders in electron-electron and electron-ion interaction, the latter serves as an external field. Using the concept of a weak PP we can expand S into powers of the PP:

$$S = S' + \sum_{n=1}^{\infty} \frac{(-1)^n}{n!} \int_0^{1/k_B T} \mathrm{d}\tau_1 \ldots \int_0^{1/k_B T} \mathrm{d}\tau_n \, T_\tau \left[H_i''(\tau_1) \ldots H_i''(\tau_n) \, S' \right], \tag{5.26}$$

where

$$S' = T_\tau \exp\,[-\int_0^{1/k_\mathrm{B}T} \mathrm{d}\tau\, H_i'(\tau)]\,. \tag{5.27}$$

Finally we obtain:

$$\Delta\Omega = \Omega_0 - k_\mathrm{B}T[\langle S'\rangle_\mathrm{c} - 1] + \sum_{n=1}^{\infty} \Omega_n\,. \tag{5.28}$$

Here

$$\Omega_n = -k_\mathrm{B}T\,\frac{(-1)^n}{n!}\int \mathrm{d}\tau_1\,\mathrm{d}^3r_1 \ldots \int \mathrm{d}\tau_n\,\mathrm{d}^3r_n \sum_{\nu_1} w_0(\boldsymbol{r}_1 - \boldsymbol{R}_{\nu_1})$$
$$\ldots \sum_{\nu_n} w_0\,(\boldsymbol{r}_n - \boldsymbol{R}_{\nu_n}) \sum_{\sigma_1\ldots\sigma_n} \langle T_\tau\,[\Psi^+_{\sigma_1}(\boldsymbol{r}_1,\tau_1)\,\Psi_{\sigma_1}(\boldsymbol{r}_i,\tau_i)$$
$$\ldots \Psi^+_{\sigma_n}(\boldsymbol{r}_n,\tau_n)\,\Psi^+_{\sigma_n}(\boldsymbol{r}_n,\tau_n)\,S']\rangle_\mathrm{c}\,. \tag{5.29}$$

As was stated above, Ω_0 is the thermodynamic potential of the ideal electron gas, the second term in (5.28) is the change of the thermodynamic potential of electrons on account of their interaction. The sum of these two terms

$$\Omega_e = \Omega_0 - k_\mathrm{B}T[\langle S'\rangle_\mathrm{c} - 1] \tag{5.30}$$

is the thermodynamic potential of the interacting gas of electrons, placed into a neutralizing background. The third term in (5.28) presented by the sum describes the changes of the thermodynamic potential of the electrons on account of their interaction with the ion subsystem of the metal. The structure of separate terms of this sum will be described in detail.

The first term contains the integral over the function $\langle T_\tau[\Psi^+_{\sigma_1}(\boldsymbol{r}_1,\tau_1)\,\Psi_{\sigma_1}(\boldsymbol{r}_1,\tau_1)\,S']\rangle$ which, due to homogeneity of space, does not depend on coordinates. Therefore the integration on coordinates will refer only to the PP and will be equal to zero because of electro-neutrality of the system.

The second term includes the so called polarization Green's function (which can be obtained from two-particle Green's function by omitting disconnected diagrams)

$$F_2(\boldsymbol{r}_1 - \boldsymbol{r}_2, \tau_1 - \tau_2) \equiv -\sum_{\sigma_1\sigma_2} \langle T_\tau[\Psi^+_{\sigma_1}(\boldsymbol{r}_1,\tau_1)\,\Psi_{\sigma_1}(\boldsymbol{r}_1,\tau_1)$$
$$\times\,\Psi^+_{\sigma_2}(\boldsymbol{r}_2,\tau_2)\,\Psi_{\sigma_2}(\boldsymbol{r}_2,\tau_2)\,S']\rangle_\mathrm{c}\,. \tag{5.31}$$

Fourier representation of this function is determined by the equation

$$F_2(r,\tau) = k_\mathrm{B}T \sum_{\omega_n} \mathrm{e}^{\mathrm{i}\omega_n\tau}\,\frac{1}{V}\sum_q \mathrm{e}^{\mathrm{i}qr}\,F_2(q,\omega_n) \tag{5.32}$$

where $\omega_n = 2\pi nk_\mathrm{B}T$) and may be represented with the help of the diagram expansion shown in figure 5.1. According to [16] the following general rules are valid:

$-f_2(q,\omega_n) \equiv$ 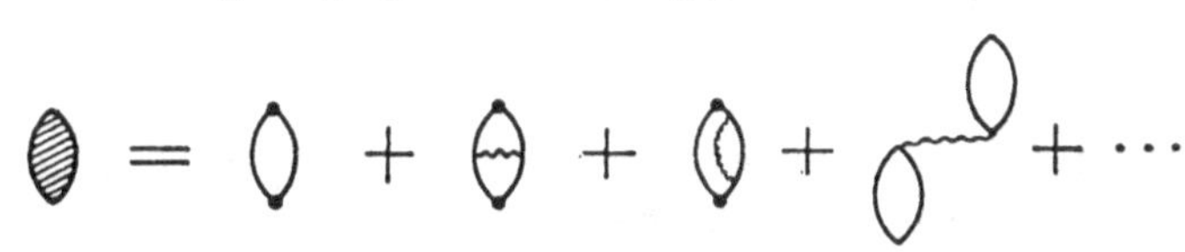 (5.33)

1. Every diagram vertex (where only two Fermi lines meet) is assigned with momentum q and frequency $\omega_n = 2\pi nk_\mathrm{B}T$. Every diagram line is assigned by the momentum p and the frequency $\varepsilon_n = (2n+1)\,\pi k_\mathrm{B}T$. Momenta and frequencies of all the lines must satisfy the laws of conservation at every vertex.

2. Every solid inner line is assigned with the value $[i\omega_n - \varepsilon_0(p) + \mu]^{-1}$, $\varepsilon_0(p) = p^2/2m$, every wavy line with the value $4\pi e^2/q^2$.

3. All the independent momenta are either summed up $\frac{1}{V} \sum_p$ or integrated $(2\pi)^{-3} \int d^3p$; all independent frequencies are summed up $k_B T \sum_{\omega_n}$ (with the exception of momenta and frequencies of external vertices).

4. The factor $(2S + 1)^F (-1)^F$ must be put in front of the expression obtained for every diagram, where F is the number of closed loops formed by Fermi lines in the diagram, S the spin of particles.

In the described functional expansion of $F_2(q, \omega_n)$ there are two types of diagrams: irreducible and reducible diagrams. Reducible diagrams may be cut into two disconnected ones by omitting one wavy line. If this cannot be done the diagram is called irreducible. Let us sum up all irreducible diagrams of $F_2(q, \omega_n)$. This sum gives the polarization operator $\pi(\boldsymbol{q}, \omega_n)$

(5.34)

Then the function $F_2(\boldsymbol{q}, \omega_n)$ may be represented in the form of a series that can be summed up:

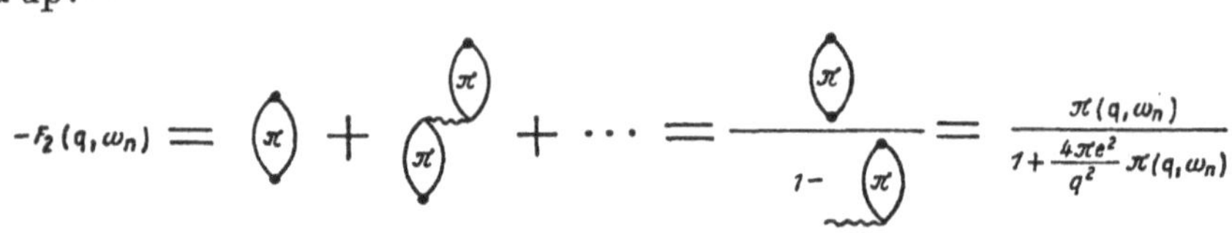

(5.35)

Using (5.31), (5.32), (5.33), the Ω_2 value may be transformed into

$$\Omega_2 = -\frac{V}{2} \sum_{\boldsymbol{q}}' \frac{\pi(\boldsymbol{q})}{1 + 4\pi e^2/q^2 \, \pi(\boldsymbol{q})} |\varrho(\boldsymbol{q}) \, w_0(q)|^2 , \tag{5.36}$$

where $\pi(\boldsymbol{q}) \equiv \pi(\boldsymbol{q}, 0)$. The appearance of zero frequency in (5.36) is caused by the static external field (adiabatic approximation). With the dielectric function of the electron gas

$$\varepsilon(\boldsymbol{q}, \omega) = 1 + \frac{4\pi e^2}{q^2} \pi(\boldsymbol{q}, \omega) \tag{5.37}$$

the quantity Ω_2 may be expressed according to

$$\Omega_2 = -\frac{V}{2} \sum_{\boldsymbol{q}}' \pi(\boldsymbol{q}) \, \varepsilon(\boldsymbol{q}) \, |\varrho(\boldsymbol{q}) \, w(\boldsymbol{q})|^2 , \tag{5.38}$$

where $\varepsilon(\boldsymbol{q}) \equiv \varepsilon(\boldsymbol{q}, 0)$, and $w(\boldsymbol{q}) = w_0(\boldsymbol{q})/\varepsilon(\boldsymbol{q})$ is the screened PP.

Any term of the sum Ω_n with $n > 2$ may be analysed in a similar way. On Fourier representation (5.29) may be expressed as follows:

$$\Omega_n = V \frac{(-1)^n}{n} \sum_{\boldsymbol{k}_1 \ldots \boldsymbol{k}_n}' F_n(\boldsymbol{k}_1, \boldsymbol{k}_2, \ldots, \boldsymbol{k}_n) \, \varrho(\boldsymbol{k}_1) \, w_0(\boldsymbol{k}_1) \ldots$$
$$\ldots \varrho(\boldsymbol{k}_n) \, w_0(\boldsymbol{k}_n) \, \Delta(\boldsymbol{k}_1 + \ldots + \boldsymbol{k}) , \tag{5.39}$$

where $F_n(\boldsymbol{k}_1, \dots, \boldsymbol{k}_n)$ is the Fourier transformation of the chronological product shown in (5.29) (with the coefficient $-1/(n-1)!$ at zero frequency. The function $F_n(\boldsymbol{k}_1, \dots, \boldsymbol{k}_n)$ may be called n-point polarization function which is presented as a sum of connected topologically non equivalent diagrams with n fixed external vertices. Though an expression for F_n with $n > 2$ cannot be obtained in such a simple from as for F_2, it is possible to make partial summation and single out of F_n polarization operators $\pi(k, \omega_n)$, as it was done in [13, 14] for zero temperature:

$$(-1)^n F_n(\boldsymbol{k}_1, \dots, \boldsymbol{k}_n, \omega_1, \dots, \omega_n) = \frac{\Lambda_n(\boldsymbol{k}_1, \dots, \boldsymbol{k}_n, \omega_1, \dots, \omega_n)}{\varepsilon(\boldsymbol{k}_1, \omega_1) \dots \varepsilon(\boldsymbol{k}_n, \omega_n)\,.} \tag{5.40}$$

Here Λ_n is determined by the set of diagrams with n vertices which have no contribution referring to any polarization part. The final expression for Ω_n takes the form:

$$\Omega_n = \frac{V}{n} \sum_{k_1 \dots k_n}' \Lambda_n(\boldsymbol{k}_1, \dots, \boldsymbol{k}_n)\, \varrho(\boldsymbol{k}_1)\, w(\boldsymbol{k}_1) \dots \varrho(\boldsymbol{k}_n)\, w(\boldsymbol{k}_n)\, \Delta(\boldsymbol{k}_1 + \dots + \boldsymbol{k}_n) \tag{5.41}$$

(here $\Delta(\boldsymbol{k})$ is Kronecker's symbol).
Thus we have clarified the structure of the thermodynamic potential $\Delta\Omega$ of the electron gas in the presence of ions. Within the adiabatic approximation it is possible to determine the character of inter-ion interaction, which is the subject of the next paragraph.

5.3. Inter-Ion Interaction in Liquid Metals. Many Body Forces

The presence of collectivized electrons in the metal and the adiabatic hypothesis result in the ion Hamiltonian $H_i + \Delta\Omega$. If terms having no ion co-ordinates are singled out

$$H_i + \Delta\Omega = \Omega_e + N\,\Delta E + H^*(\boldsymbol{R}_1, \dots, \boldsymbol{R}_N)\,, \tag{5.42}$$

the thermodynamic potential of the whole system may be expressed in the following way:

$$\Omega = \Omega_e + N\,\Delta E + \Omega\,, \tag{5.43}$$

where

$$\Delta E = -\frac{v_0^2}{2(2\pi)^3} \int \frac{\pi(\boldsymbol{q})\, |w_0(q)|^2}{\varepsilon(\boldsymbol{q})}\, \mathrm{d}^3\boldsymbol{q}\,, \tag{5.44}$$

$$\Omega_i = -k_B T \ln \mathrm{Sp}^{(i)} \left[\exp \frac{-H^* + \mu_i N}{k_B T}\right]. \tag{5.45}$$

The last term in (5.43) is the thermodynamic potential of the system of classical particles (ions) the interaction of which may be described by some effective potential, i.e.

$$H^* = \sum_\nu \frac{P_\nu^2}{2M} + U^*(\boldsymbol{R}_1, \dots, \boldsymbol{R}_N)\,, \tag{5.46}$$

where

$$U^*(\boldsymbol{R}_1, \dots, \boldsymbol{R}_N) = U_2(\boldsymbol{R}_1, \dots, \boldsymbol{R}_N) + U_3(\boldsymbol{R}_1, \dots, \boldsymbol{R}_N) + \dots, \tag{5.47}$$

$$U_2(\boldsymbol{R}_1, \dots, \boldsymbol{R}_N) = \frac{N}{2v_0} \sum_q \left(|\varrho(\boldsymbol{q})|^2 - \frac{1}{N}\right)\left[\frac{4\pi z^2 e^2}{q^2} - \frac{v_0^2 \pi(\boldsymbol{q})\,|w_0(q)|^2}{\varepsilon(\boldsymbol{q})}\right]$$
$$\equiv \frac{1}{2} \sum_{\nu\nu'} \Phi_2^{(2)}(\boldsymbol{R}_v - \boldsymbol{R})_{\nu'}, \tag{5.48}$$

$$U_n(\boldsymbol{R}_1, \dots, \boldsymbol{R}_N) \doteq \Omega_n(\boldsymbol{R}_1, \dots, \boldsymbol{R}_N), \qquad n > 2. \tag{5.49}$$

Expression (5.47) shows the way in which electrons influence the indirect ion interaction. As may be seen from (5.48) the term U_2 describes the central interaction between pairs of ions which consists of the direct Coulomb repulsion and the indirect ion-electron-ion interaction. The terms U_n with $n > 2$ describe indirect interactions between ions which cannot be reduced to pure pair ones. The latter fact is of principal importance as the traditional theory of metals [17] considers intrinsically the presence of the pair central interaction between ions which results in the additivity of forces. As it was first shown in [13, 14, 18] there are interactions connecting three and more particles in addition to the pair forces in solid metals. Therefore inter-ion interactions are not additive and there appear specific non-pair forces of covalent type in metals. The analysis shows an analogous phenomenon in liquid metals. The consequences of the existence of nonadditive many-particle forces in solid metals are well-known and are obtained by standard methods [17] by analyzing the dynamic matrix of the system. For instance, calculations of dynamic modula of elasticity show that the contribution of many-ion forces to the longitudinal sound velocity is of the same order as the contribution resulting from the indirect pair interaction. The fact of nonvalidity of Cauchy relations for elastic modula which was experimentally established may also be explained by many-ion interaction, leading to non-pair forces [19]. Finally let us point out the direct experimental evidence of non-pair inter-ion interaction in the aluminium phonon spectrum [20]. The analysis of the many particle interaction in solid metals is facilitated by the crystal symmetry and is usually carried out in momentum representation without the immediate use of an inter-particle interaction potential in coordinate space. In the case of liquid metals any physical property is a functional of the inner-particle interaction potential that is why the analysis of the inter-ion interaction's influence on any characteristic may be carried out only determining the potential in coordinate space. Let us use the formulae (5.47) up to (5.49). The effective potential of interaction between the pairs of ions in second PP approximation according to (5.48) is

$$\Phi_2^{(2)}(R) = \frac{1}{2\pi^2 R} \int_0^\infty \sin(qR)\left[\frac{4\pi z^2 e^2}{q^2} - \frac{v_0^2 \pi(q)\,|w_0(q)|^2}{\varepsilon(q)}\right] q\,\mathrm{d}q. \tag{5.50}$$

However expression (5.50) does not account for all contributions to the effective binary interaction. Indeed, as is seen from U_n (5.49) the term of order n with respect to PP contains contributions coming not only from the interaction of n, but also from a smaller number of ions as it also takes into account the possibility of repeated scattering by the same ion. In particular, Ω_n $(n > 2)$ contains also indirect pair interactions as a result of $(n-1)$-fold scattering by one ion and one-fold scattering by another. Therefore the effective binary interaction potential may be described in general as a series in powers of PP:

$$\Phi_2(R) = \Phi_2^{(2)}(R) + \Phi_2^{(3)}(R) + \dots, \tag{5.51}$$

where the upper index shows the order of PP. Similarly if the potential $\Phi_n(\boldsymbol{R}_1, \dots, \boldsymbol{R}_n)$ of the irreducible n-particle interaction is introduced it can also be described as a series of type (5.51), the first term of which will be of nth-order of PP. The role of terms of different orders of PP is not the same. In the case of solid metals members of the 3L order appear to be singled out, while calculating the energy of a crystal structure. It is connected with a considerable compensation between the ion lattice and the part of the electron energy of second order in PP. The terms of higher order are as a rule weak [22]. Since densities of solid and liquid metals do not much differ (at least near melting temperatures) it may be expected that for liquid metals their thermodynamic properties are determined mainly by considering terms up to the 3rd order of the PP. Accordingly in (5.47) only the terms written explicitly remain where U_2 is determined from (5.48), and U_3 may be represented as a sum of homogenous, two-particle and three-particle terms:

$$U_3 = U_3^{(0)} + U_3^{(2)} + U_3^{(3)}, \tag{5.52}$$

where

$$U_3^{(0)} = \frac{v_0}{3N} \sum_{\boldsymbol{k}_1\boldsymbol{k}_2} \Lambda_3(\boldsymbol{k}_1, \boldsymbol{k}_2, -\boldsymbol{k}_1 - \boldsymbol{k}_2)\, w(\boldsymbol{k}_1)\, w(\boldsymbol{k}_2)\, w(-\boldsymbol{k}_1 - \boldsymbol{k}_2), \tag{5.53}$$

$$U_3^{(2)} = \frac{1}{2!} \sum_{\substack{\nu\nu' \\ (\nu \neq \nu')}} \Phi_2^{(3)}(|\boldsymbol{R}_\nu - \boldsymbol{R}_{\nu'}|), \tag{5.54}$$

$$U_3^{(3)} = \frac{1}{3!} \sum_{\substack{\nu_1\nu_2\nu_3 \\ (\nu_1 \neq \nu_2 \neq \nu_3)}} \Phi_3^{(3)}(\boldsymbol{R}_{\nu_1}, \boldsymbol{R}_{\nu_2}, \boldsymbol{R}_{\nu_3}), \tag{5.55}$$

$$\Phi_2^{(3)}(R) = \frac{2!}{3} \frac{v_0}{N^2} \sum_{\boldsymbol{k}_1\boldsymbol{k}_2} \Lambda_3(\boldsymbol{k}_1, \boldsymbol{k}_2, -\boldsymbol{k}_1 - \boldsymbol{k}_2)\, w(\boldsymbol{k}_1)\, w(\boldsymbol{k}_2) \cdot \\ \cdot w(-\boldsymbol{k}_1 - \boldsymbol{k}_2) \left[e^{-i\boldsymbol{k}_1\boldsymbol{R}} + e^{-i\boldsymbol{k}_2\boldsymbol{R}} + e^{i(\boldsymbol{k}_1+\boldsymbol{k}_2)\boldsymbol{R}}\right], \tag{5.56}$$

$$\Phi_3^{(3)}(\boldsymbol{R}_1, \boldsymbol{R}_2, \boldsymbol{R}_3) = \frac{3!}{3} \frac{v_0}{N^2} \sum_{\boldsymbol{k}_1\boldsymbol{k}_2} \Lambda_3(\boldsymbol{k}_1, \boldsymbol{k}_2, -\boldsymbol{k}_1 - \boldsymbol{k}_2) \cdot \\ \cdot w(\boldsymbol{k}_1)\, w(\boldsymbol{k}_2)\, w(-\boldsymbol{k}_1 - \boldsymbol{k}_2)\, e^{-i\boldsymbol{k}_1\boldsymbol{R}_1 - i\boldsymbol{k}_2\boldsymbol{R}_2 + i(\boldsymbol{k}_1+\boldsymbol{k}_2)\boldsymbol{R}_3} \tag{5.57}$$

As it was shown above the effective binary interaction potential up to the 3rd order in the PP is a sum

$$\Phi_2(R) = \Phi_2^{(2)}(R) + \Phi_2^{(3)}(R). \tag{5.58}$$

Let us give the results of the calculation of some terms in (5.58). The calculation is based on the two-parameter model PP of Cohen (5.6) with parameters taken from [9]. As valence z the so called effective valence was used. It was introduced in the following way. Since wave functions of conduction electrons are orthogonal to wave functions of inner shell electrons, the conduction electrons do not enter the volume occupied by ions. This results in an effective raise of the electron density, and consequently in an effective increase in the ion charge. If the ion volume is equal to v_i the effective values of electron density n_e^* and valence z^* may be evaluated by the formulae:

$$n_e^* = z^*/v_0 , \qquad z^* = z/(1 - v_i/v_0) .$$

The dielectric screening function $\varepsilon(q)$ was chosen in random phase approximation with the Hubbard specification [23]. Figure 5.1 shows the results of the calculation

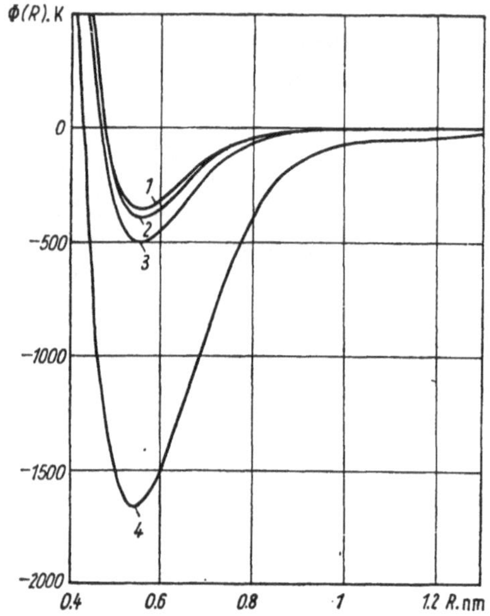

Fig. 5.1. Potential of effective inter-ion interaction in liquid Li

of $\Phi_2^{(2)}(R)$ for a typical alkaline metal — Rubidium. Curve 1 shows a pair ion-ion potential in an equilibrium lattice of a crystal at 0 K; curve 2 shows it in a liquid metal at melting temperature; curve 3 shows it at 1000 K at the saturation line from the liquid side and curve 4 shows it in the liquid metal at temperature and density corresponding to the critical point liquid-gas. The characteristic feature of the acquired potential is its strong dependence on density: the increase in temperature and the decrease in density cause an increase in the depth of the potential well, and around the critical point it increases by a factor four. The effective diameter of the ion becomes somewhat smaller but the form of the potential curve and the position of the minimum remains almost unchanged. A similar behaviour is characteristic of other alkaline metals as well.

We show also the results of $\Phi_2(R)$ calculations with terms of the 3rd order of PP taken into account [24]. Figure 5.2 shows the data for liquid sodium at melting temperature. The calculation is done with the Gursky-Krasko PP (5.7) with the same $\varepsilon(q)$. Curves 1, 2 and 3 on figure 5.2 present $\Phi_2^{(3)}$, $\Phi_2^{(2)}$ and Φ_2 correspondingly. As is seen from the picture, inclusion of terms of the 3rd order of PP leads to a decrease of the effective diameter of collision and to an almost 35% increase in the depth of the potential well. The ion-ion potential becomes more long-ranged and the position of the potential curve minimum remains practically unchanged. Such a considerable change of the potential well depth does not mean however that the inclusion of terms of the 3rd order of PP will lead to a considerable change in the thermodynamic charac-

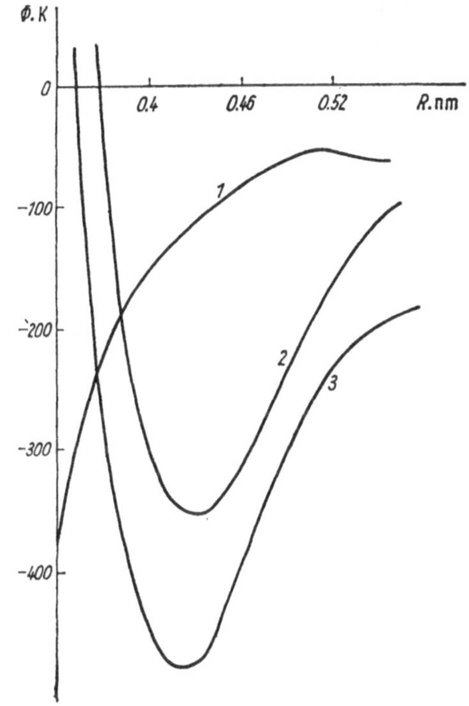

Fig. 5.2. Potential of effective inter-ion interaction in liquid Na at the melting temperature: 1 — $\Phi_2^{(3)}$; 2 — $\Phi_2^{(2)}$; 3 — $\Phi_2^{(2)} + \Phi_2^{(3)}$

seristics of liquid metals (e.g. the energy of cohesion). The matter is that the calculation of thermodynamic characteristics must take into account the irreducible 3-particle interaction, described by the potential $\Phi_3^{(3)}$, which, as calculations for solid metals thow is of the same order as $\Phi_2^{(3)}$ and compensates the $\Phi_2^{(3)}$ contribution to a high degree. In calculations for solid metals it is not necessary to single out pair and non-pair interactions from terms of a certain order in PP, but it is possible to determine their conjoint contribution at once. For liquids we cannot avoid such a separation since the presence of pair and non-pair interactions requires to use many-particle distribution functions of various order which cannot be brought down to one another. At present the joint influence of pair and non-pair interactions of the third order in PP on thermodynamic characteristics of liquid metals was not yet investigated.

Let us discuss separately the question of inter-ion interaction in metallic hydrogen. In this case the purely Coulomb interaction of electrons and protons plays the role of a PP. As follows from calculations of the electron energy [22], the electron-proton structure in the density interval of metallic hydrogen shows a vivid tendency to form highly anisotropic structures. This result is achieved if the electron-ion interaction is taken into consideration up to the terms of the second order. However, because of oscillatory parts of the lattice energy, the cohesion energy (counted from the energy of isolated atoms) becomes close to zero or even positive, which must cause an instability of the lattice (which disintegrates into atoms). Thus, inclusion of terms of the third order in the electron-ion interaction is a matter of principle. Physically it is caused by a stronger interaction of electrons and ions at small distances in metallic hydrogen as compared with usual metals, which in turn is connected with the absence of the ion frame. In this case the spatial inhomogenity of the electron density is so important that more than the 2nd order perturbation must be taken into account, which leads to the increase of cohesion. These characteristics are corroborated of the Φ_2 investigation [25]. The dielectric permeability was chosen in the from given in [26]. Figure 5.3 shows the results for different contributions in $\Phi(R)$ with a density

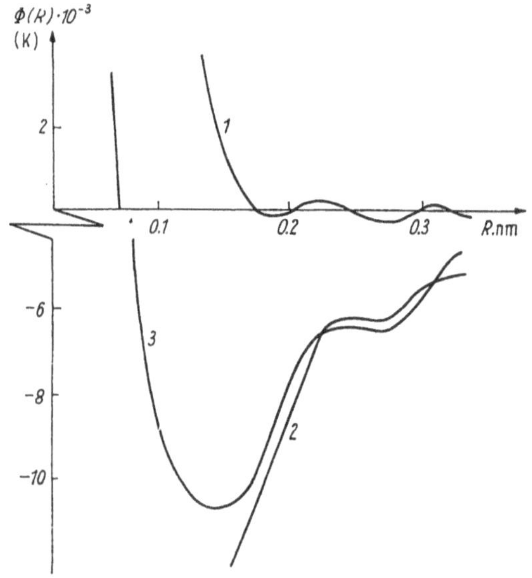

Fig. 5.3. Potential of the effective inter-ion interaction in metallic hydrogen ($r_s = 1.45$)

corresponding to $r_s = 1.45$ (r_s is the standard designation for a sphere radius corresponding to the atomic volume, expressed in Bohr radii). Inclusion of only second order terms (curve 1) does not lead to the formation of a potential well. The contribution of the 3rd order term, $\Phi_2^{(3)}(R)$ (curve 2), is decisive for the determination of $\Phi_2(R)$ and leads to the appearence of a deep minimum on the resulting curve 3.

We show as well data of the analysis of the $\Phi_2(R)$ shape depending on pressure in the interval r_s from 1.05 to 1.6. (As compared to the equation of the metallic hydrogen state [27], the given interval corresponds to pressure from zero to 10^6 MPa). Curves 1, 2, 3 in figure 5.4 refer to $r_s = 1.6$; 1.25; 1.05 respectively. As follows from the graphs a change in the pressure leads to a change in the depth of the potential well. The decrease of the minimum depth corresponds to the growth of the repulsive contribution to the energy of the direct ion-ion interaction shown in [22]. Further, in contrast to simple liquid metals, curve $\Phi_2(R)$ shows several minima. Their presence shows the change of the metallic hydrogen structure which is accompanied by the reconstruction of short-range order and the change of coordination numbers; to a certain degree it reflects the presence of different structures and a possibility of phase transitions in the crystal phase. Comparison of figures 5.3 and 5.4 shows that a change of pressure leads to the alternation of more symmetric and less symmetric structures, as is the case for a solid phase too.

Thus, if $\Phi_2(R)$ is available one can calculate all thermodynamic properties of liquid metals [28, 29], which, as a whole, show a satisfactory agreement with experimental data.

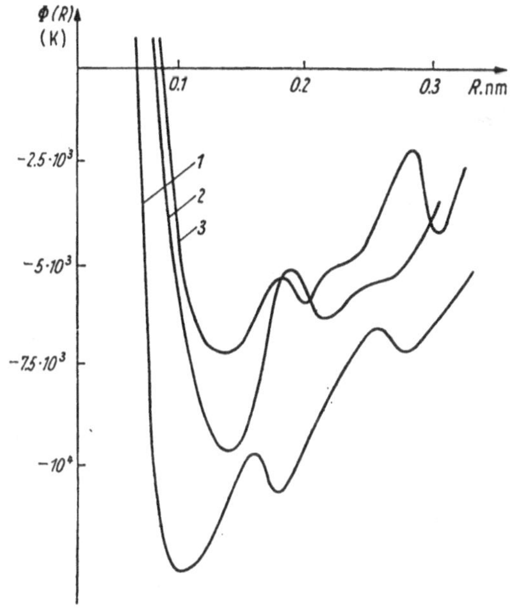

Fig. 5.4. Dependence of the potential of effective inter-ion interaction in metallic hydrogen on the density. 1 — $r_s = 1.6$; 2 — $r_s = 1.25$; 3 — $r_s = 1.05$

5.4. Transport Processes in Liquid Metals. Diffraction Model

The idea of an effective inter-ion interaction is very fruitful as with its help and the use of many-particle distribution functions one may in principle calculate any equilibrium property of metals. The smallness of the electron-ion interaction and the possibility of applying the perturbation theory makes it possible to construct a theory of kinetic phenomena in which the main role plays the electron transport and their scattering by ions (electrical conductivity, heat conductivity, thermopower, electromigration, etc.). Using the method of Green functions, in the transport theory it is possible to get not only expressions for kinetic coefficients in Born (quadratic in PP) approximation of the perturbation theory, but also to go beyond this approximation by taking into account terms of higher order of the PP. This part of the article is devoted to the theory of electron transport phenomena, based on the method of equations of motion for Green functions. There is also a separate analysis (by the method of diagrams) of the problem of electric conductivity in connection with the investigation of the problem of the applicability of Born approximation.

According to the PP theory, metals may be considered as two-component systems consisting of ions and quasi-free electrons. Therefore the connection between diffusion fluxes and thermodynamic forces in metals is determined by the usual phenomenological equations for binary liquids. For stationary processes the connection is as follows [30, 31] (see also sect. 3.2.):

$$\frac{1}{V}\boldsymbol{J}_n^{\rm d} = \sum_{m=0}^{1} L_{mn}(e_m\boldsymbol{E} - \nabla\mu_m) - L_{nq}\frac{\nabla T}{T}\,,$$
$$\frac{1}{V}\boldsymbol{J}_q = \sum_{m=0}^{1} L_{qm}(e_m\boldsymbol{E} - \nabla\mu_m) - L_{qq}\frac{\nabla T}{T}\,. \tag{5.59}$$

Here index $m = 0, 1$ marks the electron and ion subsystem, i.e. $\boldsymbol{J}_0^{\rm d} = \boldsymbol{J}_e^{\rm d}$, $e_0 = e$, $\mu_0 = \mu_{\rm e}$ and $m_0 = m$ represent correspondingly diffusion flux, charge, chemical potential and electron mass; and $\boldsymbol{J}_1^{\rm d} = \boldsymbol{J}_i^{\rm d}$, $e_1 = -ze$, $\mu_1 = \mu_i$ and $m_1 = M$ represent diffusion flux, charge, chemical potential and ion mass; $\boldsymbol{E}$ is the intensity of electric field, $\boldsymbol{J}_q$ is the heat current. Kinetic coefficients occuring in (5.59) may be found with the help of linear response theory and with the method of the nonequilibrium statistical operator [31]:

$$L_{mn} = \frac{1}{V}\int_{-\infty}^{\infty} {\rm e}^{-\varepsilon t}\left\langle\!\left\langle \int J_{mx}^{\rm d}(\boldsymbol{r}, t)\,{\rm d}^3r; \int B_{nx}(\boldsymbol{r}', 0)\,{\rm d}^3r'\right\rangle\!\right\rangle {\rm d}t\,, \tag{5.60}$$

$$L_{mq} = \frac{1}{V}\int_{-\infty}^{\infty} {\rm e}^{-\varepsilon t}\left\langle\!\left\langle \int J_{mx}^{\rm d}(\boldsymbol{r}, t)\int {\rm d}^3r; \int B_{qx}(\boldsymbol{r}', 0)\,{\rm d}^3r'\right\rangle\!\right\rangle {\rm d}t\,, \tag{5.61}$$

$$L_{qq} = \frac{1}{V}\int_{-\infty}^{\infty} {\rm e}^{-\varepsilon t}\int\left\langle\!\left\langle J_{qx}^{\rm d}(\boldsymbol{r}, t)\,{\rm d}^3r; \int B_{qx}(\boldsymbol{r}', 0)\,{\rm d}^3r'\right\rangle\!\right\rangle {\rm d}t\,, \tag{5.62}$$

where

$$B_{nx}(\boldsymbol{r}, t) = \int_{-\infty}^{t} J_{nx}^{\rm d}(\boldsymbol{r}, t')\,{\rm d}t'\,,$$
$$B_{qx}(\boldsymbol{r}, t) = \int_{-\infty}^{t} J_{qx}(\boldsymbol{r}, t')\,{\rm d}t'\,. \tag{5.63}$$

Here $J_{nx}^{\rm d}(\boldsymbol{r}, t)$ and $J_{qx}(\boldsymbol{r}, t)$ are projections of density operators of the diffusion flux of particles and the density of heat current on the x axis, double angular brackets denote the retarted Green's function

$$\langle\!\langle A(t); B(t')\rangle\!\rangle = \frac{\rm i}{\hbar}\,\theta(t - t')\,\langle[A(t), B(t')]\rangle\,, \tag{5.64}$$

where $\theta(t)$ is the Heaviside function, square brackets designate the commutator, angular brackets mean quantum-mechanical and statistical averages. As diffusion fluxes are connected by the equation

$$\sum_{n=0}^{1} m_n \boldsymbol{J}_n^{\rm d} = 0\,,$$

the kinetic coefficients L_{nm} and L_{nq} satisfy the equations

$$\sum_{n=0}^{1} m_n L_{nq} = 0\,; \qquad \sum_{n=0}^{1} m_n L_{nm} = 0\,. \tag{5.65}$$

Taking into account these correlations and introducing the total diffusion flux of the electric charge

$$\boldsymbol{I} = \sum_{n=0}^{1} e_n \boldsymbol{J}_n^d\,, \tag{5.66}$$

equations (5.59) may be written in a more usual notation (cf. 3.2.):

$$\frac{1}{V} \boldsymbol{I} = \sigma[(\boldsymbol{E} - \nabla\mu) - Q\nabla T] \,,$$
$$\frac{1}{V} \boldsymbol{J}_q = -\lambda\nabla T + \frac{1}{V} QT\boldsymbol{I} \,, \tag{5.67}$$

where

$$\sigma = L_{00}e^2\left(1 + \frac{zm}{M}\right)^2 , \tag{5.68}$$

$$\lambda = \frac{1}{T} L_{qq} - \frac{1}{TL_{00}} L_{0q}^2 \,, \tag{5.69}$$

$$Q = \frac{1}{T} \frac{L_{0q}}{L_{00}\left(1 + \dfrac{zm}{M}\right)e} \,, \tag{5.70}$$

are the coefficients of electric conductivity, heat conductivity and differential thermopower, respectively, and

$$\mu = \left(\mu_e - \frac{m}{M}\mu_i\right)e + \left(\frac{zme}{M}\right)^{-1} .$$

Thus the problem of defining kinetic coefficients is reduced to the determination of the corresponding Green functions. The largest difficulty here is to take into account the electron-electron interaction (see chap. 3.). To overcome this difficulty it is possible to use the so called diffraction model in which the electron-electron interaction is not counted in the Hamiltonian, but a screened PP is used instead of the initially unscreened one. Simultaneously the ion-ion interaction must be determined in such a way that the thermodynamic potential of the system in the diffraction model Hamiltonian, should coincide with the thermodynamic potential determined with the Hamiltonian (5.12) in every order of the perturbation theory. This program is rather easily realized in Born approximation, but gives large difficulties when higher order approximations of the perturbation theory are included. At the same time it is highly universal as it permits to obtain all kinetic coefficients of metals simultaneously. This part of the article shows how the diffraction model may help to determine transport coefficients up to the 2nd order of the PP and also how to go beyond the Born approximation by an accurate calculation of electron-ion interaction.

5.5. Kinetic Equations for Electron Green's Functions and their Solution

Kinetic coefficients are determined by expressions (5.60) up to (5.62) which comprise operators of heat and particle fluxes. These operators are determined on the basis of the equations of balance of particle numbers and energy [31], each of them contains electron and ion contributions. As the electron mobility is several orders higher than the ion mobility, the contributions of the ion subsystem to the transport and diffusion flux may be neglected. If the current operator is taken in zeroth order of the PP, the following result may be obtained:

$$J_e^d(\boldsymbol{r}) = J_e(\boldsymbol{r}) = \frac{1}{V} \sum_{\boldsymbol{k}_1 \boldsymbol{k}_2 \sigma} \frac{\hbar\,(\boldsymbol{k}_1 + \boldsymbol{k}_2)}{2m}\, a^+_{\boldsymbol{k}_1\sigma} a_{\boldsymbol{k}_2\sigma}\, \mathrm{e}^{-\mathrm{i}(\boldsymbol{k}_1 - \boldsymbol{k}_2)\boldsymbol{r}}\,, \tag{5.71}$$

$$J_q(\boldsymbol{r}) = \frac{1}{V} \sum_{\boldsymbol{k}_1 \boldsymbol{k}_2 \sigma} \left[\left(\frac{\hbar \boldsymbol{k}_2^2}{2m} - \mu_e\right) \frac{\hbar \boldsymbol{k}_1}{m} + \left(\frac{\hbar \boldsymbol{k}_1^2}{2m} - \mu_e\right) \frac{\hbar \boldsymbol{k}_2}{m}\right] \cdot$$

$$\cdot\, a^+_{\boldsymbol{k}_1\sigma} a_{\boldsymbol{k}_2\sigma} \exp\left[-\mathrm{i}(\boldsymbol{k}_1 - \boldsymbol{k}_2)\, \boldsymbol{r}\right]. \tag{5.72}$$

Thus in accordance with (5.60—61) the problem of finding kinetic coefficients is reduced to the evaluation of Green functions of the type

$$\langle\langle a^+_{\boldsymbol{k}\sigma}(t)\, a_{\boldsymbol{k}'\sigma}(t);\, B(0)\rangle\rangle \equiv \langle\langle a^+_{\boldsymbol{k}\sigma} a_{\boldsymbol{k}'\sigma};\, B\rangle\rangle = \frac{1}{2\pi} \int_{-\infty}^{\infty} \langle\langle a^{+\prime}_{\boldsymbol{k}\sigma} a_{\boldsymbol{k}'\sigma};\, B\rangle\rangle_\omega\, \mathrm{e}^{\mathrm{i}\omega t}\, \mathrm{d}\omega \tag{5.73}$$

with corresponding expressions for the operator B. Using the current operators (5.71) and formulae (5.60) up to (5.62) we get the following expressions for phenomenological coefficients in the stationary case:

$$\begin{pmatrix} L_{00} \\ L_{0q} \end{pmatrix} = \frac{1}{V} \sum_{k\sigma} \frac{\hbar k_z}{m} \left\langle\left\langle a^+_{\boldsymbol{k}\sigma} a_{\boldsymbol{k}\sigma}\,;\, \begin{pmatrix} B_{z0} \\ B_{qz} \end{pmatrix}\right\rangle\right\rangle_0, \tag{5.74}$$

$$L_{qq} = \frac{1}{V} \sum_{k\sigma} \left[E(k) - \mu_e\right] \frac{\hbar k_z}{m} \langle\langle a^+_{\boldsymbol{k}\sigma} a_{\boldsymbol{k}\sigma};\, B_{qz}\rangle\rangle_0\,,$$

$$\begin{pmatrix} B_{0z} \\ B_{qz} \end{pmatrix} = \sum_{k\sigma} \frac{\hbar k_z}{m} \begin{pmatrix} 1 \\ E(k) - \mu_e \end{pmatrix} \int_{-\infty}^{0} a^+_{\boldsymbol{k}\sigma}(t)\, a_{\boldsymbol{k}\sigma}(t)\, \mathrm{d}t\,. \tag{5.75}$$

The equations of motion for the Green's functions (5.73) are as follows [32]:

$$\mathrm{i}\hbar \frac{\partial}{\partial t} \langle\langle a^+_{\boldsymbol{k}\sigma} a_{\boldsymbol{k}\sigma};\, B\rangle\rangle = -\,\delta(t)\, \langle[a^+_{\boldsymbol{k}\sigma} a_{\boldsymbol{k}\sigma}, B]\rangle -$$

$$- \sum_{\boldsymbol{k}_1} [w(\boldsymbol{k}_1 - \boldsymbol{k})\, \langle\langle \varrho(\boldsymbol{k}_1 - \boldsymbol{k})\, a^+_{\boldsymbol{k}_1\sigma} a_{\boldsymbol{k}\sigma};\, B\rangle\rangle$$

$$- w(\boldsymbol{k} - \boldsymbol{k}_1)\, \langle\langle \varrho(\boldsymbol{k} - \boldsymbol{k}_1)\, a^+_{\boldsymbol{k}\sigma} a_{\boldsymbol{k}_1\sigma};\, B\rangle\rangle]\,.$$

$$\mathrm{i}\hbar \frac{\partial}{\partial t} \langle\langle \varrho^{(n)}(\boldsymbol{k}_1 - \boldsymbol{k})\, a^+_{\boldsymbol{k}\sigma} a_{\boldsymbol{k}\sigma};\, B\rangle\rangle = -\,\delta(t)\, \langle[\varrho^{(n)}(\boldsymbol{k}_1 - \boldsymbol{k})\, a^+_{\boldsymbol{k}_1\sigma} a_{\boldsymbol{k}\sigma}, B]\rangle$$

$$- [E(k_1) - E(k)]\, \langle\langle \varrho^{(n)}(\boldsymbol{k}_1 - \boldsymbol{k})\, a^+_{\boldsymbol{k}_1\sigma} a_{\boldsymbol{k}\sigma};\, B\rangle\rangle$$

$$- \sum_{\boldsymbol{k}_2} [w(\boldsymbol{k}_2 - \boldsymbol{k}_1)$$

$$\cdot \langle\langle \varrho^{(n)}(\boldsymbol{k}_1 - \boldsymbol{k})\, \varrho(\boldsymbol{k}_2 - \boldsymbol{k}_1)\, a^+_{\boldsymbol{k}_2\sigma} a_{\boldsymbol{k}\sigma};\, B\rangle\rangle$$

$$- w(\boldsymbol{k} - \boldsymbol{k}_2)\, \langle\langle \varrho^{(n)}(\boldsymbol{k}_1 - \boldsymbol{k}) \varrho \cdot (\boldsymbol{k} - \boldsymbol{k}_2)$$

$$a^+_{\boldsymbol{k}_1\sigma} a_{\boldsymbol{k}_2\sigma};\, B\rangle\rangle]$$

$$+ \mathrm{i}\hbar \langle\langle \varrho^{(n+1)}(\boldsymbol{k}_1 - \boldsymbol{k})\, a^+_{\boldsymbol{k}_1\sigma} a_{\boldsymbol{k}\sigma};\, B\rangle\rangle\,,$$

$$n = 0, 1, 2, \ldots \tag{5.76}$$

Here $\varrho^{(n)}(\boldsymbol{k})$ is the nth order derivative of $\varrho(\boldsymbol{k})$ with respect to the time, $\boldsymbol{E}(k) = \frac{\hbar^2 k^2}{2m}$.

Let us turn in equations (5.74) and (5.75) to the Fourier representation according to (5.73) and then with the help of equation (5.75) express the function $\langle\langle \varrho a^+ a; B\rangle\rangle_\omega$ through functions $\langle\langle \varrho^{(n)} \varrho a^+ a; B\rangle\rangle_\omega$. Putting the expressions obtained into equation (5.74) we get:

$$
\begin{aligned}
[\hbar\omega + \mathrm{i}\delta]\, \langle\langle a^+_{\boldsymbol{k}\sigma} a_{\boldsymbol{k}\sigma}; B\rangle\rangle_\omega = &- \mathrm{i}\langle [a^+_{\boldsymbol{k}\sigma} a_{\boldsymbol{k}\sigma}, B]\rangle + \\
&+ \sum_{n=0}^{\infty} (\mathrm{i}\hbar)^n \sum_{\boldsymbol{k}_1} \Bigg\{ w(\boldsymbol{k}_1 - \boldsymbol{k}) \frac{\langle [\varrho^{(n)}(\boldsymbol{k}_1 - \boldsymbol{k})\, a^+_{\boldsymbol{k}_1\sigma} a_{\boldsymbol{k}\sigma}, B]\rangle}{[E(k_1) - E(k) + \hbar\omega + \mathrm{i}\delta]^{n+1}} \\
&- w(\boldsymbol{k} - \boldsymbol{k}_1) \frac{\langle [\varrho^{(n)}(\boldsymbol{k} - \boldsymbol{k}_1)\, a^+_{\boldsymbol{k}\sigma} a_{\boldsymbol{k}_1\sigma}, B]\rangle}{[E(k) - E(k_1) + \hbar\omega + \mathrm{i}\delta]^{n+1}} \Bigg\} + \sum_{n=0}^{\infty} (\mathrm{i}\hbar)^n \sum_{\boldsymbol{k}_1 \boldsymbol{k}_2} \\
&\cdot \Bigg\{ w(\boldsymbol{k}_2 - \boldsymbol{k}_1)\, w(\boldsymbol{k}_1 - \boldsymbol{k}) \frac{\langle\langle \varrho^{(n)}(\boldsymbol{k}_1 - \boldsymbol{k})\, \varrho(\boldsymbol{k}_2 - \boldsymbol{k}_1)\, a^+_{\boldsymbol{k}_2\sigma} a_{\boldsymbol{k}\sigma}; B\rangle\rangle_\omega}{[E(k_1) - E(k) + \hbar\omega + \mathrm{i}\delta]^{n+1}} \\
&+ w(\boldsymbol{k} - \boldsymbol{k}_1)\, w(\boldsymbol{k}_1 - \boldsymbol{k}_2) \frac{\langle\langle \varrho^{(n)}(\boldsymbol{k} - \boldsymbol{k}_1)\, \varrho(\boldsymbol{k}_1 - \boldsymbol{k}_2)\, a^+_{\boldsymbol{k}\sigma} a_{\boldsymbol{k}_2\sigma}; B\rangle\rangle_\omega}{[E(k) - E(k_1) + \hbar\omega + \mathrm{i}\delta]^{n+1}} \\
&- w(\boldsymbol{k} - \boldsymbol{k}_2)\, w(\boldsymbol{k}_1 - \boldsymbol{k}) \frac{\langle\langle \varrho^{(n)}(\boldsymbol{k}_1 - \boldsymbol{k})\, \varrho(\boldsymbol{k} - \boldsymbol{k}_2)\, a^+_{\boldsymbol{k}_1\sigma} a_{\boldsymbol{k}_2\sigma}; B\rangle\rangle_\omega}{[E(k_1) - E(k) + \hbar\omega + \mathrm{i}\delta]^{n+1}} \\
&- w(\boldsymbol{k} - \boldsymbol{k}_1)\, w(\boldsymbol{k}_2 - \boldsymbol{k}) \frac{\langle\langle \varrho^{(n)}(\boldsymbol{k} - \boldsymbol{k}_1)\, \varrho(\boldsymbol{k}_2 - \boldsymbol{k})\, a^+_{\boldsymbol{k}_2\sigma} a_{\boldsymbol{k}_1\sigma}; B\rangle\rangle_\omega}{[E(k) - E(k_1) + \hbar\omega + \mathrm{i}\delta]^{n+1}} \Bigg\}.
\end{aligned}
\tag{5.77}
$$

This equation cannot be solved immediately because of the presence of Green's functions of higher order. To get a closed equation let us make the following approximation:

$$
\begin{aligned}
\langle [\varrho^{(n)}(\boldsymbol{k} - \boldsymbol{k}_1)\, a^+_{\boldsymbol{k}\sigma} a_{\boldsymbol{k}_1\sigma}, B]\rangle \cong{}& \langle \varrho^{(n)}(\boldsymbol{k} - \boldsymbol{k}_1)\rangle \langle [a^+_{\boldsymbol{k}\sigma} a_{\boldsymbol{k}_1\sigma}, B]\rangle + \\
&+ \langle a^+_{\boldsymbol{k}\sigma} a_{\boldsymbol{k}_1\sigma}\rangle \langle [\varrho^{(n)}(\boldsymbol{k} - \boldsymbol{k}_1), B]\rangle ,
\end{aligned}
\tag{5.78}
$$

$$
\begin{aligned}
\langle\langle \varrho^{(n)}(\boldsymbol{k} - \boldsymbol{k}_1)\, \varrho(\boldsymbol{k}_1 - \boldsymbol{k}_2)\, a^+_{\boldsymbol{k}\sigma} a_{\boldsymbol{k}_2\sigma}; B\rangle\rangle_\omega \cong{}& \langle \varrho^{(n)}(\boldsymbol{k} - \boldsymbol{k}_1)\, \varrho(\boldsymbol{k}_1 - \boldsymbol{k}_2)\rangle \cdot \\
&\cdot \langle\langle a^+_{\boldsymbol{k}\sigma} a_{\boldsymbol{k}_2\sigma}; B\rangle\rangle_\omega + \langle a^+_{\boldsymbol{k}\sigma} a_{\boldsymbol{k}_2\sigma}\rangle \cdot \\
&\cdot \langle\langle \varrho^{(n)}(\boldsymbol{k} - \boldsymbol{k}_1)\, \varrho(\boldsymbol{k}_1 - \boldsymbol{k}_2); B\rangle\rangle_\omega .
\end{aligned}
\tag{5.79}
$$

If we restrict us to the second order with respect to the PP, the formal possibility of the approximation (5.79) follows from the fact that it is made in those terms of equation (5.77) which contain squares of PP as a factor. Owing to this we may omit in the right hand side of (5.79) the second term which is nonzero only in the first order of the PP. Approximation (5.78) is also permissible as it refers to terms which are free terms in equation (5.77) with respect to the electron function and here it is sufficient to take into account terms of zeroth order in the PP. Furtheron restricting ourselves to the static case, we obtain instead of (5.77):

$$
\begin{aligned}
&\frac{2\pi \mathrm{i}}{N\hbar} \sum_{\boldsymbol{k}_1} |w(\boldsymbol{k} - \boldsymbol{k}_1)|^2\, S\left(\boldsymbol{k} - \boldsymbol{k}_1; \frac{E(k) - E(k_1)}{\hbar}\right) \cdot \\
&\cdot \{\langle\langle a^+_{\boldsymbol{k}\sigma} a_{\boldsymbol{k}\sigma}; B\rangle\rangle_0 - \langle\langle a^+_{\boldsymbol{k}_1\sigma} a_{\boldsymbol{k}_1\sigma}; B\rangle\rangle_0\} = - \langle [a^+_{\boldsymbol{k}\sigma} a_{\boldsymbol{k}\sigma}, B]\rangle .
\end{aligned}
\tag{5.80}
$$

Here

$$S(\boldsymbol{k}, \omega) = \frac{N}{2\pi} \int_{-\infty}^{\infty} \langle \varrho(\boldsymbol{k}, t)\, \varrho(-\boldsymbol{k}, 0)\rangle\, e^{i\omega t}\, dt \tag{5.81}$$

is the dynamic structure factor of the liquid metal (Van-Hove's correlation function) Thus the integral equations for the Green's functions are of the type of Boltzmann's equation. The kernels of these equations depend on the ion dynamic structure factor which is determined by all possible ion collective motions, and by the PP electron-ion interaction.

The equations (5.80) for L_{00} on the one hand and L_{0q}, L_{qq} on the other hand differ only by a free term and are solved by expanding $w^2 S(\boldsymbol{k}, \omega)$ into Legendre polynomial series and the Green's functions — into spherical harmonics:

$$\left\langle\left\langle a^+_{\boldsymbol{k}\sigma} a_{\boldsymbol{k}\sigma};\begin{pmatrix} B_{0z} \\ B_{qz}\end{pmatrix}\right\rangle\right\rangle_0 = \sum_{lm} \begin{pmatrix} f_{lm}(E) \\ g_{lm}(E)\end{pmatrix} Y_{lm}(\theta, \varphi)\,. \tag{5.82}$$

Inserting this expansion into (5.80) and taking equal coefficients at similar spherical harmonics we can easily obtain equations for f_{lm} and g_{lm}. However, as it follows from (5.74) in order to obtain the desired transport coefficients we must know only the functions $f_{10}(E) \equiv f(E)$ and $g_{10}(E) \equiv g(E)$. They satisfy the following equations:

$$-i\sqrt{\frac{4\pi}{3}}\left(-\frac{\partial n}{\partial E}\right)\frac{\hbar k}{m} = \frac{v_0}{2\pi\hbar^2}\int_0^{\infty} k_1^2\, dk_1 \int_0^{\pi} d\gamma \cdot \sin\gamma\, |w(\boldsymbol{k}-\boldsymbol{k}_1)|^2 \cdot \cdot S\left(\boldsymbol{k}-\boldsymbol{k}_1; \frac{E-E_1}{\hbar}\right)[f(E) - f(E_1)\cos\gamma]\,, \tag{5.83}$$

$$-i\sqrt{\frac{4\pi}{3}}\left(-\frac{\partial n}{\partial E}\right)\frac{\hbar k}{m}(E-\mu_e) = \frac{v_0}{2\pi\hbar^2}\int_0^{\infty} k_1^2 dk_1 \int_0^{\pi} d\gamma \cdot \sin\gamma\, |w(\boldsymbol{k}-\boldsymbol{k}_1)|^2 \cdot \cdot S(\boldsymbol{k}-\boldsymbol{k}_1; \frac{E-E_1}{\hbar}\,[g(E) - g(E_1)\cos\gamma]\,, \tag{5.84}$$

where $n(E)$ is the Fermi distribution function, γ is the angle between $\boldsymbol{k}$ and $\boldsymbol{k}_1$, and $E = E(k)$, $E_1 = E(k_1)$. As the electron gas is highly degenerate at the discussed temperatures, only the electrons with energies of the order of the Fermi energy $E \approx E_F$ take part in the transport process. The electron energy loss due to interaction with the ion equals $E - E_i$. In metals this energy is of the order of $k_B T_D$, where T_D is the Debye temperature. If $E_F \gg k_B T_D$, then at high temperatures the functions $f(E_i)$ and $g(E_i)$ may be expanded into a Taylor power series near the point E. Retaining the terms of the lowest order in $k_B T_D/E_F$ we get the solution of equations (5.83), (5.84) in the form:

$$f(E) = -i\sqrt{\frac{4\pi}{3}}\left(-\frac{\partial n}{\partial E}\right)\sqrt{\frac{2E}{m}}\,\tau(E)\,, \tag{5.85}$$

$$g(E) = -i\sqrt{\frac{4\pi}{3}}\left(-\frac{\partial n}{\partial E}\right)\sqrt{\frac{2E}{m}}\,(E-\mu_e)\,\tau(E)\,, \tag{5.86}$$

where

$$\tau^{-1}(E) = \frac{v_0}{2\pi\hbar^2} \int_0^\infty k_1^2 \mathrm{d}k_1 \int_0^\pi \mathrm{d}\gamma \cdot \sin\gamma \, |w(\boldsymbol{k} - \boldsymbol{k}_1)|^2 \, S\left(\boldsymbol{k} - \boldsymbol{k}_1; \frac{E - E_1}{\hbar}\right) \cdot$$

$$\cdot (1 - \cos\gamma) . \tag{5.87}$$

Inserting now (5.82) into (5.74) and using the expressions (5.85) and (5.86) we may obtain the expressions for phenomenological coefficients thus restoring the transport coefficients (5.68) up to (5.70):

$$\sigma = \frac{e^2 z\tau}{mv_0}, \tag{5.88}$$

$$\lambda = \frac{\pi^2}{3} \frac{k_{\mathrm{B}}^2 T z\tau}{mv_0}, \tag{5.89}$$

$$Q = \frac{\pi^2 k_{\mathrm{B}}^2 T}{3eE_{\mathrm{F}}} \left[\frac{3}{2} + \frac{E_{\mathrm{F}}}{\tau} \left(\frac{\partial\tau}{\partial E}\right)_{E=E_{\mathrm{F}}}\right]. \tag{5.90}$$

(While deriving (5.88) up to (5.90) we used the well-known approximate indentities for integrals containing the Fermi distribution and used $\tau \equiv \tau(E_{\mathrm{F}})$).

The formulae (5.88) up to (5.90) are valid for both liquid and solid metals with crystalline structure of rather high symmetry. The relaxation time τ, given by formula (5.87), differs from the usually used Ziman relaxation time [33] because it contains a dynamic structure factor instead of a static one thus depending on the ion subsystem dynamics (see also sect. 3.3.3.). In the case of solid metals of cubic symmetry the dynamic structure factor takes the form [34]

$$S(\varkappa, \omega) = \frac{1}{2\pi N} \sum_{n_1 n_2} \mathrm{e}^{-\mathrm{i}\varkappa(n_1 - n_2)} \int_{-\infty}^{\infty} \mathrm{d}t \, \mathrm{e}^{\mathrm{i}\omega t} \langle \mathrm{e}^{-\mathrm{i}\varkappa U_{n_1}(t)} \mathrm{e}^{\mathrm{i}\varkappa U_{n_2}(0)} \rangle$$

where $\boldsymbol{U}_{\boldsymbol{n}}(t)$ is the ion desplacement from the equilibrium position, $\boldsymbol{n}$ is the radius vector of the ion. If $\boldsymbol{U}_{\boldsymbol{n}}(t)$ is expanded into a series with respect to normal ion lattice oscillations $S(\varkappa, \omega)$ may be presented in a series

$$S(\varkappa, \omega) = S_0(\varkappa, \omega) + S_1(\varkappa, \omega) + \dots \tag{5.91}$$

where the term $S_0(\varkappa, \omega)$ describes the elastic electron scattering on a crystal lattice, $S_1(\varkappa, \omega)$ — the inelastic one phonon — electron scattering (where the number of phonons in the system changes by unity) and so on. It may be shown for temperatures higher than the Debye temperature [35], that

$$S_0(\varkappa, \omega) = \frac{1}{N} \sum_{n_1 n_2} \mathrm{e}^{-\mathrm{i}\varkappa(n_1 - n_2)} \, \delta(\omega) ,$$

$$S_1(\varkappa, \omega) \cong \frac{k_{\mathrm{B}} T}{Ms^2} \, \delta(\omega)$$

and so on. (Here s is the sound velocity). Inserting (5.91) into (5.87) and retaining only the two first terms (one-phonon approximation) we come to a well-known expression [36]

$$\tau^{-1} = \frac{v_0 k_B T m k_F}{2\pi\hbar M s^2 \hbar^2} \int_0^\pi d\gamma \cdot \sin\gamma \, |w(2k_F \sin\gamma/2)|^2 \, (1 - \cos\gamma) \, . \tag{5.92}$$

Retaining in (5.91) the following expansion terms, we may take into account the dependence of transport coefficients on many-phonon processes.

Let us consider now the case of liquid metals. If we restrict ourselves again to "one-phonon" electron scattering, then, in order to determine $S(\varkappa, \omega)$, it is sufficient to use the hydrodynamic approximation ($\varkappa, \omega \to 0$). If we ignore sound wave attenuation, then [37]

$$S(\varkappa, \omega) = \frac{k_B T}{M s^2} \left\{ \left(\frac{c_p}{c_v} - 1 \right) \right\} \delta(\omega) + \frac{1}{2} [\delta(\omega + s\varkappa) + \delta(\omega - s\varkappa)] \Bigg\} \tag{5.93}$$

where c_p and c_v are specific heat capacities at constant pressure and volume. The first term in (5.93) emerges in the result of ion disorder, the second-describes sound waves. For a crystal we have $c_p = c_v$, and we obtain automatically the result (5.92). In liquid metals we have $c_p \neq c_v$ and there appears a contribution in the kinetic coefficient, caused by a structure scattering. Inserting (5.93) into (5.87) and taking into account that $E \approx E_F \gg s\varkappa$, we obtain the expression found earlier in [38]:

$$\tau^{-1} = \frac{k_B T \beta_T m k_F}{2\pi\hbar^3} \int_0^\pi d\gamma \cdot \sin\gamma \left| w \left((2k_F \sin \frac{\gamma}{2} \right) \right|^2 (1 - \cos\gamma) \tag{5.94}$$

where β_T is the isothermal compressibility. This expression gives a correct qualitative result. For instance, it follows from (5.94) and (5.92) that at melting point the electroresistivity $\varrho = \sigma^{-1}$ has a jump, the ratio of the electroresistivity at the melting point for solid and liquid phases being

$$\frac{\varrho_1}{\varrho_{sol}} = \beta_T \left(\frac{M s^2}{v_0} \right)_{sol} . \tag{5.95}$$

If we take experimental compressibility data, sound velocity and density, then for Na, K, Al, Pb the ratio in (5.95) equals 1.38, 1.38; 2.61; 1.63, respectively while the experiment gives 1.49; 1.45; 2.85; 1.9.

Since the hydrodynamic approximation gives a good representation of $S(\varkappa, \omega)$ only at small $\varkappa$ and ω, (5.94) and (5.95) give a qualitative description only. As other exact models for $S(\varkappa, \omega)$ do not exist, a quantitative estimation of (5.87) may be received only by means of sum rules for $S(\varkappa, \omega)$. In this way we obtain the Ziman formula and its correction [39]:

$$\tau^{-1} = \frac{m v_0}{4\pi\hbar^3 k_F^3} \int_0^{2k_F} |w(\varkappa)|^3 \, \varkappa^3 \left[S(\varkappa) + \frac{m k_B T}{M E_F} \frac{3\varkappa^2 - 2k_F^2}{4k_F^2} + \dots \right] d\varkappa \, . \tag{5.96}$$

We conclude with the remark that the above discussed theory of transport coefficients was applied also for viscosity calculations [40], optical properties [41], sound absorption coefficients in liquid metals [42], electrotransport in alloys [43].

5.6. Theory of Electroconductivity of Liquid Metals with Account of Many Particle Effects

One of the main approximations of the given theory of transport coefficients is the use of the Born approximation in the perturbation theory. The calculations of the electroresistivity in this approximation with the Ziman relaxation time (5.96) lead to a qualitative agreement with the experiment, however, the quantitative disagreement for a number of metals appears to be large. In the work [44—47] the corrections are obtained to the Born approximation for the electroconductivity in higher orders of the perturbation theory in the PP. But both estimations of terms of the 3rd order and approximate summation of the whole perturbation series lead in some cases to consequences which show the Born approximation to be insufficient. It is essential here that the basis of the calculation is the diffraction model of the metal, where the electron-electron interaction is taken into account implicitly by means of the use of the screened PP in the Hamiltonian of the system. However, passing over from the second order (Born) to higher approximations of the perturbation theory, the diffraction model cannot be applied, as conduction electron scattering at pairs correlated in space and at triplets of ions does not coincide with independent scattering at pairs of ions. Therefore it is necessary to determine anew the screening in higher orders of perturbation theory in accordance with the range of inter-ion correlation. Let us show the possibility of obtaining a formally rigorous expansion of the specific electroresistivity into a power series of the PP, while the electron-electron interaction (i.e. screening) being exactly accounted for [48—50].

For the calculation of the electroconductivity tensor $\sigma_{\mu\nu}(\omega)$ for the frequency ω let us use the linear response theory [31]:

$$\sigma_{\mu\nu}(\omega) = \frac{ine^2}{m\omega}\,\delta_{\mu\nu} + \frac{e^2}{\omega}\int\limits_0^\infty e^{i\omega t}\,\langle[J_\mu(t), J_\nu(0)]\rangle\,dt\,, \tag{5.97}$$

where n is the electron density, J_μ is an operator of the μ-th current component (here we consider the volume of the system and Planck's constant to be equal to unity). The static electroconductivity in question may be obtained by the limiting procedure $\omega \to 0$. The Hamiltonian of the system may be split up as follows:

$$H = H_i + H_e + H_{ee} + H_{ei}\,, \tag{5.98}$$

where H_i refers to the ion subsystem, H_e and H_{ee} are kinetic and potential electron energy operators and H_{ei} — the Hamiltonian of the interaction between electrons and ions. The latter may be represented in the form

$$H_{ei} = \sum_{\boldsymbol q} w_0(\boldsymbol q)\,\varrho(\boldsymbol q)\,\varrho_e(-\boldsymbol q) \equiv \sum_{\boldsymbol q} \tilde\varrho(\boldsymbol q)\,\varrho_e(-\boldsymbol q)\,, \tag{5.99}$$

where

$$\varrho_e(\boldsymbol q) = \sum_{\boldsymbol k\sigma} a^+_{\boldsymbol k-\boldsymbol q\sigma} a_{\boldsymbol k\sigma}\,.$$

According to a general structure of transport coefficients, we assume the electroconductivity to be of the following form:

$$\sigma(\omega) \equiv \frac{1}{3}\,\sigma_{\mu\mu}(\omega) = \frac{ne^2}{m}\,\frac{\tau(\omega)}{1 - i\omega\tau(\omega)}\,, \tag{5.100}$$

where $\tau(\omega)$ is the "relaxation time" to be determined. The inverse relaxation time may be presented by a power series with respect to the PP. Correspondingly for $\delta(\omega)$ we obtain the expansion

$$\sigma(\omega) = \frac{ine^2}{m\omega} + \frac{ne^2}{m\omega^2}\left[\frac{1}{\tau_2} + \frac{1}{\tau_3} + \frac{1}{\tau_4} + \frac{1}{i\omega}\left(\frac{1}{\tau_2}\right)^2 + \dots\right], \tag{5.101}$$

where the indices at τ determine the order in the PP and the argument ω was not written out for brevity. After performing an analogous expansion in (5.97), the comparison of (5.97) and (5.101) gives the microscopic definition $\tau^{-1}(\omega)$ in arbitrary order in w_0.

Let us transform (5.97), i.e. twice perform an integration by parts in the 2nd term- The appearing time derivatives of the operators J_μ are easily defined as the corresponding commutators with the Hamiltonian. For the real part of $\tau^{-1}(\omega)$ which determines the conductivity we obtain:

$$\operatorname{Re}\left[\frac{1}{\tau_2(\omega)} + \frac{1}{\tau_3(\omega)} + \dots\right] = -\frac{1}{6mn\omega}\sum_{\boldsymbol{k}_1\boldsymbol{k}_2}(\boldsymbol{k}_1\cdot\boldsymbol{k}_2)$$
$$\cdot\int_0^\infty e^{i\omega t}\langle[\tilde{\varrho}(\boldsymbol{k}_1, t)\,\varrho_e(-\boldsymbol{k}_1, t), \tilde{\varrho}_e(\boldsymbol{k}_2, 0)\,\varrho(-\boldsymbol{k}_2, 0)]\rangle\, dt\,. \tag{5.102}$$

As before, we consider ions as classical particles, the motion of which may be ignored while calculating the electronic characteristics. Then we may regard the ion variables belonging to one time argument, $t = 0$, and the whole dynamics of the system will be reflected in retarded Green's functions

$$G_{\mathrm{R}}(\boldsymbol{k}_1, \boldsymbol{k}_2, t) = -i\theta(t)\,\langle[\varrho_e(-\boldsymbol{k}_1, t), \varrho_e(-\boldsymbol{k}_2, 0)]\rangle_e\,, \tag{5.103}$$

and averaging with respect to the "e" index is performed with the Hamiltonian (5.98), in which H_i is ommitted. After calculating (5.103) an additional averaging over the ion variables should be done.

In order to apply the diagram technique, let us proceed from the retarded Green's functions to causal Green's functions

$$G(\boldsymbol{k}_1, \boldsymbol{k}_2, t) = -i\,\langle T\{\varrho_e(-\boldsymbol{k}_1, t)\,\varrho_e(-\boldsymbol{k}_2, 0)\}\rangle_e\,. \tag{5.104}$$

As well G_{R} as G depend parametrically on ion coordinates. Using a well-known theorem about the connection between the retarded and the causal Green's functions [16] we may write instead of (5.102):

$$\operatorname{Re}\left[\frac{1}{\tau_2(\omega)} + \frac{1}{\tau_3(\omega)} + \dots\right] = -\frac{i}{6mn\omega}\sum_{\boldsymbol{k}_1\boldsymbol{k}_2}(\boldsymbol{k}_1\cdot\boldsymbol{k}_2)$$
$$\cdot\langle\varrho(\boldsymbol{k})_1\,\tilde{\varrho}(\boldsymbol{k}_2)\,[G(\boldsymbol{k}_1, \boldsymbol{k}_2, \omega) - G^*(\boldsymbol{k}_1, \boldsymbol{k}_2, \omega)]\rangle_i\,\operatorname{sign}\omega\,, \tag{5.105}$$

where the parenthesis with the index "i" means averaging over ion variables. The expression (5.105) is the starting point for obtaining a perturbation expansion in powers of the PP.

Let us change in the function G to the interaction representation for the operators $\varrho_e(\boldsymbol{k}, t)$, choosing as a perturbation the Hamiltonian $H_{ee} + H_{ei}$. Then, applying the standard methods of quantum field theory, G may be presented in the form:

$$G(\boldsymbol{k}_1, \boldsymbol{k}_2, t) = -\mathrm{i}\langle T\{\tilde{\varrho}_e(-\boldsymbol{k}_1, t)\, \tilde{\varrho}_e(-\boldsymbol{k}_2, 0)\, S\}\rangle\,, \tag{5.106}$$

where

$$S = T \exp\{-\mathrm{i}\int_{-\infty}^{\infty} [H_{ei}(t) + H_{ee}(t)]\, \mathrm{d}t\}\,, \tag{5.107}$$

and $\tilde{\varrho}_e(\boldsymbol{k}, t)$ is an operator in the interaction representation. To extract in (5.106) explicitly PP powers let us expand S into a power series in w_0. This leads to the following expansion for G:

$$G(\boldsymbol{k}_1, \boldsymbol{k}_2, t) = \sum_{n=0}^{\infty} \frac{(-\mathrm{i})^n}{n!} G_{n+2}(\boldsymbol{k}_1, \boldsymbol{k}_2, t)\,. \tag{5.108}$$

Here

$$G_2(\boldsymbol{k}_1, \boldsymbol{k}_2, t) = -\mathrm{i}\,\langle T\{\tilde{\varrho}_e(-\boldsymbol{k}_1, t)\, \tilde{\varrho}_e(-\boldsymbol{k}_2, 0)\, S_{ee}\}\rangle \tag{5.109}$$

is a Green's function for electrons, placed into a homogenous neutralizing background, and

$$G_{n+2}(\boldsymbol{k}_1, \boldsymbol{k}_2, t) = -\mathrm{i}\int_{-\infty}^{\infty} \dots \int_{-\infty}^{\infty} \mathrm{d}t_1 \dots$$
$$\dots \mathrm{d}t_n \langle T\{\tilde{\varrho}_e(-\boldsymbol{k}_1, t)\, \tilde{\varrho}_e(-\boldsymbol{k}_2, 0)\, H_{ei}(t_1) \dots H_{ei}(t_n)\, S_{ee}\}\rangle \tag{5.110}$$

is n^{th} order correction to G_2, accounting for spatial ion distribution structure. S_{ee} in (5.109) and (5.110) is defined by the expression of (5.107) type at $H_{ei} = 0$ and formally exactly accounts for the interaction between electrons.

It is easy to be seen that, in the quadratic approximation, (5.105) leads to the Ziman formula. In fact in this approximation electron and ion variables are separated in (5.105). Summing up a diagram series for G_2 is performed directly and gives:

$$G_2(\boldsymbol{k}_1, \boldsymbol{k}_2, \omega) = -\frac{\pi(\boldsymbol{k}_1, \omega)}{\varepsilon(\boldsymbol{k}_1, \omega)} \Delta(\boldsymbol{k}_1 + \boldsymbol{k}_2); \qquad \varepsilon(\boldsymbol{k}, \omega) = 1 + v_0(\boldsymbol{k})\, \pi(\boldsymbol{k}, \omega)\,, \tag{5.111}$$

where $\varepsilon(\boldsymbol{k}, \omega)$ is the electron gas dielectric function, $\pi(\boldsymbol{k}, \omega)$ is the polarization operator, $v_0(\boldsymbol{k})$ is the Fourier component of the interelectron interaction potential and $\Delta(\boldsymbol{k})$ is the Kronecker symbol. Inserting (5.111) into (5.105) we obtain:

$$\operatorname{Re} \frac{1}{\tau_2(\omega)} = \frac{1}{3mn\omega} \sum_{\boldsymbol{k}} \boldsymbol{k}^2 \left| \frac{w_0(k)}{\varepsilon(\boldsymbol{k}, \omega)} \right|^2 \operatorname{Im} \pi(\boldsymbol{k}, \omega) \langle |\varrho(\boldsymbol{k})|^2\rangle \operatorname{sign} \omega\,. \tag{5.112}$$

In the limit $\omega \to 0$ we have $\operatorname{Im} \varepsilon(\boldsymbol{k}, \omega) \to 0$, and $\varepsilon(\boldsymbol{k}, \omega)$ remains finite. Therefore concerning the static case it is sufficient to preserve the ω-dependence in the numerator of (5.112). Then from (5.112) we obtain:

$$\operatorname{Re} \frac{1}{\tau_2(0)} = \frac{1}{3mn} \sum_{\boldsymbol{k}} \boldsymbol{k}^2\, |w(k)|^2 \langle |\varrho(\boldsymbol{k})|^2\rangle f(\boldsymbol{k})\,, \tag{5.113}$$

where

$$f(\boldsymbol{k}) = \lim_{\omega \to 0} \frac{\operatorname{Im} \pi(\boldsymbol{k}, \omega) \operatorname{sign} \omega}{\omega}\,. \tag{5.114}$$

For instance, in the random phase approximation follows

$$f(\boldsymbol{k}) \equiv f_0(\boldsymbol{k}) = \frac{m^2}{2\pi^2 k}\,\theta(2P_F - k)\,,$$

and (5.113) gives the Ziman result. However, in general, $f(k)$ differs from $f_0(\boldsymbol{k})$. If exchange and correlation effects are taken into account by writing $\pi(\boldsymbol{k}, \omega)$ in the form [18]

$$\pi(\boldsymbol{k}, \omega) = \frac{\pi_0(\boldsymbol{k}, \omega)}{1 - v_0(\boldsymbol{k})\, g(\boldsymbol{k})\, \pi_0(\boldsymbol{k}, \omega)}\,, \tag{5.115}$$

where $\pi_0(\boldsymbol{k}, \omega)$ is the polarization operator in the random phase approximation and $g(\boldsymbol{k})$ plays the role of an effective electron-electron interaction, then

$$\mathrm{Re}\,\frac{1}{\tau_2(0)} = \frac{1}{3mn} \sum_{\boldsymbol{k}} \boldsymbol{k}^2 \left|\frac{w_0(\boldsymbol{k})}{\tilde{\varepsilon}(\boldsymbol{k})}\right|^2 \langle|\varrho(\boldsymbol{k})|^2\rangle\, f_0(\boldsymbol{k})\,, \tag{5.116}$$

where the "effective" dielectric function is

$$\tilde{\varepsilon}(\boldsymbol{k}) = \varepsilon(\boldsymbol{k})\,[1 - v_0(\boldsymbol{k})\, g(\boldsymbol{k})\, \pi_0(\boldsymbol{k})]\,. \tag{5.117}$$

Thus the screening of the pseudo-potential in the Ziman formula is given not by $\varepsilon(\boldsymbol{k})$, but by $\tilde{\varepsilon}(\boldsymbol{k})$. In numerical calculations in the Ziman formula the functions $\pi(\boldsymbol{k})$ and $f_0(\boldsymbol{k})$ can not be chosen independently, they must be consistent in accordance with (5.113) and (5.114).

Let us proceed to find corrections to the Born approximation. We restrict ourselves here to find the corrections of the 3rd order of the pseudopotential, higher corrections could be found analogously.

Inserting (5.110) into (5.105) and extracting the 3rd order we get

$$\mathrm{Re}\,\frac{1}{\tau_3(\omega)} = -\frac{1}{3mn\omega} \sum_{\boldsymbol{k}_1\boldsymbol{k}_2\boldsymbol{k}_3} (\boldsymbol{k}_1 \cdot \boldsymbol{k}_2)\, \cdot$$
$$\cdot \langle \tilde{\varrho}(\boldsymbol{k}_1)\, \tilde{\varrho}(\boldsymbol{k}_2)\, \tilde{\varrho}(\boldsymbol{k}_3)\rangle\, \mathrm{Re}\, \Gamma_3(\boldsymbol{k}_1, \boldsymbol{k}_2, \boldsymbol{k}_3, \omega)\, \mathrm{sign}\, \omega\,. \tag{5.118}$$

Here

$$\Gamma_3(\boldsymbol{k}_1, \boldsymbol{k}_2, \boldsymbol{k}_3, \omega) = -\mathrm{i} \int\!\!\int_{-\infty}^{\infty} \mathrm{d}t_1\, \mathrm{d}t\, \mathrm{e}^{\mathrm{i}\omega t}\, \cdot$$
$$\cdot \langle T\{\tilde{\varrho}_e(-\boldsymbol{k}_1, t)\, \tilde{\varrho}_e(-\boldsymbol{k}_2, 0)\, \tilde{\varrho}_e(-\boldsymbol{k}_3, t_1)\, S_{ee}\}\rangle\,. \tag{5.119}$$

In the absence of the electron-electron interaction, Γ_3 is represented as a sum of two three-vertex diagrams each of them having the structure given in figure 5.5. Here straight lines represent electron Green's functions, wavy lines correspond to the elec-

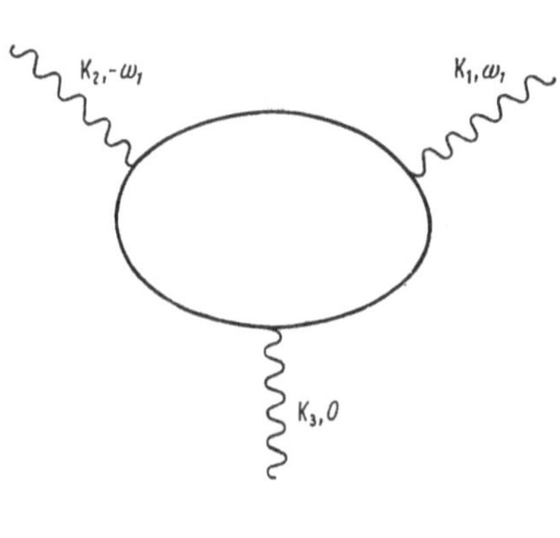

Fig. 5.5. The three-vertex part Γ_3

tron-ion interaction (which plays the role of the external field), both of them bearing the frequency ω of the electric field. The account of electron-electron interaction is taken as in [18]. After a partial summation, corresponding to a transition from "thin" to "thick" lines of the external field, we get:

$$\Gamma_3(\boldsymbol{k}_1, \boldsymbol{k}_2, \boldsymbol{k}_3, \omega) = \frac{\Lambda_3(\boldsymbol{k}_1, \boldsymbol{k}_2, \boldsymbol{k}_3, \omega)}{\varepsilon(\boldsymbol{k}_1, \omega)\, \varepsilon(\boldsymbol{k}_2, -\omega)\, \varepsilon(\boldsymbol{k}_3, \omega)}, \tag{5.120}$$

where $\Lambda_3(\boldsymbol{k}_1, \boldsymbol{k}_2, \boldsymbol{k}_3, \omega)$ is the set of diagrams with three inputs of the external field, not containing polarization parts, which might belong to "thick" external lines; two inputs have the frequency ω.

Let us insert (5.120) into (5.118) and perform the limiting procedure $\omega \to 0$. Here $\mathrm{Re}\, \varepsilon$ and $\mathrm{Im}\, \Lambda_3$ are finite and $\mathrm{Im}\, \varepsilon$, $\mathrm{Re}\, \Lambda_3 \approx \omega$. Therefore it is sufficient for small ω values to find the limiting value of $\mathrm{Im}\, \Lambda_3$ at $\omega = 0$. As in [18] we consider the ring diagram approximation (see fig. 5.5). It is easy to be seen that $\mathrm{Im}\, \Lambda_3(\boldsymbol{k}_1, \boldsymbol{k}_2, \boldsymbol{k}_3, \omega)$ coincides with the three-vertex [18]:

$$\mathrm{Im}\, \Lambda_3(\boldsymbol{k}_1, \boldsymbol{k}_2, \boldsymbol{k}_3, 0) = 6\Lambda_3(\boldsymbol{k}_1, \boldsymbol{k}_2, \boldsymbol{k}_3);$$

it is given explicitly in [18].

For the relaxation time up to the third order of the PP we obtain finally:

$$\begin{aligned}\mathrm{Re}\, \frac{1}{\tau_3(0)} = \frac{mv^2}{32\pi^7 n} \int\!\!\int \mathrm{d}\boldsymbol{k}_1\, \mathrm{d}\boldsymbol{k}_2\, w(\boldsymbol{k}_1)\, w(\boldsymbol{k}_2)\, w(-\boldsymbol{k}_1 - \boldsymbol{k}_2)\, \cdot \\ \cdot\, S_3(\boldsymbol{k}_1, \boldsymbol{k}_2, -\boldsymbol{k}_1 - \boldsymbol{k}_2) \left\{ \frac{2\theta\, (2P_F \cos\theta_{12} - k_1)}{3k_1 k_2 [k_1 \operatorname{sign} \cos\theta_{12} + k_2 |\cos\theta_{12}|]} - \right. \\ \left. - \frac{v_0(k_1)\, \Lambda_3^{(0)}(\boldsymbol{k}_1, \boldsymbol{k}_2, -\boldsymbol{k}_1 - \boldsymbol{k}_2)}{k_1 \varepsilon(k_1)\, [1 - v_0(k_1)\, g(k_1)\, \pi_0(k_1)]^2} \right\} (\boldsymbol{k}_1 \cdot \boldsymbol{k}_2)\, \theta(2P_F - k_1)\end{aligned} \tag{5.121}$$

where $S_3(\boldsymbol{k}_1, \boldsymbol{k}_2, \boldsymbol{k}_3)$ is the 3-particle structure factor (irreducible part of the correlation function $N^2\langle \varrho(\boldsymbol{k}_1)\, \varrho(\boldsymbol{k}_2)\, \varrho(\boldsymbol{k}_3)\rangle$), and θ_{12} is the angle between $\boldsymbol{k}_1$ and $\boldsymbol{k}_2$.

The formula (5.121) solves the problem of finding the corrections to the Born approximation of the 3rd order of the PP. The lack of experimental data and good theoretical models for $S_3(\boldsymbol{k}_1, \boldsymbol{k}_2, \boldsymbol{k}_3)$ make it difficult to estimate correction (5.121) numerically, but it is possible to draw important qualitative conclusions. In all variants of the theory of electroconductivity already at the Hamiltonian level a screened pseudopotential is introduced, what means in fact to use some "superposition" approximation for the screening properties of the electron gas, in order to give corrections to the Born approximation.

As a more exact result (5.121) shows, that the error arising here may be considerable, since, in contrast to the diffraction model, in general not only one, but all terms of (5.121) contribute to the electro-conductivity having the same order of magnitude. As estimations show, the terms in curly brackets of (5.121) partially compensate each other. We give the results of the calculation [50] of the correction to (5.121) in the "gas" approximation of the 3-particle structure factor

$$S_3(\boldsymbol{k}_1, \boldsymbol{k}_2, \boldsymbol{k}_3) \cong S(\boldsymbol{k}_1)\, S(\boldsymbol{k}_2)\, S(\boldsymbol{k}_3),$$

where $S(\boldsymbol{k})$ is the two-particle structure factor. We have chosen as $S(\boldsymbol{k})$ a structure factor of solid spheres in the Percus-Yevick approximation [21], the model PP — in the form of Gursky-Krasko (5.7) and the dielectric function according to [26]. Figure 5.6 gives the curve of the dependence of the specific electro-resistivity of liquid metals in the interval from the melting temperature to 1000 K along the saturation curve.

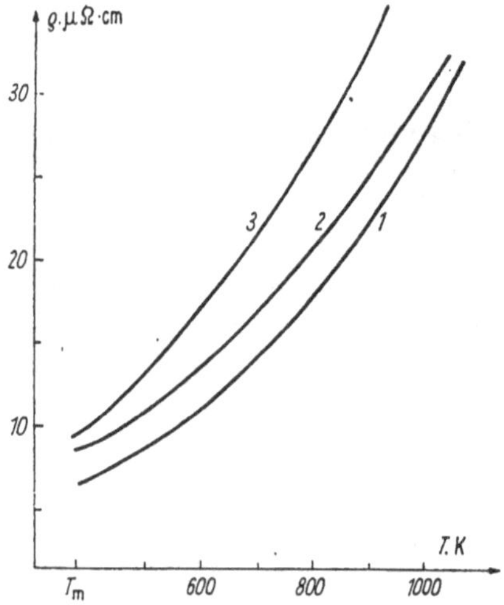

Fig. 5.6. Dependence of the resistivity of liquid Na on the temperature along the saturation curve: 1 — calculated according to the Ziman formula; 2 — with the correction after (5.121); 3 — experimental curve

As it follows from figure 5.6, at temperatures close to the melting point the contribution of terms of the 3rd order improves the Born approximation, shifting the calculated curve to the experimental one. With increasing temperature 3-particle corrections become less important and at temperatures close to the boiling point, calculated values approach the results of the Born approximation. Such a behaviour of the correction of 3rd order is naturally connected with the diminished role of many-particle effects with decreasing metal density.

To conclude, the mutual compensation of different terms of the 3rd order correction to the electroconductivity explains the relative success of the Born approximation, on the one hand, but on the other hand it requires a careful estimation of all conclusions concerning the exactness of the Born approximation and the convergence of the entire perturbation series in the case that the screening properties of the electron gas are not properly accounted for.

5.7. Remarks

In conclusion of this Chapter we point out some particularities of the theory of liquid metals with respect to the general frame of plasma theory.

Compared with usual plasmas, liquid metals are characterized by relatively high densities and low temperature values so that electrons are strongly degenerate whereas the ion motion remains classical. The kinetic energy of electrons (Fermi energy) is large compared with the potential energy of the electron-ion interaction so that liquid metals may be considered as weakly coupled nonideal plasmas. Therefore, perturbation theory works quite well in the evaluation of the thermodynamic and transport properties.

More rigorously, because of the high density we have to introduce a quasiparticle representation (see 3.3.6., 3.3.7.) in which electrons with an effective mass m_{eff} interact with ions via a relatively weak pseudopotential. One of the first attempts to introduce an effective electron-ion potential is the Hellmann potential [51] which may be generalized to include the Krasko-Gurski potential (see chap. 3. and [52]). The effects of the pseudopotential are increasing with density, and it is one of the most important problems in liquid metal theory to introduce a good pseudopotential. For instance pseudopotentials for liquid metals may be found in form of muffin-tin potentials [53].

Onother marginal point in highly dense plasmas and liquid metals is the coherent scattering of electrons by different ions as described by the structure factor. As pointed out in chapter 1. and chapter 3. (see 3.3.3.), structure factor effects in plasmas also increase with density and are of importance for the theory of highly dense plasmas (see [54, 55]). In contrast to crystal-like metals, in liquid metals and highly dense plasmas the structure factor represents only short range order in the ion subsystem; however, a more general theory should work with the dynamical structure factor $S(q, \omega)$, (see sects. 3.3.3., 5.6.) to include the effects of ion dynamics (phonons etc.).

A systematic extension of the theory of liquid metals and highly dense plasmas is found by improving the Born approximation introducing a t-matrix approach and scattering phase shifts, (see chaps. 1., 3. and [55—57]) for the case of dense plasmas. In liquid metal theory, the generalization of the Ziman formula has been considered in the last decade especially in connection with the description of liquid transition metals (see [58]). Using the Green's function technique, third order effects of the t-matrix on the conductivity in liquid metals were considered by Popielawski et al. [59].

The correct description of electron correlation, especially the dynamical screening, remains a general problem in liquid metal theory as well as in highly dense plasmas (see also sects. 3.3.5., 3.4. and [52]).

6. The Electrical Conductivity for the Partially Ionized Hydrogen Plasma

D. Kremp, G. Röpke, M. Schlanges

6.1. Expressions for the Electrical Conductivity

The electrical conductivity of a hydrogen plasma for parameter values of temperature and pressure where the plasma is partially ionized is determined not only by scattering between free carriers, but is influenced also by the occurrence of bound states. The general methods to calculate the electrical conductivity of a plasma were described in chapters 3. and 4. In this chapter, we give explicit results for the partially ionized hydrogen plasma which take into account bound states as well as the scattering of the free particles by the bound states. Especially the influence of bound states on the conductivity was discussed within the frame of linear response theory in section 3.3.7. The main results of this approach are summarized below. In chapter 4. generalized quantum kinetic equations are given, which include the effects of the formation, the decay and the scattering of H-atoms in the three-particle collisions.

For simplification we consider the partially ionized hydrogen plasma in such a density-temperature range, in which three-body scattering, inelastic and reaction processes can be neglected approximately. In the stationary case with a homogeneous electrical field E (E in z-direction) the following kinetic equation for the electrons is obtained

$$eE \frac{\partial}{\partial p_{e,z}} f(\boldsymbol{p}_e) = I_{ee} + I_{ep} + I_{e\mathrm{H}} \,. \tag{6.1}$$

The electron distribution function $f(\boldsymbol{p}_e)$ is normalized to the free electron density ($\boldsymbol{p}_e$: spin and momentum).

The two particle collision terms on the right hand side of equation (6.1) for the electron-electron and electron-proton scattering are given by

$$I_{ej} = \frac{(2\pi\hbar)^3}{\hbar\Omega} \int \mathrm{d}\boldsymbol{p}_j\, \mathrm{d}\boldsymbol{p}_j'\, \mathrm{d}\boldsymbol{p}_e'\, |\langle \boldsymbol{p}_e\boldsymbol{p}_j|\, T_{ej}\, |\boldsymbol{p}_j'\boldsymbol{p}_e'\rangle|^2$$
$$\cdot\, 2\pi\delta(E_{p_ep_j} - E'_{p_ep_j})\, \{f(\boldsymbol{p}_e')\, f(\boldsymbol{p}_j') - f(\boldsymbol{p}_e)\, f(\boldsymbol{p}_j)\} \,. \tag{6.2}$$

($j = e, p$)

From the many-channel representation of the three-particle term, given in chapter 4., we get for the collision integral of the elastic electron-atom scattering

$$I_{e\mathrm{H}} = \frac{(2\pi\hbar)^3}{\hbar\Omega} \sum_n \int \mathrm{d}\boldsymbol{P}\, \mathrm{d}\boldsymbol{P}'\, \mathrm{d}\boldsymbol{p}_e'\, |\langle \boldsymbol{p}_e\boldsymbol{P}n|\, T_{e\mathrm{H}}\, |n\boldsymbol{P}'\boldsymbol{p}_e'\rangle|^2$$
$$\cdot\, 2\pi\delta(E_{p_ePn} - E'_{p_eP'n})\, \{f(\boldsymbol{p}_e')\, f_n(\boldsymbol{P}') - f(\boldsymbol{p}_e)\, f_n(\boldsymbol{P})\} \,. \tag{6.3}$$

(Ω — system volume)
Here $E_{p_e p_j}$ and $E_{p_e p_n}$ are the energies of the particles in the corresponding scattering channels and $f_n(\boldsymbol{P})$ is the distribution function of H-atoms in the state with the internal quantum number n and total momentum $\boldsymbol{P}$.

The known expressions in terms of the differential cross sections follow immediately from the given T-matrix representations using the relations of scattering theory.

In the case of small deviations from the thermodynamic equilibrium different methods are known to calculate the electrical conductivity from kinetic equations [1] In the adiabatic approximation and if the electron-electron collision term is neglected ("Lorentz-Plasma") one can derive in a known manner the following explicit expression, see chapter 3.,

$$\sigma = \frac{4\pi e^2}{3m_e k_B T} \int\limits_0^\infty \frac{\mathrm{d}p_e}{(2\pi\hbar)^3} \frac{p_e^3 f^0(p_e)}{[n_p^* Q_{ep} + n_H^* Q_{eH}]} \tag{6.4}$$

where Q_{ep} denotes the electron-proton and Q_{eH} the electron-atom transport cross section. Here f_0 is the equilibrium distribution function

$$f^0(p_e) = n_e \Lambda_e^3 \exp\left(-\frac{p_e^2}{2m_e k_B T}\right), \qquad (\Lambda_e^2 = 2\pi\hbar^2/m_e k_B T)\,.$$

Because of the chemical picture, in (6.4) appear the number densities of the free electrons (n_e^*), protons (n_p^*) and H-atoms (n_H^*). They are determined by the balance equation

$$n_e = n_e^* + n_H^* \tag{6.5}$$

where n_e is the total electron density. A mass action law can be derived for the equilibrium case to calculate the number densities [2, 3].

The contributions of the two- and three-particle collision terms are given in formula (6.4) by the transport cross sections Q_{ep} and Q_{eH}. The relation between these cross sections and the T-matrix elements are given in section 6.2.

As shown in chapter 3., expressions for the electric conductivity of a partially ionized plasma can also be obtained from linear response theory.

Starting with the free particle states $|\boldsymbol{p}_e\rangle$ and the two particle bound states $|n\boldsymbol{P}\rangle$, the following generalized linearized Boltzmann equation is derived [4]:

$$E\{(\dot{R}, n_\nu) + \langle \dot{R}(\varepsilon), \dot{n}_\nu\rangle\} = \sum_{\nu'} B_{\nu'}\langle \dot{n}_{\nu'}(\varepsilon), \dot{n}_\nu\rangle\,, \tag{6.6}$$

where

$$\{|\nu\rangle\} = \{|\boldsymbol{p}_e\rangle, |\boldsymbol{P}n\rangle\}\,,$$

$$R^z = \sum_{i=1}^N e_i r_i^z \quad \text{describes the } z\text{-coordinate of the centre of charge,}$$

$$(A, B) = \int\limits_0^\beta \mathrm{d}\tau\, \mathrm{Tr}\,\{\varrho_s A(-i\hbar\tau)\, B\}\,,$$

$$\langle A(\varepsilon), B\rangle = \int\limits_{-\infty}^0 \mathrm{d}t^{\varepsilon t}\, e\big(A(t), B\big)\,,$$

$$\varrho_s = Z_s^{-1} \exp\big(-\beta(H_s - \mu N)\big)\,,$$

and $n_\nu = a_\nu^+ a_\nu$ being the occupation numbers of the state $|\nu\rangle$. The B_ν are related to the corresponding distribution function $f_\nu = \big(f(\boldsymbol{p}_e), f_n(\boldsymbol{P})\big)$ of free particles and two particle bound states according to

$$\langle n_\nu \rangle = \mathrm{Tr}\,\{\varrho_q n_\nu\} = f_\nu^0 + \sum_{\nu'} B_{\nu'}(n_{\nu'}, n_\nu) = f_\nu^0 + \delta f_\nu^0 \,. \tag{6.7}$$

The equations (6.6) for the parameters B_ν can be solved using the Kohler variational principle. The most simple variational ansatz is

$$B_\nu = \begin{cases} \alpha \boldsymbol{p} \cdot \boldsymbol{E} & \text{for} \quad \nu = |\boldsymbol{p}\rangle \\ 0 & \text{for} \quad \nu = |n\boldsymbol{P}\rangle \,. \end{cases} \tag{6.8}$$

Then, the following expression for the electric conductivity σ is obtained ($n^{\text{free}} = n_e^* = n_p^*$ — pair density of free particles):

$$\sigma = \frac{e^2 n^{\text{free}}}{\hbar^2} \Omega \,\{\sum_{p'p} p_z' p_z \langle \dot{n}_p(\varepsilon), \dot{n}_{p'} \rangle\}^{-1} \,. \tag{6.9}$$

Improvements of this approach can easily be reached by using further moments in the variational ansatz (6.8), see [5].

This very general expression (6.9) includes many body effects like three-particle scattering or screening, which can be treated by an excellent many body theory (for instance Green's function technique, see chap. 3.), and further transport mechanism like, for instance, hopping conductivity. These correlation functions are also immediately related to the transport cross sections (see 6.2.).

6.2. The Calculation of the Transport Cross Sections

The contribution of the particle interaction for the electrical conductivity can be expressed by transport cross sections, which appear in formula (6.8) explicitely. For the electron-proton and the electron-atom scattering they are given by

$$Q_{ea} = 2\pi \int_0^\pi \mathrm{d}\theta (1 - \cos\theta) \sin\theta \, \sigma_{ea}(\theta) \,; \; (a = p, \mathrm{H}) \tag{6.10}$$

Here the σ_{ep} and σ_{eH} are the differential cross sections, which can be calculated from the T-matrices of the corresponding scattering channels:

$$\sigma_{ep}(\theta) = (2\pi)^4 \hbar^2 m_e^2 \,|\langle \boldsymbol{p}|\, T_{ep} |\boldsymbol{p}_p' \rangle|^2 \,, \tag{6.10a}$$

$$\sigma_{eH}(\theta) = (2\pi)^4 \hbar^2 m_e^2 \,|\langle \boldsymbol{p}n|\, T_{e\mathrm{H}} \,|n'\boldsymbol{p}'\rangle|^2 \,. \tag{6.10b}$$

In order to take into account the influence of the surrounding plasma on the interaction between the particles we assume (Debye) screened two-body potentials

$$V_{ab} = \frac{e_a e_b}{r_{ab}} \exp(-\varkappa r_{ab})$$

with

$$\varkappa^2 = 4\pi\beta(n_e^* + n_p^*)\, e^2 \,, \qquad (\beta = 1/k_B T) \,.$$

The electron-proton scattering with Debye potential between the particles is well investigated. Here extensive numerical calculations were made, to get the scattering phase shifts for different values of the energy and the orbital quantum number of the relative motion [6, 5]. For the investigation of the transport properties in the region of the partially ionized hydrogen plasma we can use the perturbation expansion method, to calculate the T-matrix of the electron-proton scattering. So one gets in a simple way an analytic expression up to the second Born approximation [7]. For the $e-p$ transport cross section we can get approximately the following expression up to the order e^6

$$Q_{ep} = \frac{2\pi a_0^2}{K^4}\left\{\ln(1+4y) - \frac{4y}{1+4y} - \frac{2}{\lambda(3y+1)}\left[(y^{-1}+4)\ln\left(1+\frac{4y}{y^{-1}+4}\right) - \ln(1+4y)\right]\right\}. \tag{6.11}$$

The dimensionless variables are given by

$$K = \frac{p_e a_0}{\hbar}; \qquad \lambda = \varkappa a_0; \qquad y = \frac{K^2}{\lambda^2}.$$

(a_0 — Bohr radius)
The transport cross section for the elastic scattering of an electron on a H-atom can be calculated with (6.10) and (6.10b) from the T-matrix of the corresponding three particle problem. The T-operator of the e-H channel is given by the Lippmann-Schwinger equation ($T_3 = T_{eH}$)

$$T_{eH} = V_{eH} + V_{eH}G^0_{eH}T_{eH}. \tag{6.12}$$

Here $V_{eH} = V_{ee} + V_{ep}$ is the potential and $G^0_{eH} = (z - H^0_{eH})^{-1}$ is the free Green function in the channel. For the Hamiltonian of the three particle system we have $H_3 = H^0_{eH} + V_{eH}$ where in H^0_{eH} the interaction is included, leading to the two-body subcluster (H-atom). V_{ee} and V_{ep} are then the two body potentials between the free electron and the bound particles.

The elastic e-H scattering was investigated in the literature with different methods, which are all based on the two-body Coulomb forces between the elementary particles [8]. Assuming (Debye) screened potentials in the three particle interaction we will calculate the T-matrix with the perturbation expansion method. Up to the second Born approximation we have

$$\langle \boldsymbol{pP}n|\, T_{eH}\,|n'\boldsymbol{P}'\boldsymbol{p}'\rangle = \langle \boldsymbol{pP}n|\, V_{eH}\,|n'\boldsymbol{P}'\boldsymbol{p}'\rangle + \sum_{\bar n}\int \mathrm{d}\bar{\boldsymbol{P}}\,\mathrm{d}\bar{\boldsymbol{p}}\,\frac{\langle \boldsymbol{pP}n\,|V_{eH}|\bar n\bar{\boldsymbol{P}}\bar{\boldsymbol{p}}\rangle\langle \bar{\boldsymbol{p}}\bar{\boldsymbol{P}}\bar n\, V_{eH}|\, n'\boldsymbol{P}'\boldsymbol{p}'\rangle}{E_{nPp} - E_{\bar n\bar P\bar p} + i\varepsilon}. \tag{6.13}$$

Here $|n\boldsymbol{Pp}\rangle$ is the asymptotic state in the e-H scattering channel with the electron momentum $\boldsymbol{p}$, the total momentum $\boldsymbol{P}$ and the quantum number n of the H-atom. The energy is

$$E_{nPp} = \tilde{E}_n + \frac{P^2}{2M} + \frac{p^2}{2m_e}.$$

The screening of the atom we can treat in first order of the perturbation theory with the shifted binding energies $\tilde{E}_n = E_n + \varkappa e^2$ and the unscreened wave functions $\langle \boldsymbol{r} \mid n\rangle$.

As it is seen from (6.13), different effects of the three particle interaction appear in the terms of the T-matrix expansion. The first Born approximation describes electron scattering on a static charge distribution of the H-atom. The interaction due to the polarization of the H-atom by the colliding electron is included in first order by the second Born approximation. Neglecting exchange effects, after a straightforward calculation for the first Born approximation follows

$$\langle \boldsymbol{pP}n|\ T^{(1)}_{e\mathrm{H}}\ |n'\boldsymbol{P}'\boldsymbol{p}'\rangle = \langle \boldsymbol{pP}n|\ V_{e\mathrm{H}}\ |n'\boldsymbol{P}'\boldsymbol{p}'\rangle =$$

$$= \frac{4\pi\hbar^2}{(2\pi\hbar)^3}\,\frac{e^2}{q^2+\hbar^2\varkappa^2}\,\delta(\boldsymbol{P}-\boldsymbol{P}'+\boldsymbol{q})\,P_{nn'}(\boldsymbol{q})\,, \tag{6.14}$$

where

$$P_{nn'}(\boldsymbol{q}) = \left\langle n \left| \exp\left(-\frac{i}{\hbar}\frac{m_p}{M}\boldsymbol{qr}\right) - \exp\left(\frac{i}{\hbar}\frac{m_e}{M}\boldsymbol{qr}\right)\right| n'\right\rangle = F_{nn'}(\boldsymbol{q}) - 1$$

with

$$\boldsymbol{q} = \boldsymbol{p} - \boldsymbol{p}'\,, \qquad M = m_p + m_e\,.$$

Because of the difference of masses $m_e/m_p \ll 1$ the atomic form factor $F_{nn'}(\boldsymbol{q})$ for $n = n' = 1$ can be given by (adiabatic approximation)

$$F_{11}(q) = \left[1 + \left(\frac{qa_0}{2\hbar}\right)^2\right]^{-2}. \tag{6.15}$$

Here the unscreened atomic wave function

$$\langle \boldsymbol{r} \mid n\rangle|_{n=1} = (\pi a_0^3)^{-1/2}\exp\left(-r/a_0\right)$$

was used. If we treat the screening of the atom in a better approximation, we get screened atomic form factors. For $\varkappa = 0$ (no screening) from (6.14) and (6.15) the well known result for the e-H scattering with bare Coulomb interactions follows.

Now we consider the second term on the right hand side of (6.13). That gives in addition to the first term the second Born approximation of the T-matrix.

In adiabatic approximation with (6.14) and (6.15) the integration over the total momentum $\overline{\boldsymbol{P}}$ yields

$$\langle \boldsymbol{pP}n \mid T^{(2)}_{e\mathrm{H}} \mid n'\boldsymbol{P}'\boldsymbol{p}'\rangle = \frac{(4\pi\hbar^2e^2)^2}{(2\pi\hbar)^6}\,\delta(\boldsymbol{P}-\boldsymbol{P}')\sum_{\bar{n}}\int \mathrm{d}\overline{\boldsymbol{p}}\,\cdot$$

$$\cdot\left\{\frac{P_{n\bar{n}}(\boldsymbol{p}-\overline{\boldsymbol{p}})}{(\boldsymbol{p}-\overline{\boldsymbol{p}})^2+\hbar^2\varkappa^2}\,\frac{P_{\bar{n}n'}(\overline{\boldsymbol{p}}-\boldsymbol{p}')}{(\overline{\boldsymbol{p}}-\boldsymbol{p}')^2+\hbar^2\varkappa^2}\cdot\frac{1}{\tilde{E}_n+\dfrac{p^2}{2m_e}-\tilde{E}_{\bar{n}}-\dfrac{\overline{p}^2}{2m_e}+i\varepsilon}\right\}. \tag{6.16}$$

A convenient expression for (6.16) is obtained by using the model of the polarization potential. Neglecting the kinetic energies of the electrons in the Green function and using the $P_{nn'}(\boldsymbol{q})$ in adiabatic approximation yields:

$$\langle \boldsymbol{pP}n|\, T_{\mathrm{eH}}^{(2)}\, |n'\boldsymbol{P}'\boldsymbol{p}'\rangle = \frac{(4\pi\hbar^2 e^2)^2}{(2\pi\hbar)^6}\,\delta(\boldsymbol{P}-\boldsymbol{P}') \sum_{\bar{n}} \int \mathrm{d}\bar{\boldsymbol{q}}\,\cdot$$

$$\cdot\left\{\frac{1}{E_n - E_{\bar{n}}} \cdot \frac{\left\langle n\right| \mathrm{e}^{-\frac{i}{\hbar}(\boldsymbol{p}-\bar{\boldsymbol{p}})\boldsymbol{r}} - 1\left|\bar{n}\right\rangle}{(\boldsymbol{p}-\bar{\boldsymbol{p}})^2 + \hbar^2\varkappa^2}\, \frac{\left\langle \bar{n}\right| \mathrm{e}^{-\frac{i}{\hbar}(\bar{\boldsymbol{p}}-\boldsymbol{p}')\boldsymbol{r}} - 1\left|n'\right\rangle}{(\bar{\boldsymbol{p}}-\boldsymbol{p}')^2 + \hbar^2\varkappa^2}\right\}. \tag{6.17}$$

For elastic scattering processes at the ground state ($n = n' = 1$, $p = p'$) for instance, we introduce the polarization potential according to

$$\langle \boldsymbol{pP}n|\, T_{\mathrm{eH}}^{(2)}\, |n\boldsymbol{P}'\boldsymbol{p}\rangle = \frac{(4\pi\hbar^2)^2}{(2\pi\hbar)^6}\,\delta(\boldsymbol{P}-\boldsymbol{P}') \int \mathrm{d}\bar{\boldsymbol{p}}\; v_p(\boldsymbol{p}-\bar{\boldsymbol{p}})\,. \tag{6.18}$$

The behaviour of the polarization potential $v_p(r) = \int \frac{\mathrm{d}\boldsymbol{q}}{(2\pi)^3}\, \mathrm{e}^{i\boldsymbol{qr}}\, v(\boldsymbol{q})$ for large distances ($r \to \infty$) is found from the matrix elements

$$\frac{\left\langle n\right| \mathrm{e}^{-\frac{i}{\hbar}(\boldsymbol{p}-\bar{\boldsymbol{p}})\boldsymbol{r}} - 1\left|\bar{n}\right\rangle}{(\boldsymbol{p}-\bar{\boldsymbol{p}})^2 + \hbar^2\varkappa^2} = \int \mathrm{d}\boldsymbol{r}\, \mathrm{e}^{-\frac{i}{\hbar}(\boldsymbol{p}-\bar{\boldsymbol{p}})\boldsymbol{r}} \int \mathrm{d}\boldsymbol{r}'\, \psi_n^*(\boldsymbol{r}')\, \psi_{\bar{n}}(\boldsymbol{r}')\,\cdot \left\{\frac{\mathrm{e}^{-\varkappa|\boldsymbol{r}-\boldsymbol{r}'|}}{|\boldsymbol{r}-\boldsymbol{r}'|} - \frac{\mathrm{e}^{-\varkappa|\boldsymbol{r}|}}{|\boldsymbol{r}|}\right\} \tag{6.19}$$

by using the dipole approximation and expanding the exponential function with respect to $\varkappa r'$. In the opposite case, for $r = 0$, the polarization potential has, according to (6.18), the finite value

$$v_p(r)|_{r=0} = e^4 \sum_{n>1} (E_1 - E_n)^{-1} \left\{\int \mathrm{d}\boldsymbol{r}'\, \frac{\mathrm{e}^{-\varkappa r'}}{r'}\, \psi_1^*(\boldsymbol{r}')\, \psi_n(\boldsymbol{r}')\right\}^2. \tag{6.20}$$

The general behaviour of the polarization potential for arbitrary r is interpolated by the following expression

$$v_p(r) = -\frac{e^2\alpha_p}{2}\left(\frac{1+\varkappa r}{r^2 + r_0^2}\right)^2 e^{-2\varkappa r} \tag{6.21}$$

where α_p is the ground state polarizability

$$\alpha_p = -2e^2 \sum_n \frac{|\langle 1|\, Z\, |n\rangle|^2}{E_1 - E_n}$$

and the parameter r_0 is determined by the value of $v_p(0)$ (see, for instance, [9]). For the Coulomb case ($\varkappa = 0$) one finds the value $r_0 = 1{,}6\, a_0$. Notice that the polarization potential can also be introduced for excited states.

To avoid the approximations of the polarization potential, we must evaluate the expression (6.16). The problem is then the summation over the states $|\bar{n}\rangle$.

From the sattering theory it is known that the differences of the atomic energies can be replaced by a constant term [10]. For the further calculation we replace these differences $(\tilde{E}_1 - \tilde{E}_{\bar{n}})$ in the denominator of (6.16) by the effective energy difference $\Delta\tilde{E} = \Delta E + \varkappa e^2$ where the value for ΔE can be fitted.

With this approximation and using

$$1 = \sum_{\bar{n}} |\bar{n}\rangle \langle\bar{n}|$$

from (6.16) the following expression can be obtained for the case of elastic scattering

$$\langle \boldsymbol{pP}n| \, T_{e\mathrm{H}}^{(2)} \, |n\boldsymbol{P}'\boldsymbol{p}'\rangle = \frac{4a_0^2e^2}{\pi(2\pi\hbar)^3} \, \delta(\boldsymbol{P} - \boldsymbol{P}') \int \mathrm{d}\bar{\boldsymbol{K}}$$

$$\cdot \left\{ \frac{1 - F_{nn}(\boldsymbol{K} - \bar{\boldsymbol{K}}) - F_{nn}(\bar{\boldsymbol{K}} - \boldsymbol{K}') + F_{nn}(\boldsymbol{K} - \boldsymbol{K}')}{[\lambda^2 + (\bar{\boldsymbol{K}} - \boldsymbol{K})^2]\,[\lambda^2 + (\bar{\boldsymbol{K}} - \boldsymbol{K}')^2]} \, \frac{1}{\tilde{\alpha} + K^2 - \bar{K}^2 + \mathrm{i}\varepsilon} \right\} \tag{6.22}$$

where the dimensionless variables are given by

$$K = \frac{pa_0}{\hbar}; \qquad \lambda = \varkappa a_0; \qquad \frac{2m_e a_0^2}{\hbar^2} \Delta\tilde{E} = \tilde{\alpha}\,.$$

For the ground state case follows that the evaluation of (6.22) reduces to integral terms of the type

$$I = \int \mathrm{d}\bar{\boldsymbol{K}} \, \frac{1}{[\lambda^2 + (\boldsymbol{K} - \bar{\boldsymbol{K}})^2]\,[\eta + (\bar{\boldsymbol{K}} - \boldsymbol{K}')^2]} \, \frac{1}{\tilde{\alpha} + K^2 - \bar{K}^2 + \mathrm{i}\varepsilon}\,. \tag{6.23}$$

The two-angle problem in (6.18) can be solved by using the Feynman formulae [7]

$$\frac{1}{ab} = \frac{1}{2} \int_{-1}^{+1} \mathrm{d}t \left(a \, \frac{1+t}{2} + b \, \frac{1-t}{2} \right)^{-2}. \tag{6.24}$$

Introducing spherical polar co-ordinates, the K-integration then is carried out with the help of the residue theory for a path of integration, enclosing all poles in the upper complex half plane. After integration over the angles the remaining parameter integration of (6.24) is done with Euler substitution methods.

For the second term of the perturbation expansion of the T-matrix then the following analytic expression can be derived

$$\langle \boldsymbol{pP}n| \, T_{e\mathrm{H}}^{(2)} \, |n\boldsymbol{Pp}\rangle|_{n=1} = \frac{4a_0^2e^2}{\pi(2\pi\hbar)^3} \, \delta(\boldsymbol{P} - \boldsymbol{P}') \cdot$$

$$\cdot \left\{ \left[1 - \frac{32}{(4 - \lambda^2)^2} + \frac{16}{(4 + \tau^2)^2} \right] I(\lambda^2) - \frac{\partial}{\partial\eta} \left[\frac{8}{(\eta - \lambda^2)} \, I(\eta) \right]_{\eta=4} \right\} \tag{6.25}$$

with

$$I(\eta) = \frac{2\pi^2}{\tau f}\left\{\arctan\frac{K_0\omega - \eta^{1/2}Rv_1}{f} - \arctan\frac{K_0\omega - \lambda Rv_2}{f}\right\},$$

$$\omega^2 = \tau^2 + 2(\eta + \lambda^2) + \frac{1}{\tau^2}(\eta - \lambda^2)^2,$$

$$v_{1/2} = 2\left(\omega + \frac{1}{\tau}(\eta - \lambda^2) \pm\right)\tau^{-1},$$

$$R = \tilde{\alpha} - \lambda^2 - \frac{\lambda(\eta - \lambda^2)}{\tau v_1},$$

$$f^2 = \left[\tilde{\alpha} - \frac{1}{2}(\eta + \lambda^2)\right]^2 - \frac{(\eta - \lambda^2)^2}{\tau^2}\left(\tilde{\alpha} + \frac{\tau^2}{4}\right) - K_0^2\omega,$$

$$K_0^2 = -\tilde{\alpha} - K^2,$$

$$\tau = 2K\sin\frac{\theta}{2} \qquad (\theta - \text{scattering angle}).$$

It follows that our result (6.25) is real for the case $K^2 < |\tilde{\alpha}| \left(\frac{p^2}{2m_e} < |\Delta\tilde{E}|\right)$, where the parameter $\tilde{\alpha}$ represents approximately the atomic energy differences in (6.16). A real and an imaginary part exist, if the energy of the colliding electron is greater than the value of the branch point $|\tilde{\alpha}|$.

The presented calculation of the second Born term can be improved. So an exact treatment of the first contribution ($\bar{n} = 1$) in the sum of the virtual states in (6.16) is possible by an addition of corresponding corrections. This first contributions ($\bar{n} = 1$) which gives the static interaction part of the second Born term (6.16) becomes essential in the case of higher screening ($\lambda > 0.2$). In figure 6.1 numerical results for the transport cross section of the elastic electron-H-atom scattering are shown. The results on the basis of the second Born approximation for the T-matrix (6.13) with (6.14), (6.15) and (6.25) for different values of the screening parameter λ are given by the solid curves. Here for the effective energy difference we took $\Delta E = E_1$, so that $\tilde{\alpha} = -1 + 2\lambda$. The imaginary part of the T-matrix for $K^2 > |\tilde{\alpha}|$ was neglected.

The dotted curve in figure 6.1 was obtained by using the polarization potential for $\lambda = 0$.

The dashed line, which results from a fit formula form Meyer and Bartoli [11] combining calculations for different energy regions, was also presented, to give a comparison with out results for the limiting (Coulomb) case $\lambda = 0$. For energies $K^2 = |\tilde{\alpha}|$ the curves of the T-matrix calculation have peaks. They follow from the atomic energy differences in (6.16) which we have approximately replaced by a (screened) effective energy difference.

The behaviour of the curves $\lambda \neq 0$ for $K^2 < |\tilde{\alpha}|$ clearly shows the influence of the screening and polarization effects. So a strong decreasing of the polarization interaction for low energies is observed.

As expected, we get for all cases in figure 6.1 a high energy behaviour of Q_{eH}, which is well described, using the first Born approximations for the T-matrix.

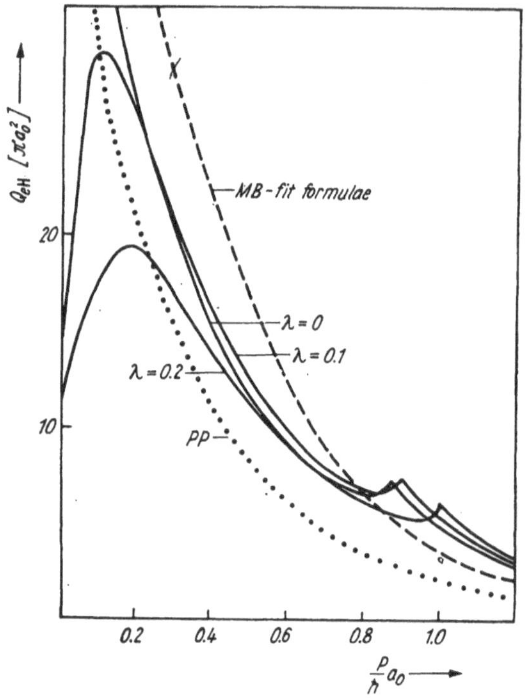

Fig. 6.1. The transport cross section Q_{eH} for the elastic scattering of an electron on a H-atom in the ground state.
——— second Born approximation (for different screening parameters $\lambda = \varkappa a_0$),
········· polarization potential for $\lambda = 0$,
– – – fit curve from [11] for $\lambda = 0$

6.3. Discussion

Now we will calculate the electrical conductivity on the basis of the given particle scattering contributions. As it is seen from (6.4) and (6.9) the influence of the different interaction processes on the electrical conductivity is connected with the composition of the partially ionized hydrogen plasma. For the calculation of the particle densities we have used a mass action law in the following approximation [3]

$$\frac{n_{\mathrm{H}}^*}{n_{\mathrm{e}}^* n_{\mathrm{p}}^*} = \Lambda_e^3 \exp\left[-\beta(E_1 - \varkappa e^2)\right] \tag{6.26}$$

which can be derived from the linear response theory and from the kinetic theory in the equilibrium case. With this special form it is possible to describe the pressure ionization (Mott effect) in a dense plasma. In figure 6.2 the degree of ionization $\alpha = n_e^*/n_e$ is given as a function of the total electron density n_e for different temperatures. In the ideal region we have $\alpha \approx 1$ which corresponds to the case of the fully ionized hydrogen plasma. With higher densities the α-curves show at first the typical decreasing because of the formation of H-atoms. Near the Mott-density ($\varkappa \approx a_0^{-1}$) we obtain a strong increasing of α, which describes the changing into the fully ionized dense plasma state (Mott transition).

With the employed mass action law (6.26) the behaviour of the plasma composition can be described of course only in an approximate way. For a better description a mass action law must be derived on the basis of a consistent many-particle theory, which

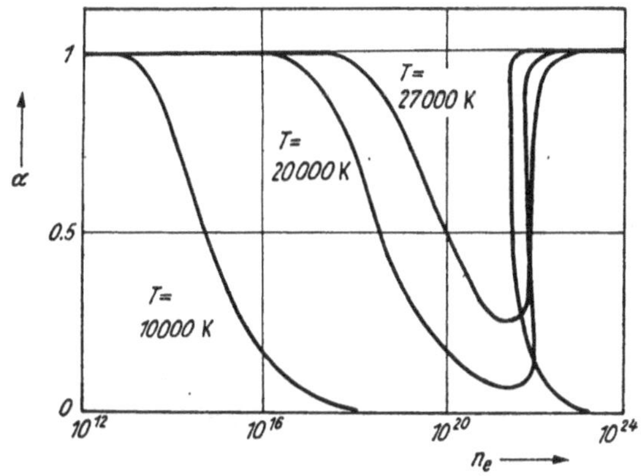

Fig. 6.2. Degree of ionization as a function of the total electron density for different temperatures

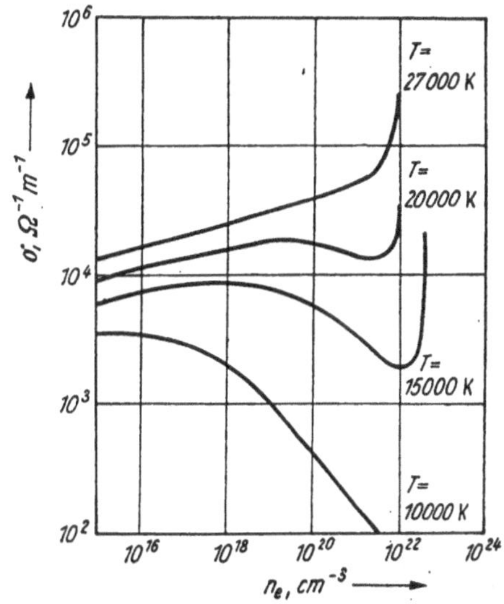

Fig. 6.3. The electrical conductivity as a function of the total electron density for different temperatures

also takes into account the influence of bound states in excited states, exchange effects collective plasma effects and so on.

Finally we will discuss our results for the electrical conductivity. In figure 6.3 the behaviour of the conductivity is shown as a function of the total electron density for different temperatures. For the calculation we have used (6.26) and the transport cross sections (6.10) in second Born approximation [12]. A characteristic feature is the minimum and the following increasing of the conductivity. This effect is produced first by the increasing importance of the e-H scattering contribution because of the formation of bound states with higher densities, and on the other hand by the pressure ionization of the atoms in the region of the Mott density. As expected, the influence of the H-atom formation on the conductivity is strong at lower temperatures.

In figure 6.4 the influence of different approximations of the e-H scattering contribution on the electrical conductivity for a temperature $T = 15000$ K is presented. One curve was shwown, for which the minimum is absent. Here the existence of bound states was included only by the mass action law (6.26). The minimum appears if we take into account the cross section Q_{eH}, what is done in different manner. The change in the

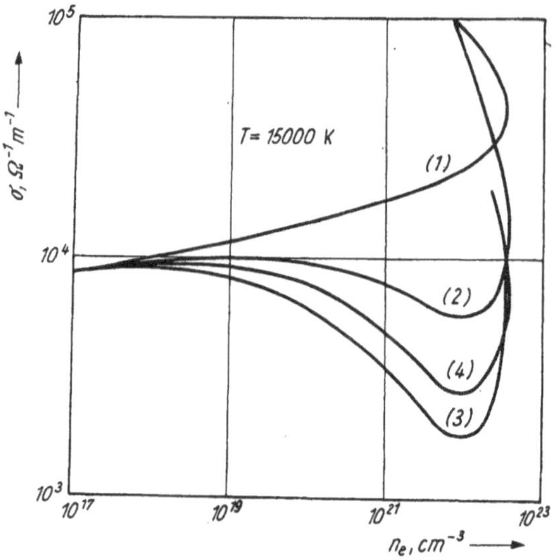

Fig. 6.4. The electrical conductivity as a function of the total electron density from different approximations for the e-H scattering; (1) without Q_{eH}; (2) Q_{eH} in first Born approximation; (4) from the polarization potential; (3) Q_{eH} in second Born approximation

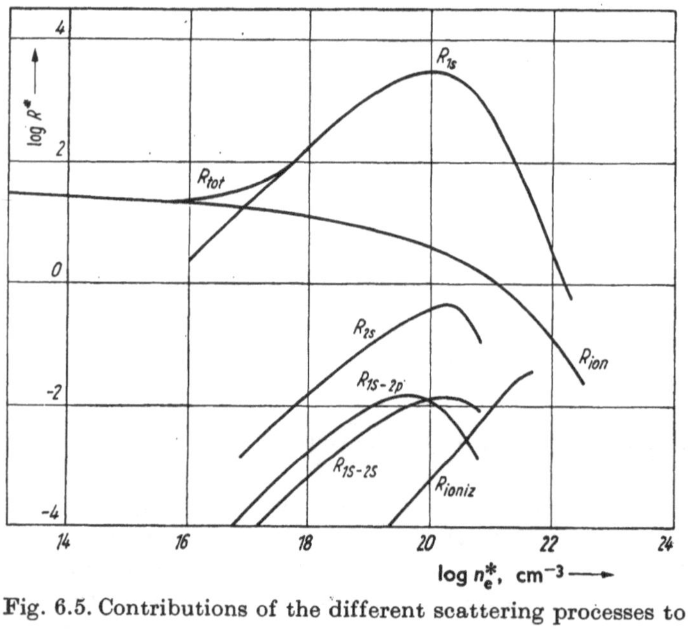

Fig. 6.5. Contributions of the different scattering processes to the total specific resistivity

behaviour of the electrical conductivity is essential, if the polarization interaction in the e-H scattering contribution is taken into account. This can be seen from the curves which we have obtained using the polarization potential and the second Born approximation in the corresponding T-matrix calculation.

With the most simple variational ansatz (6.8) a formulae follows for the total specific resistivity R^{tot}, which contains the contributions of all scattering processes in an additive way. They are shown in figure 6.5 (see also [13]).

In the low density region the electron-ion-contribution is the most important one because the plasma is fully ionized. With higher densities bound states are formed according to the mass action law and the e-H scattering process is the essential contribution to the total resistivity. In the high density region the e-H scattering contribution decreases due to the pressure ionization of the bound states (Mott).

Other e-H scattering contributions (scattering on the excited $2s$-state, inelastic scattering with the $1s$-$2p$ and the $1s$-$2s$ transition, ionization processes) in first Born approximation are small so that the total resistivity is essentially given by the collision terms of the e-i — and the elastic e-H scattering at the ground state.

One can expect that the contribution of the second Born approximation for the elastic e-H scattering at excited states can be essential due to the much higher polarizability of these states as compared to that of the ground state.

7. References

Citation of journals from Soviet Union are performed using the following abbreviations:

Dokl. Akad. Nauk SSSR: Доклады Академии Наук СССР — ДАН СССР
Fiz. tverd. tela: Физика твердого тела — ФТТ
Ing. fiz. zh.: Инжениер-физический журнал — ИФЖ
Teor. matem. fiz.: Теоретическая и математическая физика — ТМФ
Teplofiz. vyssok. temp.: Теплофизика высоких температур — ТВТ
Ukrain. fiz. zh.: Украйнский физический журнал — УФЖ
Usp. fiz. nauk: Успехи физических наук — УФН
Zh. eksp. teor. fiz.: Журнал экспериментальной и теоретической физики — ЖЭТФ

7.1. References to Chapter 1

[1] Ebeling, W., Kraeft, W. D., Kremp, D.: Theory of Bound States and Ionization Equilibrium in Plasmas and Solids. Akademie-Verlag, Berlin 1976; Russian translation: Мир Moscow 1979.

[2] Norman, G. E., Starostin, A. N. (Г. Э. Норман, А. Н. Старостин): ТВТ 8 (1970) 413.

[3] Fortov, V. E. (В. Е. Фортов): Модели уравнения состояния вещества. Р.И.О. ОИХФ АН СССР, Chernogolovka 1979.

[4] Gryaznov, V. K., et al. (В. К. Грязнов и др.): Теплофизические свойства рабочих сред ядерной энергетической установки с газофазным реактором. Атомиздат. Moscow 1980.

[5] Research of Uranium Plasmas and their Technological Applications. Proc. of Symposium in Gainessuill, Florida. Ed. Thom, K., Schneider, R. T., NASA, Washington 1971; 2nd Symposium on Uranium Plasmas, Atlanta, Georgia, 1971.

[6] Fortov, V. E. (В. Е. Фортов): Динамические методы в физике неидеальной плазмы, т. 1, 2. Р. И. О. ОИХФ АН СССР, Chernogolovska 1981.

[7] Ragan, C. E., Silbert, M. G., Diven, B. C.: J. Appl. Phys. 48 (1977) 2860; Proc. VII. Internat. AIRAPT Conference, Le Cresot 1978.

[8] Linhart, J. G.: Nucl. Fusion **10** (1970) 211.

[9] Physics of High Energy Density. Ed. Caldirola, P., Knöpfel, H., Academic Press, New York 1971; Russian translation: Мир, Moscow 1974.

[10] Юпитер, т. I, II., Ed. Zharkov, V. N. (В. Н. Жарков) Мир, Moscow 1978.

[11] Prokhorov, A. M., Anisimov, S. I., Pashinin, P. P. (А. М. Прохоров, С. И. Анисимов, П. П. Пашинин): УФН **119** (1976) 401.

[12] Zamalin, V. M., Norman, G. E., Filinov, V. S. (В. М. Замалин, Г. Э. Норман, В. С. Филинов): Метод Монте-Карло в статистической термодинамике. Наука, Moscow 1977.

[13] Balescu, R.: Statistical Mechanics of Charged Particles. Wiley Interscience, London 1963; Russian translation: Мир, Moscow 1965.

[14] Ermokhin, N. N., Kovaliov, B. N., Kulik, P. P., Riabii, V. A.: J. Physique (Paris) **39 C** (1978) 200.
[15] Alekseev, V. A. (В. А. Алексеев): ТВТ **8** (1970) 641; УФН **10** (1972) 393.
[16] Asinovskii, E. I., Zaigarnik, V. A. (Э. И. Асиновский, В. А. Зейгарник): ТВТ **12** (1974) 1278.
[17] Ross, R. G., Greenwood, D. A.: Progr. Mater. Sci. **14** (1971) 173.
[18] Al'tshuler, L. V. (Л. В. Альтшулер): УФН **85** (1965) 147.
[19] Kormer, S. B. (С. Б. Кормер): УФН **94** (1968) 641.
[20] Mineev, V. N., Ivanov, A. G. (В. Н. Минеев, А. Г. Иванов): УФН **119** (1976) 75.
[21] Lomakin, B. N., Fortov, V. E. (Б. Н. Ломакин, В. Е. Фортов): ЖЭТФ **63** (1972) 42.
[22] Bushman, A. V., et al. (А. В. Бушман, Б. Н. Ломакин, В. А. Сеченов, В. Е. Фортов, О. Е. Щекотов): ЖЭТФ **69** (1975) 1624.
[23] Tkachenko, B. K., et al. (Б. К. Ткаченко, и др.): Физика горения и взрыва, № 5, 763, 1976; IV Конференция по теплофизическим свойствам веществ. Minsk 1978; VI Конференция по динамике излучающего газа. Тезисы, Moscow 1980; A.J.J.J. Journal **14** (1976) 371; ТВТ **16** (1978) 411.
[24] Fortov, V. E., Ivanov, Yu. V., et al. (В. Е. Фортов, Ю. В. Иванов, и др.): ДАН СССР **221** (1975) 1307.
[25] Bespalov, V. E., et al. (В. Е. Беспалов, В. К. Грязнов, А. Н. Дремин, В. Е. Фортов): ЖЭТФ **69** (1975) 2059.
[26] Ivanov, Yu. V., et al. (Ю. В. Иванов, В. Б. Минцев, А. Н. Дремин, В. Е. Фортов): ЖЭТФ **71** (1976) 216; Письма в ЖЭТФ **2** (1976) 97.
[27] Mintsev, V. B., Fortov, V. E., Gryaznov, V. K. (В. Б. Минцев, В. Е. Фортов, В. К. Грязнов): ЖЭТФ **79** (1980) 116.
[28] Fortov, V. E., et al. (В. Е. Фортов, А. А. Леонтьев, В. К. Грязнов, А. Н. Дремин): ЖЭТФ **71** (1976) 225.
[29] Gryaznov, V. K., et al. (В. К. Грязнов, М. В. Жерноклетов, В. Н. Зубарев, И. Л. Иосилевский, В. Е. Фортов): ЖЭТФ **78** (1980) 573.
[30] Al'tshuler, L. V., et al. (Л. В. Альтшулер, А. В. Бушман, А. А. Леонтьев, М. В. Жерноклетов, В. Е. Фортов): ЖЭТФ **78** (1980) 741.
[31] Bespalov, V. E., Gryaznov, V. K., Fortov, V. E. (В. Е. Беспалов, В. К. Грязнов, В. Е. Фортов): ЖЭТФ **76** (1979) 141.
[32] Mintsev, V. B., Fortov, V. E. (В. Б. Минцев, В. Е. Фортов): Письма в ЖЭТФ **30** (1979) 401.
[33] Trainor, R. J., Shaner, J. W., et al.: Phys. Rev. Letters **42** (1979) 1154; Vessel, L., Solem, J.: Phys. Rev. Letters **40** (1978) 1391.
[34] Kunavin, A. G., Kirillin, A. V., Korshunov, Yu. S. (А. Г. Кунавин, А. В. Кириллин, Ю. С. Коршунов): ТВТ **13** (1974) 1304; **11** (1973) 261.
[35] Isakov, I. M., Lomakin, B. N. (И. М. Исаков, Б. Н. Ломакин): ТВТ **17** (1979) 262.
[36] Al'tshuler, L. V., et al. (Л. В. Альтшуле р, Б. Н. Моисеев, Л. В. Попов, Г. В. Симаков, Р. Ф. Трунин): ЖЭТФ **54** (1968) 785.
[37] Trunin, R. F., et al. (Р. Ф. Трунин, М. А. Подурец, Г. В. Симаков, Л. В. Попов, Б. Н. Моисеев): ЖЭТФ **62** (1972) 1044.
[38] Trunin, R. F., et al. (Р. Ф. Трунин, М. А. Подурец, Б. Н. Моисеев, Г. В. Симаков, Л. В. Попов): ЖЭТФ **56** (1969) 1172.
[39] Zubarev, V. N., et al. (В. Н. Зубарев, М. А. Подурец, Л. В. Попов, Г. В. Симаков, Р. Ф. Трунин): Сб. Детонация, p. 61, Chernogolovka 1978.
[40] Al'tshuler, L. V., et al. (Л. В. Альтшулер, Н. Н. Калиткин, Л. В. Кузьмина, Б. С. Чекин): ЖЭТФ **77** (1979) 1410.
[41] Avronin, E. N., et al. (Е. Н. Аврорин, Б. К. Водолага, А. С. Владимиров, В. А. Симоненко, Б. Т. Черноволюк): Письма в ЖЭТФ **31** (1980) 727; Volkov, L. P., et al. (Л. П. Волков, Н. П. Волош, Р. А. Мангасаров, В. А. Симоненко, Г. В. Синько, В. Л. Сорокин): Письма в ЖЭТФ **31** (1980) 546.
[42] Kirzhnits, D. A., Lozovik, Yu. E., Shpatakovskaya, G. V. (Д. А. Киржниц, Ю. Е. Ло́зовик, Г. В. Шпа́таковская): УФН **117** (1975) 3.

[43] Zel'dovich, Ya. B., Raizer, Yu. P. (Я. Б. Зельдович, Ю. П. Райзер): Физика ударных волн и высокотемпературных гидродинамических явлений. Nauka, Moscow 1966.
[44] Fortov, V. E., Krasnikov, Yu. G. (В. Е. Фортов, Ю. Г. Красников): ЖЭТФ 59 (1970) 1645.
[45] Lomakin, B. N., Fortov, V. E., Shchekotov, O. E. (Б. Н. Ломакин, В. Е. Фортов, О. Е. Щекотов): ТВТ 9 (1971) 869.
[45a] Gryaznov, V. K., Iosilevskiĭ, I. L., Fortov, V. E. (В. К. Грязнов, И. Л. Иосилевский, В. Е. Фортов): Zhurnal Prikladnoĭ Mekhaniki i Tekhnicheskoi Fiziki 3 (1973) 70.
[46] Zelener, B. V. (Б. В. Зеленер): ТВТ 15 (1977) 893.
[47] Fortov, V. E., Gryaznov, V. K., Ivanova, A. N.: XIII. Internat. Conf. on phenomena in ionized gazes, Berlin 1977, p. 218.
[48] Alder, B. J., Wainwright, T. E.: J. Chem. Phys. 33 (1960) 1439.
[49] Hooveretal, W. J.: J. Chem. Phys. 55 (1971) 1128; 49 (1968) 3609; 52 (1970) 4931.
[50] Gryaznov, V. K., Iosilevskii, I. L. (В. К. Грязнов, И. Л. Иосилевский): Chislennye metody mekhaniki sploshnoĭ sredy (Численные методы механики сплошной среды) 4 (1973) 166; Collection "Teplofizicheskie scoistva nizkotemperaturnoi plazmy" (Сб. „Теплофизические свойства низкотемпературной плазмы"). Ed. Ievlev, V. M. (В. М. Иевлев), Наука, Moscow 1976, p. 25.
[51] Stillinger, F. H., Lovett, R.: J. Chem. Phys. 49 (1968) 1991.
[52] Volkov, L. P., Vladimirov, N. P., et al. (Л. П. Волков, Н. П. Владимиров и др.): Письма в ЖЭТФ 31 (1980) 623.
[53] Kalitkin, N. N. (Н. Н. Калиткин): ЖЭТФ 38 (1960) 1534; Preprint Inst. Appl. Math. Acad. Sci. USSR (Препринт Института прикладной математики АН СССР) No. 35, Moscow 1975; Fizika Plaszmy 2 (1976) 858.
[54] Kirzhnits, D. A., Shpatakovskaya, G. V. (Д. А. Киржниц, Г. В. Шпатаковская): ЖЭТФ 62 (1972) 2082; Preprint Inst. Appl. Math. Acad. Sci. USSR (Препринт Института прикладной математики АН СССР) No. 54, Moscow 1975.
[55] McMahan, A. K., Ross, M.: Proc. VI Internat. AIRAPT Conf., vol. 2, p. 920, Colorado 1978.
[56] Nikoforov, A. F., et al. (А. Ф. Никофоров, В. Г. Новиков, Ю. Ю. Орлов, В. Б. Уваров): Preprint Inst. Appl. Math. Acad. Sci. USSR (Препринт Института прикладной математики АН СССР) No. 172, Moscow 1979.
[57] Sin'ko, G. V. (Г. В. Синько): Chislennye metody mekhaniki sploshnoi sredy (Численные методы механики сплошной среды) 10 (1979) 124.
[58] Gryaznov, V. K., Iosilevskiĭ, I. L., Fortov, V. E. (В. К. Грязнов, И. Л. Иосилевский, В. Е. Фортов): Fizika Plazmy (1981).
[59] Mason, E., Spurling, T.: The virial equation of state. In: The International Encyclopedia of Physical Chemistry and Chemical Physics, Topic 10, Vol. 2. (Ed. E. S. Rowlinson); Russian translation. Мир, Moscow 1972.
[60] Kormer, E. B., et al. (С. Б. Кормер, А. И. Фунтиков, В. Д. Урлин, А. Н. Колесникова): ЖЭТФ 42 (1962) 686.
[61] Fortov, V. E., et al. (В. Е. Фортов, А. Н. Дремин, А. А. Леонтьев, С. В. Першин): Письма в ЖЭТФ 20 (1974) 30.
[62] Scidmore, C., Morris, E.: Proc. Sympos. on Thermodynamics of Nuclear Materials, I.A.E.A. Vienna, Springer-Verlag, p. 173, 1982.
[63] Glushak, B. A., Zhernoklyetov, M. V., Zubarev, V. N. (Б. А. Глушак, М. В. Жерноклетов, В. Н. Зубарев): Доклады I, Всесоюзного симпозиума по импульсным давлениям ВНИИФТРИ, Наука, Moscow 1974, p. 87.
[64] Al'tshuler, L. V., et al. (Л. В. Альтшулер, А. А. Баканова, А. В. Бушман, И. П. Додуладов, В. Н. Зубарев): ЖЭТФ 73 (1977) 1866.
[65] Landau, L. D., Zel'dovich, Ya. B. (Л. Д. Ландаи, Я. Б. Зельдович): ЖЭТФ 14 (1944) 32.
[66] Norman, G. E.: (Г. Э. Норман): ЖЭТФ 60 (1971) 1686.
[67] Очерки физики и химии низкотемпературной плазмы. Ed. Polak, L. S., (Л. С. Полак); Наука, Moscow 1971.

[68] Bushman, A. V., Fortov, V. E., Sharipdzhanov, I. I. (А. В. Бушман, В. Е. Фортов, И. И. Шарипджанов): ТВТ **15** (1977) № 1.
[69] Voropinov, A. I., Gandel'man, G. M., Podval'nyi, I. I. (А. И. Воропинов, Г. М. Гандельман, И. И. Подвальный): УФН **100** (1970) 93.
[70] Fortov, V. E. (В. Е. Фортов): Гидродинамические эффекты в неидеальной плазме **10** (1972) 86.
[71] Taylor, J. W.: J. Appl. Phys. **34** (1962) 2727.
[72] Spitzer, L.: Physics of Fully Ionized Gases. Wiley, New York u. a. 1962; Russian translation: Мир, Moscow 1965.
[73] Kihara, T., Aono, O.: J. Phys. Soc. Japan **18** (1963) 837; Itakawa, J.: J. Phys. Soc. Japan **18** (1963) 1499.
[74] Blatt, F. J.: Physics of Electronic Conduction in Solids. McGraw-Hill Book Company, New York 1968; Russian translation: Мир, Moscow 1971.
[75] Conwell, E. M., Weisskopf, V. E.: Phys. Rev. **77** (1950) 388.
[76] Sechenov, V. A., Son, E. E., Shchekotov, O. E., (В. А. Сеченов, Э. Е. Сон, О. Е. Щекотов): Письма в ЖЭТФ **1** (1975) 819; ТВТ **15** (1977) 411.
[77] Dikhter, I. Ya., Zaigarnik, V. A (И. Я. Дихтер, В. А. Зейгарник): ТВТ 15 (1977) 471.
[78] Barol'ski, S. G., Yermokhin, N. V., Kulik, P. P., et al. (С. Г. Барольский, Н. В. Ермохин, П. П. Кулик, и др.): ЖЭТФ **62** (1972) 176; ТВТ **15** (1977) 415, 659; ТВТ **17** (1979) 693.
[79] Khomkin, A. L. (А. Л. Хомкин): ТВТ **12** (1974) 870.
[80] Klimontovich, Yu. L., Ebeling, W. (Ю. Л. Климонтович, В. Эбелинг): ЖЭТФ **63** (1972) 905.
[81] Gould, H. A., DeWitt, H. E.: Phys. Rev. **155** (1967) 68; Williams, R. H. DeWitt, H. E.: Phys. Fluids **12** (1969) 2326.
[82] Valuev, A. A., Norman, G. E. (А. А. Валуев, Г. Э. Норман): ТВТ **145** (1977) 689.
[83] Vorob'ev, V. S., Khomkin, A. L. (В. С. Воробьев, А. Л. Хомкин): ТВТ **14** (1976) 204.
[84] Gryaznov, V. K., et al. (В. К. Грязнов, Ю. В. Иванов, А. Н. Старостин, В. Е. Фортов): ТВТ **14** (1976) 643.
[85] Keeler, R. N., van Thiel, M., Alder, B. J.: Physica **31** (1965) 1437.
[86] Pavlov, G. A., Kucherenko, V. I. (Г. А. Павлов, В. И. Кучеренко): ТВТ **15** (1977) 409.
[87] Kovaliov, B. M. et al. (Б. М. Ковалев, П. П. Кулик, Б. Н. Ломакин, В. Е. Фортов): Физика горения и взрыва **10** (1974) 289.
[88] Askerov, B. M. (Б. М. Аскеров): Кинетические явления в полупроводниках. Наука, Moscow 1970.
[89] Yermokhin, N. V., et al. (Н. В. Ермохин, Б. М. Ковалев, П. П. Кулик, В. А. Рябый): ТВТ **15** (1977) 695.
[90] Kalitkin, N. N., Kuz'mina, L. V., Rogov, V. S. (Н. Н. Калиткин, Л. В. Кузьмина, В. С. Рогов): Физика плазмы **2** (1976) 858: Численные методы механики сплошной среды **4** (1973) 88; ТВТ **6** (1968) 801; 8 (1970) 689.
[91] Mayev, S. A. (С. А. Маев): ЖЭТФ **40** (1970) 567.
[92] Podlubnyi, L. N., Rostovskii, V. S. (Л. Н. Подлубный, В. С. Ростовский): VI Всесоюзная конференция по теплофизическим свойствам веществ. Тезисы докладов, Minsk 1978 p. 201.
[93] Landau, L. D., Lifschitz, E. M. (Л. Д. Ландау, Е. М. Лифшиц): Квантовая механика. Наука, Moscow 1974; German translation: Lehrbuch der Theoretischen Physik, Bd. III, Quantenmechanik. Akademie-Verlag, 6. Auflage, Berlin 1979.
[94] Shkarofskii, I., Johnston, T., Bachinskii, M. (И. Шкаровский, Т. Джонстон, М. Бачинский): Кинетика частиц в плазме. Атомиздат, Moscow 1969
[95] Harm, H., Manson, E. A., Smith, F. J.: Phys. Fluids **14** (1971) 278.
[96] Biberman, L. M., Norman, G. E. (Л. М. Биберман, Г. Э. Норман): УФН **91** (1967) 193.

[97] Batenin, V. M., Minayev, P. V.: (В. М. Батенин, П. В. Минаев): ТВТ **15** (1977) 647.

[98] Kobzev, G. A. Kurilenkov, Yu. K,. Norman, G. E. (Г. А. Кобзев, Ю. К. Куриленков, Г. Э. Норман): ТВТ **15** (1977) 193.

[99] Kurilenkov, Yu. K., Minayev, P. V. Ю. К. Куриленков, П. В. Минаев): ЖЭТФ **74** (1978) 563.

[100] Shor, B. W.: J. Phys. **B8** (1975) 2023.

[101] Lisitsa, V. S.: (В. С. Лисица): УФН **122** (1977) 449.

[102] Koval'skaya, G. A., Sevost'yanenko, V. G (Г. А. Ковальская, В. Г. Севостьяненко). Сб. ,,Свойства низкотемпературной плазмы и методы ее диагностики''. Ed. Zhukov, M. F. (М. Ф. Жуков), Наука, Novosibirsk 1977, p. 11.

[103] Andreev, S. I., Gavrilova, T. V.: (С. И. Андреев, Т. В. Гаврилова): ТВТ **13** (1975) 647.

[104] Kon'kov, A. A., Razin, A. P., Rudnev, V. S. (А. А. Коньков, А. П. Разин, В. С. Руднев): Quant. Spec. Rad. Trans. **7** (1967) 345.

[104a] Model', I. T. (И. Т. Модель): ЖЭТФ **32** (1957) 714.

[105] Bespalov, V. E., et al. (В. Е. Беспалов, Л. Г. Дьячков, Г. А. Кобзев, В. Е. Фортов): ТВТ **17** (1979) 266.

[106] Avilova, I. A., Biberman, L. M., et al. (И. А. Авилова, Л. М. Биберман и др.): Оптические свойства горячего воздуха. Наука, Moscow 1970

[107] Asinovskii, E. I., et al. (Э. И. Асиновский, и др.): ТВТ **8** (1970) 905.

[108] Kon'kov, A. A., Rezin, A. P., Sokolov, A. I. (А. А. Коньков, А. П. Резин, А. И. Соколов): ТВТ **12** (1974) 806.

[109] Schlüter, D.: Z. Phys. **210** (1968) 80.

[110] Tsikulin, M. A., Popov, E. G. (М. А. Цикулин, Е. Г. Попов): Излучатедьные свойства ударных волн в газах. Наука, Moscow 1977.

[111] Roth, J.: J. Appl. Phys. **35** (1964) 1429.

[112] Heitler, W.: Quantum theory of radiation. Oxford 1954; Russian translation: ГИТТЛ 1940.

[113] Bekefi, J.: Radiative processes in plasmas. Wiley, New York 1966; Russian translation: Мир, Moscow 1971.

[114] Gryaznov, V. K., Ivanova, A. N., Kozyrev, A. A. (В. К. Грязнов, А. Н. Иванова, А. А. Козырев): VI Всесоюзная конференция по теории атомов и атомных спектров. Voronezh 1980.

[115] Sobel'man, I. I. (И. И. Собельман): Введение в теорию атомных спектров. Наука, Moscow 1977.

[116] Höhne, F., Zimmermann, R., J. Phys. B **15** (1982) 2551.

[117] Röpke, G., Höhne, F.: phys. stat. sol. (b) **107** (1981) 603.

[118] Röpke, G., Kilimann, K., Kremp, D., Kraeft, W. D., Zimmermann, R.: phys. stat. sol. (b) **88** (1978) K59.

[119] Zimmermann, R., Kilimann, K., Kraeft, W. D., Kremp, D., Röpke, G.: phys. stat. sol. (b) **90** (1978) 175.

[120] Röpke, G., Kilimann, K., Kraeft, W. D., Kremp, D.: Phys. Lett. **A68** (1978) 329.

[121] Röpke, G., Seifert, T., Stolz, H., Zimmermann, R.: phys. stat. sol. (b) **100** (1980) 215.

[122] Röpke, G., Der, R.: phys. stat. sol. (b) **92** (1979) 501.

[123] Röpke, G., Meister, C.-V., Kollmorgen, K., Kraeft, W. D.: Ann. Physik (Leipzig) **36** (1979) 377.

[124] Röpke, G., Seifert, T., Kilimann, K.: Ann. Physik (Leipzig), **38** (1981) 381.

[125] Griem, H. R.: Spectral Line Broadening by Plasmas. Academic Press, New York 1974.

[126] Sobel'man, I. I., Vainshtein, L. A., Yukov, E. A.: Excitation of Atoms and Boradening of Spectral Lines. Springer, Berlin 1981.

7.2. References to Chapter 2

[1] Spitzer, L., Härm, R.: Transport Phenomena in a Completely Ionized Plasma. Phys. Rev., 89 (1953) 977.

[2] Kovaliov, B. M., Kulik, P. P., Riaby, V. A.: (Б. М. Ковалев, П. П. Кулик, Рябый: В. А. 22, (1972) 92.

[3] Kovaliov, B. M., Kulik, P. P., Riaby, V. A.: Contr. papers X ICPIG, Oxford. Donald Parsons and Co. Publ., 1971, p. 266.

[4] Günther, K., Popovič, M. M., Popovič, S. S., Radtke, R.: J. Phys., D.: Applied Physics, 9 (1976) 1139.

[5] Günther, K., Radtke, R.: Beiträge aus der Plasmaphysik, 12 (1972) 63.

[6] Klimontovich, Yu. L. (Ю. Л. Климонтович): Кинетическая теория электромагнитных процессов. Наука, Moscow 1980.

[7] Kaklyugin, A. S., Norman, G. E. (А. С. Каклюгин, Г. Э. Норман): ТВТ 11 (1973) 238.

[8] Vorobjov, V. S., Khomkin, A. D. (В. С. Воробьев, А. Д. Хомкин): ТВТ 15 (1977) 188.

[9] Norman, G. E., Valuev, A. A.: Contr. papers XII ICPIG, Eindhoven. North Holland Publ. Comp., 1975, p. 257.

[10] Valuev, A. A., Norman, G. E. (А. А. Валуев, Г. Э. Норман): ТВТ 15 (1977) 689.

[11] Batanova, T. N., Kovaliov, B. M., Kulik, P. P., Riaby, V. A.: Contr. papers XIII ICPIG, Berlin. Phys. Soc. GDR, 1977, p. 583.

[12] Valuev, A. A., Norman, G. E. (А. А. Валуев, Г. Э. Норман): ТВТ 15 (1977) 191.

[13] Norman, G. E., Shchurova, L. Yu. (Г. Э. Норман, Л. Ю. Шурова): Физика и техника полупроводников. 13, (1979) 1409.

[14] Röpke, G., Ebeling, W., Kraeft, D. W.: Physica 101A (1980) 243.

[15] Röpke, G., Meister, C. V., Kollmorgen, K., Kraeft, W. D.: Annalen der Physik 36 (1979) 377.

[16] Mintsev, V. B., Fortov, V. E. (В. Б. Минцев, В. Е. Фортов): Письма в ЖЭТФ 30 (1979) 401.

[17] Kobzev, G. A., Kurilenko, Yu, K., Norman, G. E. (Г. А. Кобзев, Ю. К. Куриленков, Г. Э. Норман): ТВТ 15 (1977) 193.

[18] Bespalov, V. E., Gryaznov, IU. K., Fortov, V. E. (В. Е. Беспалов, В. К. Грязнов, В. Е. Фортов): ЖЭТФ 76 (1979) 140.

[19] Valuev, A. A., Kurilenko, Yu. K. (А. А. Валуев, Ю. К. Куриленков): ТВТ 18 (1980) 897.

[20] Barolskii, S. G., et al. (С. Г. Барольский, Н. В. Ермохин, П. П. Кулик, В. А. Рябый): ТВТ 14 (1976) 702.

[21] Vorobiov, V. V., Kulik, P. P., Pallo, A. V., Rakitin, A. A., Rozanov, E. K., Riaby, V. A.: J. Phys. 40 (1979) suppl. No. 7, coll. C7, pp. C7-693 and C7-694 (XIV ICPIG, Grenoble 1979).

[22] Vorobiov, V. V., Kulik, P. P., Pallo, A. V., Rozanov, E. K., Riaby, V. A.: Contr. papers XV ICPIG, Minsk USSR Academy of Sciencens 1981, part I, p. 361.

[23] Schlichting, G. (Г. Шлихтинг): Теория пограничного слоя. Наука, Moscow 1974. p. 522, 545.

[24] Kulikov, I. S. (И. С. Куликов): Термическая диссоциация соединений. Металлургия, Moscow 1969, p. 223.

[25] Ermokhin, N. V., et al. (Н. В. Ермохин, Б. М. Ковалев, П. П. Кулик, В. А. Рябый): ТВТ 15 (1977) 695.

[26] Atrazhev, V. M., Yakubov, I. T. (В. М. Атражев, И. Т. Якубов): ТВТ 18 (1980) 16.

[27] Fjodorenko, S. P. (С. П. Федоренко): Приближенное решение задач оптимального управления. Наука, Moscow 1978.

[28] Kuznetsova, N. I., Lappo, G. B. (Н. И. Кузнецова, Г. Б. Лаппо): ТВТ 17 (1979) 37.

[29] Nazarenko, I. P., Panevin, I. G. (И. П. Назаренко, И. Г. Паневин): Моделирование и методы расчета. физико-химических процессов в низкотемпературной плазме. Ed. Polaka, L. S., (Л. С. Полака), Наука, Moscow 1974, p. 86.
[30] Rozanov, E. K. (Е. К. Розанов): ТВТ 17 (1979) 639.
[31] Sechenov, V. A., Son, E. E., Shchekotov, O. E. (В. А. Сеченов, Э. Е. Сон, О. Е. Щекотов): ТВТ 15 (1977) 411.
[32] Isakov, I. M., Lomakin, B. N. (И. М. Исаков, Б. Н. Ломакин): ТВТ 17 (1979) 262.
[33] Yermokhin, N. V., Kovaliov, B. M., Kulik, P. P.: Contr. papers XII ICPIG, Eindhoven. North Holland Publ. Comp., Amsterdam 1975, p. 184.
[34] Kulik, V. Ya., Kulik, P. P., Riaby, V. A.: (В. Я. Кулик, П. П. Кулик, В. А. Рябый): ТВТ 10(1972) 715.

7.3. References to Chapter 3

[1] Gryaznov, V. K., et al. (В. К. Грязнов, И. Л. Иосилевский, Ю. Г. Красников, Н. И. Кузнецова, В. И. Кучеренко, Г. В. Лаппо, Б. Н. Ломакин, Г. А. Павлов, Э. Е. Сон, В. Е. Фортов): Теплофизические свойства, Атомиздат, Moscow 1980.
[2] Kalman, G., Carini, P. (Eds.): Strongly Coupled Plasmas. Plenum Press, New York u. a. 1978; Rice, T. M., Hensel, J. C., Philips, T. G., Thomas, G. A.: The Electron-Hole Liquid in Semiconductors. Academic Press, New York u. a. 1977.
[3] Ebeling, W., Kraeft, W. D., Kremp, D.: Theory of Bound States and Ionization Equilibrium in Plasmas and Solids. Akademie-Verlag, Berlin 1976; Russian translation: Мир, Moscow 1979.
[4] Baus, M., Hansen, P.: Physics Reports 59 (1980) 2.
[5] Gurov, K. P. (К. П. Гуров): Основания Кинетической Теории. Наука, Moscow 1966.
[6] Klimontovich, Yu., L., Ebeling, W.: Zh. eksp. teor. fiz. 43 (1962) 146, 62 (1972) 905.
[7] Klimontovich, Yu. L., Ebeling, W.: Wiss. Z. Univ. Rostock 11 (1962) 355.
[8] Klimontovich, Yu. L. (Ю. Л. Климонтович): Кинетическая Теория Неидеального Газа и Неидеальной Плазмы. Наука, Moscow 1975.
[9] Klimontovich, Yu. L., (Ю. Л. Климонтович): Кинетическая Теория Электромагнитных Процессов. Наука, Moscow 1980.
[10] Silin, V. P. (В. П. Силин): Введение в Кинетическую Теорию Газов, Наука, Moscow 1971.
[11] Balescu, R.: Statistical Mechanics of Charged Particles. Interscience Publ., London u. a. 1963; Ichimaru, S.: Basic Principles of Plasma Physics. Benjamin, New York 1973; Russian translation: Мир ,Moscow 1975.
[12] Ebeling, W.: Ann. Phys. (Leipzig) 33 (1976) 350.
[13] Klimontovich, Yu. L., Kremp, D.:, Physica 109A (1981) 517; Chapter 4. of this volume.
[14] Ebeling, W., Kraeft, W. D., Kremp, D.: 13th International Conference on Phenomena in Ionized Gases. Invited Papers. Berlin 1977.
[15] Spitzer, L.: Physics of Fully Ionized Gases. Wiley, New York u. a. 1962; Montgomery, D. C., Tidman, D. A.: Plasma Kinetic Theory. McGraw Hill, New York u. a. 1964; Liboff, R.: Introduction to the Theory of Kinetic Equations. Wiley, New York u. a. 1969; Ecker, G.: Theory of Fully Ionized Plasmas. Academic Press, New York u. a. 1972.
[16] Prigogine, I.: Introduction to Thermodynamics of Irreversible Processes. Springfield 1955.
[17] De Groot, S. R., Mazur, P.: Non-Equilibrium Thermodynamics. North Holland Publ. Comp., Amsterdam 1962.

[18] GRAD, H.: Principles of the Kinetic Theory of Gases (Handbuch der Physik Vol. 12). Springer-Verlag, Berlin u. a. 1957.

[19] KHRAPAK, A. G., YAKUBOV, I. T. (А. Г. Храпак, И. Т. Якубов): Электроны в плотных газах и в плазме. Наука, Moscow 1981; SMIRNOW, B. M. (Б. М. Смирнов): Физика слабоионизованного газа. Наука, Moscow 1978.

[20] EBELING, W., RÖPKE, G.: Ann. Phys. (Leipzig) **36** (1979) 429.

[21] RÖPKE, G., EBELING, W., KRAEFT, W. D.: Physica **101A** (1980) 243.

[22] KADOMTSEV, B. B. (Б. Б. Кадомцев): ЖЭТФ **33** (1957) 151.

[23] NERNST, W.: Theoretische Chemie. Teubner-Verlag, Stuttgart 1926; EBELING, W.: Beitr. Plasmaphysik 7 (1967) 11.

[24] FALKENHAGEN, H.: Theorie der Elektrolyte. Hirzel-Verlag, Leipzig 1971; EBELING, W., FEISTEL, R., KELBG, G., SÄNDIG, R.: J. Non-Equil. Thermodyn. 3 (1978); J. Solution Chem. 8 (1979) 53; JUSTICE, J. C., EBELING, W.: J. Solution Chem. 8 (1979) 809.

[25] EBELING, W.: 15th International Conference on Phenomena in Ionized Gases. Invited Papers. Minsk 1981 ;BELYI, V. V., KLIMONTOVICH, YU. L., NALIVAIKO, V. P.: 15th International Conference on Phenomena in Ionized Gazes. Contributed Papers. Minsk 1981; Preprint IZMIRAN 5A, Moscow 1981.

[26] KUBO, R.: J. Phys. Soc. Japan **12** (1957) 570.

[27] ZUBAREV, D. N. (Д. Н. Зубарев): Неравновесная статистическая термодинамика. Наука, Moscow 1971; German translation: SUBAREW, D. N., Statistische Thermodynamik des Nichtgleichgewichts. Akademie-Verlag Berlin 1976.

[28] KLIMONTOVICH, YU. L., (Ю. Л. Климонтович): Статистическая теория необратимых процессов MGU Publ., Moscow 1964.

[29] BOGOLYUBOV, N. N. (Н. Н. Боголюбов): Проблемы динамической теории. Гостехиздат, Moscow 1946.

[30] HUBERMANN, M., CHESTER, G. V.: Adv. Phys. **24** (1975) 489.

[31] CHRISTOPH, V., RÖPKE, G.: phys. stat. sol. (b) **80** (1977) K117.

[32] RÖPKE, G., HÖHNE, F.: phys. stat. sol. (b) **107** (1981) 603.

[33] RÖPKE, G., CHRISTOPH, V.: J. Phys. **C8** (1975) 3615.

[34] KOHLER, M.: Z. Phys. **124** (1948) 772; **125** (1949) 679.

[35] APPEL, J.: Phys. Rev. **122** (1971) 1760.

[36] MEISTER, C. V., RÖPKE, G.: Ann. Physik (Leipzig) **39** (1982) 133; MEISTER, C. V.: Dissertation (A), Wilhelm-Pieck-Universität, Rostock 1981.

[37] PALMER, R. G., WEEKS, J. D.: J. Chem. Phys. **58** (1973) 4171.

[38] KLYUCHNIKOV, N. I., TRIGER, S. A. (Н. И. Ключников, С. А. Тригер): ТМФ **26** (1976) 256; **39** (1979) 368.

[39] GOEDSCHE, F., RICHTER, R., VOJTA, G.: phys. stat. sol. (b) **91**, (1979) 581; MÖBIUS, A., GOEDSCHE, F., VOJTA, G.: Physica **95A** (1979) 294.

[40] BUSCH, G., GÜNTHERODT, H.-J., in: Solid State Physics, Vol. 29. (Ed. H. EHRENREICH, F. SEITZ, D. TURNBULL); Academic Press, New York 1974.

[41] DREIRACH, O., EVANS, R., GÜNTHERODT, M. J., KUNZI, H. U.: J. Phys. **F2** (1972) 709.

[42] ROGERS, F. J. :Phys. Rev. **A4** (1971) 1145; ROGERS, F. J., DE WITT, H. E., BOERCKER, D. B.: Phys. Lett. **82A** (1981) 331.

[43] LIE, T. J., LIBOFF, R. L.: Ann. Phys. **67** (1971) 349.

[44] TING, C. S., YING, S. C., QUINN, J. J.: Phys. Rev. **B14** (1976) 4439.

[45] KOJIMA, D. Y.: Physica **A39** (1978) 171; ALTSHULER, B. L. (Б. Л. Альтшулер): ЖЭТФ **75** (1978) 1330; FLEUROV, V. N., KOZLOV, A. N.: J. Phys. **F8** (1978) 1899.

[46] KADANOFF, L. P., BAYM, G.: Quantum Statistical Mechanics. Benjamin, New York 1982.

[47] RÖPKE, G. (Г. Рэпке): ТМФ **46** (1981) 279; Ann. Physik (Leipzig) **39** (1982) 35.

[48] GINGRICH, N. S., HEATON, L.: J. Chem. Phys. **34** (1961) 873.

[49] HUBBARD, J.: Proc. Roy. Soc. London **A261** (1961) 371.

[50] GOULD, H. A., DE WITT, H. E.: Phys. Rev. **155** (1967) 68; WILLIAMS, R. H., DE WITT, H. E.: Phys. Fluids **12** (1969) 2326.

[51] RÖPKE, G., MEISTER, C. V., KOLLMORGEN, K., KRAEFT, W.-D.: Ann. Physik (Leipzig) **36** (1979) 377.

[52] Fetter, A. L., Walecka, J. D.: Quantum Theory of Many Particle Systems, New York 1971; Stolz, H.: Einführung in die Vielelektronentheorie der Kristalle, Akademie-Verlag, Berlin 1974.
[53] Kilimann, M. K.: Doctor thesis II, Rostock 1978.
[54] Zimmermann, R., Kilimann, M. K., Kraeft, W. D., Kremp, D., Röpke, G.: phys. stat. sol. (b) **90** (1978) 175.
[55] Blümlein, J., Kraeft, W. D., Meyer, T.: Ann. Physik (Leipzig) **37** (1980) 379.
[56] Plakida, N. M. (Н. М. Плакида): ФТТ **6** (1964) 3444.
[57] Blümlein, J., in preparation.
[58] Ron, A., Tzoar, N.: Phys. Rev. **131** (1963) 12; **131** (1963) 1943.
[59] Blümlein, J., Kraeft, W. D., in preparation; Kraeft, W. D., Luft, M., Mihajlov, A. A.: Physica **120A** (1983) 263
[60] Kraeft, W. D., Blümlein, J., Meyer, T.: Beitr. Plasmaphysik **23** (1983) 9.
[61] Kohn, W.: Phys. Rev. **123** (1961) 1242.
[62] Mott, N. F.: Proc. Camb. Phil. Soc. **32** (1936) 281.
[63] Blatt, F. J.: Physics of Electronic Conduction in Solids. McGraw-Hill Book Company, New York 1968.
[64] Brooks, H.: Phys. Rev. **83** (1951) 879.
[65] Rogov, V. S. (В. С. Рогов): ТВТ 8 (1970) 689.
[66] Chapman, S., Cowling, T. G.: The Mathematical Theory of Non-Uniform Gases. University Press, Cambridge 1958.
[67] Faber, T. E.: Introduction to the Theory of Liquid Metals. University Press, Cambridge 1972, p. 330.
[68] Hart, G. A., Goodfriend, P. L.: J. Chem. Phys. **53**, (1970) 448.
[69] Günther, K., Lang, S., Radtke, R.: J. Phys. D., Appl. Phys. **16** (1983) 1235.
[70] Ivanov, Ju. V., Mintsev, V. B., Fortov, V. E., Dremin, A. N. (Ю. В. Иванов, В. Б. Минцев, В. Е. Фортов, А. Н. Дремин): ЖЭТФ71 (1976) 216.
[71] Radtke, R., Günther, K., Popovic, S. S.: J. Phys. D9 (1976) 1131.
[72] Andreev, S. I., Gavrilova, T. V. (С. И. Андреев, Т. В. Гаврилова): ТВТ **13** (1975) 176.
[73] Ogurtsova, M. H., Podmoshenskiĭ, I. V., Smirnova, V. Ja. (М. Н. Огурцова, И. В. Подмошенский, В. Я. Смирнова): ТВТ **12** (1974) 650.
[74] Bakeev, A. A., Revinskiĭ, R. E. (А. А. Бакеев, Р. Е. Ревинский): ТВТ 8 (1970) 1121.
[75] Sechenov, V. A., Son, E. E., Shchekotova, O. E. (В. А. Сеченов, Е. Е. Сон, О. Е. Щекотова): Письма в ЖЭТФ **1** (1975) 891.
[76] Bauder, U., et al.: Phys. Fluids **16** (1973) 2143; Günther, K. et al.: J. Phys. **D9** (1976) 1139; Günther, K., Radtke, R.: Ergebnisse der Plasmaphysik und der Gaselektronik, Bd. 7. Akademie-Verlag, Berlin (in press); Sechenov, V. A., et al. В. А. Сеченов и др.): ТВТ **15** (1977) 411; Kulik, P. P., et al. (П. П. Кулик и др.): ТВТ **15** (1977) 415; Proc. XIII ICPIG. Berlin 1977, p. 583; Mintsev, V. B., et al. В. Б. Минцев и др.): ЖЭТФ **79** (1980) 116; Dikhter, I. I., et al. (И. И. Дихтер и др.): ТВТ **15** (1977) 196; Ermokhin, N. V., et al. (Н. В. Ермохин и др.): ТВТ **15** (1977) 695; Bushman, A. V., Fortov, V. E. (А. В. Бушман, В. Е. Фортов): УФН **140** (1983) 177; Alekseev, V. A., Fortov, V. E. Yakubov, I. T. (В. А. Алексеев, В. Е. Фортов, И. Т. Якубов): УФН **139** (1983) 193.
[77] Langer, I. S.: Phys. Rev. **120** (1960) 714; **124** (1961) 1003; Götze, W., Wölfle, P.: Phys. Rev. **B6** (1972) 1226; Devoto, R. S.: Phys. Fluids **19** (1976) 1; Kojima, D. Y.: Physica **39A** (1978) 171; Van Vliet, K. M.: J. Math. Phys. **19** (1978) 1345; **20** (1979) 2573; Ermakov, V. V., Kalitkin, N. N. (В. В. Ермаков, Н. Н. Калиткин): Численные методы механики сплошной среды. Novosibirsk 1978; Norman, G. E., Valuev, A. A.: Plasmaphys. **21** (1979) 1609; Wilhelm, H. E.: IEEE Trans. on Plasma Science **PS9** (1981) 68.

7.4. References to Chapter 4

[1] Bogoljubow, N. N. (Н. Н. Боголюбов): Проблемы динамической теории в статистической физике. остехиздат., Moscow 1946.
[1a] Bogoljubow, N. N., Gurow, K. P. (Н. Н. Боголюбов, К. П. Гуров): ЖЭТФ 17 (1947) 614.
[2] Klimontovich, Yu. L. (Ю. Л. Климонтович): Кинетическая теория неидеального газа и Неидеальной плазмы. Наука, Moscow 1975; ; Кинетическая Теория Электромагнитных Процессов. Наука, Moscow 1980.
[2a] Klimontovich, Yu. L. (Ю. Л. Климонтович), Ebeling, W.: ЖЭТФ **63** (1972) 905.
[3] Ebeling, W.: Ann. Phys. (Leipzig) **33** (1976) 350.
[4] Klimontovich, Yu. L. (Ю. Л. Климонтович): ЖЭТФ **52** (1967) 1233; **54** (1967) 136.
[5] Peletminski, S. W. (С. В. Пелетминский): ТМФ **6** (1971) 123.
[6] Vashukov, S. I., Marusin, V. V. (С. И. Вашуков, В. В. Марусин): ТМФ **29** (1976) 255.
[7] Ivanov, C.: Dissertation (A), Wilhelm-Pieck-Universität, Rostock 1978.
[8] Taylor, J. R.: Scattering Theory. Wiley, New York 1972.
[9] Kremp, D., Kraeft, W. D., Ebeling, W.: Physica **51** (1971) 146.
[10] Ebeling, W., Kraeft, W. D., Kremp, D.: Theory of Bound States and Ionization Equilibrium in Plasma and Solids. Akademie-Verlag, Berlin 1976.
[11] Cohen, E. G. D.: Fundamental Problems in Statistical Mechanics, Vol. II. North Holland Publ. Comp., Amsterdam 1968.
[12] Kadanoff, K., Baym, G.: Quantum Statistical Mechanics, Benjamin, New York 1962.
[13] Baerwinkel, K.: Z. Naturforsch. **24a** (1968) 22.
[14] Palzew, L. A. (Л. А. Пальцев): ТМФ **11** (1972) 259.
[15] Kolesnishenko, E. G. (Е. Г. Колеснишенко): ТМФ **30** (1977) 114, 382.

7.5. References to Chapter 5

[1] Landau, L. D., Lifshits, E. M. (Л, Д. Ландау, Е. М. Лифшиц): Статистическая физика, part 2. Наука, Moscow 1978; German translation: Lehrbuch der Theoretischen Physik, Bd. IX, Statistische Physik, Teil 2. Akademie-Verlag, Berlin 1980.
[2] Herring, C.: Phys. Rev. **57** (1940) 1169.
[3] Harrison, W. A.: Pseudopotentials in the Theory of Metals. Benjamin, New York 1963: Russian translation: У. Харрисон: Псевдопотенцалы в теории металлов. Мир, Moscow 1968.
[4] Heine, V., Abarenkov, J.: Phil. Mag. **9** (1964) 451.
[5] Animalu, A. O. E., Heine, V.: Phil. Mag. **12** (1965) 1249.
[6] Heine, V., Cohen, M. L., Weaire, D.: Solid State Physics **24**, Academic Press, New York 1970; Russian translation: В. Хейне, М. Коэн, Д. Узйр: Теория псевдопотенциала. Мир, Moscow 1973.
[7] Ashcroft, N. W.: J. Phys. **C** I, ser. 2 (1968) 232.
[8] Cohen, M. H.: J. Phys. et Radium **23** (1962) 643.
[9] Brovman, E. G., Kagan, Yu., Holas, A. (Е. Г. Бровман, Ю. Каган, А. Холас): ФТТ **12** (1970) 1001.
[10] Krasko, G. L., Gurskiî, Z. A. (Г. Л. Краско, З. А. Гурский): Письма в ЖТЭФ **9** (1969) 596.
[11] Gurskiî, Z. A., Krasko, G. L. (З. А. Гурский, Г. Л. Краско): **11** (1969) 3016.
[12] Gurskiî, Z. A., Krasko, G. L. (З. А. Гурский, Г. Л. Краско): ДАН СССР **197** (1971) 810.
[13] Brovman, E. G., Kagan, Yu. (Е. Г. Бровман, Ю. Каган): ЖЭТФ **52** (1967) 557.

[14] BROVMAN, E. G., KAGAN, YU.: Neutron inelastic scattering (Proc. Symp. Copenhagen), **1**, 3, Vienna 1968.

[15] KRASNYI, YU. P., ONISHCHENKO, V. P. (Ю. П. Красный, В. П. Онищенко): УФЖ **17** (1972) 1705.

[16] ABRIKOSOV, A. A., GOR'KOV, L. P., DZYALOSHINSKÎ, I. E. (А. А. Абрикосов, Л. П. Горьков, И. Е. Дзялошинский): Методы квантовой теории поля в статистической физике. Наука, Moscow 1962.

[17] BORN, M., HUANG, K.: Dynamical Theory of Crystal Lattices. Clarendon Press, Oxford 1954; Russian translation:
М. Борн, Хуан Кунь: Динамическая теория кристаллических решеток. Мир, Moscow 1958.

[18] BROVMAN, E. G., KAGAN, YU. M. (Е. Г. Бровман, Ю. М. Каган): УФН **112** (1974) 369.

[19] BROVMAN, E. G., KAGAN, YU. M., HOLAS, A. (Е. Г. Бровман, Ю. М. Каган, А. Холас): ЖЭТФ **57** (1969) 1635.

[20] RUMJANZEV, A. YU., et al. (А. Ю. Румянцев и др.): ЖЭТФ **75** (1978) 712.

[21] KOVALENKO, N. P., FISCHER, I. Z. (Н. П. Коваленко, И. З. Фишер): УФН **108** (1972) 2097.

[22] BROVMAN, E. G., KAGAN, YU., HOLAS, A. (Е. Г. Бровман, Ю. Каган, А. Холас): ЖЭТФ **61** (1971)2429.

[23] HUBBARD, J. H.: Proc. Roy. Soc. **240** (1957) 539.

[24] KOVALENKO, N. P., KUZ'MINA, L. M. (Н. П. Коваленко, Л. М. Кузьмина): Физика жидкого состояния **7** (1979) 60.

[25] KOVALENKO, N. P., KUZ'MINA, L. M., MAIER, G. (Н. П. Коваленко, Л. М. Кузьмина, Г. Майер): Письма в ЖЭТФ **28** (1978) 239.

[26] GELDART, D., VOSKO, S. H.: Can. J. Phys. **44** (1966) 2137.

[27] KAGAN, YU., PUSCHKAREV, V. V., HOLAS, A. (Ю. Каган, В. В. Пушкарев, А. Холас): ЖЭТФ **73** (1977) 967.

[28] ONISHCHENKO, V. P., KRASNYI, YU. P. (В. П. Онищенко, Ю. П. Красный): УФЖ **18** (1973) 1194.

[29] KESSEL'MAN, P. M., KRASNYI, YU. P., ONISHCHENKO, V. P. (П. М. Кессельман, Ю. П. Красный, В. П. Онищенко): ТВТ **12** (1974) 70.

[30] DE GROOT, S., MAZUR, P.: Non-Equilibrium Thermodynamics. North Holland, Publ. Comp., Amsterdam 1962; Russian translation: С. де Гроот, П. Мазур: Неравновесная термодинамика. Мир, Moscow. 1964.

[31] ZUBAREV, D. N. (Д. Н. Зубарев): Неравновесная статистическая термодинамика. Наука, Moscow 1971; German translation: D. N. SUBAREW, Statistische Thermodynamik des Nichtgleichgewichts. Akademie-Verlag, Berlin 1976.

[32] KRASNYI, YU. P., KOSTENKO, V. M. (Ю. П. Красный, В. М. Костенко): ТМФ **14** (1973) 521.

[33] ZIMAN, J. M.: Phil. Mag **6**,68 (1961) 1013.

[34] TURTSCHIN, B. F. (Б. Ф. Туричин): Медленные нейтрон. Наука, Moscow 1963.

[35] GUREVITSCH, I. I., TARASOV, L. V. (И. И. Гуревич, Л. В. Тарасов): Физика нейтронов оизких зннргий. Наука, Moscow 1965.

[36] ZIMAN, J.: Principles of the Theory of Solids. University Press, Cambridge 1972; Russian translation: Дж. Займан: Принципы теории твердого тела. Мир, Moscow 1974.

[37] Collected papers: Рассеяние тепловых нейтронов. Атомиздат, Moscow 1970.

[38] TAKEUCHI, S., ENDO, H.: Trans. Japan. Inst. Met. **3** (1962) 30.

[39] KRASNYI, YU. P., KOVALENKO, N. P. (Ю. П. Красный, Н. П. Коваленко): ЖЭТФ **62** (1972) 829.

[40] KRASNYI, YU. P., SHVETS, V. T. (Ю. П. Красный, В. Т. Швец): УФЖ **21** (1976) 681.

[41] KOVALENKO, N. P., KRASNYI, YU. P. (Н. П. Коваленко, Ю. П. Красный): Оптика и спектр. **35** (1973) 681.

[42] KRASNYI, YU. P., SHVETS, V. T. (Ю. П. Красный, В. Т. Швец): УФЖ **20** (1975) 1885.

[43] KRASNYI, YU. P., KOSTENKO, V. P. (Ю. П. Красный, В. П. Костенко): УФЖ 20 (1975) 537, 546.
[44] SPRINGER, B.: Phys. Rev. 154 (1967) 614.
[45] NEAL, T.: Phys. Fluids 13 (1970) 249.
[46] BRINGER, A., WAGNER, D.: Zs. Phys. 241 (1971) 295.
[47] ASHCROFT, N. W., SCHAICH, W.: Phys. Rev. B1 (1970) 1370.
[48] KOVALENKO, N. P., SHVETS, V. T. (Н. П. Коваленко, В. Т. Швец): УФЖ 23 (1978) 440.
[49] KOVANLENKO, N. P., KUZ'MINA, L. M. (Н. П. Коваленко,Л. М. Кузьмина): УФЖ 23 (1978) 2023.
[50] KOVALENKO, N. P., KUZ'MINA, L. M. (Н. П. Коваленко, Л. М. Кузьмина): УФЖ 24 (1979) 1220.
[51] HELLMANN, H.: Acta phys. chim. USSR 1 (1935) 913; 4 (1936) 255.
[52] RÖPKE, G., EBELING, W., KRAEFT, W. D.: Physica 101 A (1980) 243.
[53] MUKOHOPADHAY, G., JAIN, A., RATTI, V. K.: Solid State Comm. 13 (1973) 1623.
[54] BAUS, M., HANSEN, J.-P., SJÖGREN, L.: Phys. Lett. A 82 (1981) 180.
[55] MEISTER, C.-V., RÖPKE, G.: Ann. Physik (Leipzig) 39 (1982) 133.
[56] ROGERS, F. J., DEWITT, H. E., BOERCKER, D. B.: Phys. Lett. A, 82 (1981) 331.
[57] RÖPKE, G., HÖHNE, F.: phys. stat. sol. (b) 107 (1981) 603.
[58] DREIRACH, O., EVANS, R., GÜNTHERODT, H. J., KUNZI, H. U.: J. Phys. F: Metal Phys. 2 (1972) 709; BUSCH, G., GÜNTHERODT, H. J.: Solid State Phys. 29 (1974) 235; DUNLEAVY, H. N., JONES, W.: J. Phys. F: Metal Phys. 8 (1978) 1477.
[59] GRYKO, J., POPIELAWSKI, J.: Phys. Rev. 21 A (1980) 1717; GORECKI, J., POPIELAWSKI, J.: preprint 1980, to be published.

7.6. References to Chapter 6

[1] CHAPMAN, S. and COWLING, T. G.: The mathematical Theory of non-uniform Gases. University Press, Cambridge 1953.
[2] KLIMONTOVICH, Y. L. and KREMP, D.: Physica 109A (1981) 517.
[3] EBELING, W., KRAEFT, W. D. and KREMP, D.: Theory of Bound States and Ionization Equilibrium, Akademie-Verlag, Berlin 1976.
[4] RÖPKE, G.: Ann. Phys. (Leipzig) 39 (1982) 35; Teor. matem. fiz. (USSR) 46 (1981) 279; Physica A, to be published (1983).
[5] MEISTER, C. V. and RÖPKE, G.: Ann. Phys. (Leipzig) 39 (1982) 133. RÖPKE, G. and HÖHNE, F. E.: phys. stat. sol. (b) 107 (1981) 603.
[6] ROGERS, F. J.: Phys. Rev. A4 (1971) 1145.
[7] DALITZ, R. H.: Proc. Roy. Soc. A206 (1951) 509.
[8] MOTT, N. F. and MASSEY, H. S.: The Theory of Atomic Collisions. Clarendon Press, Oxford 1965
[9] GRYAZNOV, V. K. et al. (Грязнов, В. К., и. др.): Теплофизические свойства рабочих сред газофазного ядерново реактора. Атомиздат, Moscow 1980.
[10] HOLT, A. R., MOISEIWITSCH, B. L.: J. Phys. B1 (1968) 36.
[11] MEYER, J. R and BARLOTI, F. J.: Phys. Rev. B24 (1982) 2089.
[12] KREMP, D., SCHLANGES, M. KÜHN, K. H.: Phys. Lett., in press.
[13] RÖPKE, G., SCHMIDT, M., REDMER, R.: Wiss. Z. der Wilhelm-Pieck-Universität Rostock 31 (1982) 55.

8. Subject Index

The volume is devoted to a new field of plasma physics which is in full development and promises important practical implications e.g. for modern technical projects with high local concentration of energy as nuclear fusion, gas-phase nuclear reactors, plasma-chemical technology, detonations etc. In order to satisfy the urgent need in the physical characteristics of dense (strongly coupled) plasmas a survey of recent experimental and theoretical developments is given. The authors which are well-known specialists in the field present in the first two chapters important experimental methods as shock compression and the impulse capillary isobaric discharge and the interpretation of experimental data. The chapters 3 up to 5 are devoted to the theoretical methods as kinetic equations, linear response theory, bound state theory structure factors pseudopotential and diffraction models. Transport quantities for a wide range of densities and temperatures including liquid metals are presented.

Der Band ist einem neuen Gebiet der Plasmaphysik gewidmet, das sich gegenwärtig schnell entwickelt und wichtige praktische Anwendungen, z. B. für moderne technische Projekte mit hohen lokalen Energiekonzentrationen wie Kernfusion, Gasphasen-, Kernreaktoren, plasmachemische Technologie, Detonationen etc., verspricht. Um dem dringenden Bedarf an physikalischen Charakteristika für dichte (stark gekoppelte) Plasmen entgegenzukommen, wird ein Überblick über moderne experimentelle und theoretische Entwicklungen gegeben. Die Autoren, die als Spezialisten auf dem Gebiet bekannt sind, präsentieren in den ersten beiden Kapiteln wichtige experimentelle Methoden wie die Schockkompression und die kapillare isobare Impulsentladung sowie die Deutung der experimentellen Daten. Die Kapitel 3 bis 5 sind theoretischen Methoden wie kinetischen Gleichungen, der linearen Antworttheorie, der Theorie der Bindungszustände, den Strukturfaktor-, Pseudopotential- und Diffraktionsmodellen gewidmet. Für weite Dichte- und Temperaturbereiche — mit Einschluß flüssiger Metalle — werden Transportgrößen angegeben.

FSC
www.fsc.org
MIX
Papier | Fördert
gute Waldnutzung
FSC® C083411

Kolibri 360 GmbH
Ferdinand-Jühlke-Straße 7
99095 Erfurt, Deutschland
produktsicherheit@kolibri360.de